Environment, People and Development
Experiences from Desert Ecosystems

NIPA® GENX ELECTRONIC RESOURCES & SOLUTIONS P. LTD.
New Delhi-110 034

Environment, People and Development

Experiences from Desert Ecosystems

Edited by

Mahesh Kumar Gaur

Senior Scientist – Geography
Cartography & Land Use Section (Div - I)
Central Arid Zone Research Institute (CAZRI/ICAR)
Jodhpur - 342003, Rajasthan, India

P.C. Moharana

Principal Scientist
Central Arid Zone Research Institute (CAZRI/ICAR)
Jodhpur - 342003, Rajasthan, India

NIPA® GENX ELECTRONIC RESOURCES & SOLUTIONS P. LTD.

New Delhi-110 034

NIPA® GENX ELECTRONIC
RESOURCES & SOLUTIONS P. LTD.

101,103, Vikas Surya Plaza, CU Block
L.S.C. Market, Pitam Pura, New Delhi-110 034
Ph : +91 11 27341616, 27341717, 27341718
E-mail: newindiapublishingagency@gmail.com
www: www.nipabooks.com

For customer assistance, please contact
Phone: + 91-11-27 34 17 17
Fax: + 91-11- 27 34 16 16
E-Mail: feedbacks@nipabooks.com

ISBN: 978-81-19103-32-4

Composed and Designed by NIPA®.

Foreword

The year was 2002, the locality was spread out across two countries from Cape Town to Gibeon to Gobabeb, the conference topic involved Connecting Science with Community Action to Combat Desertification, the participants included a very enthusiastic and participatory Dr Mahesh Gaur. I am happy that Dr Gaur had compiled and edited a book entitled 'Environment, People and Development - Experiences from Desert Ecosystems'. Considering his enquiring mind, this book is not unexpected but nevertheless much appreciated.

Considering Dr Gaur's wide-ranging interests and scientific networks, it is not surprising that this book covers an eclectic selection of topics presented in 21 chapters. The chapter titles themselves offer an intriguing array from rural to urban settings, from Asia and Africa, from geography to development to engineering to remote sensing, from soils to water - and everything in between. The subjects also include a broad variety of approaches ranging from more traditional academic expositions to more personal overviews of the evolution of developmental tactics. There are not many aspects of arid lands that are not touched upon in one way or another.

While each chapter is very interesting in itself, almost all of the chapters end up begging questions and asking for more. Many of the chapters are very focused on a theme or a geographical area but leave one wondering if there are not comparative examples or experiences from elsewhere. This compilation thus provides the starting place for future research and scholarly analysis. Many of these chapters provide baselines for future studies when geographic, demographic, political, development or climate conditions have changed significantly. As the globe becomes more integrated, these papers will provide a valuable component of the reference archives covering arid lands. For this assemblage of interesting papers Dr Gaur is to be seriously thanked.

Although not specifically in the titles of individual chapters, many of the papers included in this collection include information of wider application than is indicated in the papers themselves. Movement of coastal dunes is well covered in some of the papers while changes in the approach to rural development are covered in others. A chapter on urban planning provides fascinating experience of the evolution and intensification of an unintended heat island. Other chapters provide a close connection between degradation of land and water both addressing negative impacts on the dependent population. Each chapter requires attention to gain the maximum value and overview from the materials that Dr Gaur has assembled.

For bringing together this interesting array of papers, Dr Gaur is to be heartily congratulated. Experiences from desert ecosystems are always welcome and this collection will take its places amongst the valuable additions to our knowledge of desert lands.

Dr Mary Seely
Gobabeb Research and Training Centre
Desert Research Foundation of Namibia

Preface

Desert regions are familiar faces in many pages of history of the world. The earliest civilizations and two of the world's major religions were born here. Now, deserts are themselves making news. Some of the cultures have made hot deserts their home for thousands of years. Modern technologies, including innovative irrigation systems and agricultural patterns have made deserts much more hospitable now. Despite such improvements, arid zones remain vulnerable to climate change and extremes of climate due to not-so-hospitable climate, poor soil and vegetation. In the wake of global climate changes, the desert ecosystem has again come under scrutiny to become one of the heart-throbs of research. Therefore, desert ecosystems seem to be dwindling between development and degradation.

Many of the desert ecosystems like Thar desert in India is well studied by researchers from many parts of the world including its own inhabitants. Information is now available about this region for the rest of the world to explore it more for mankind and its development. There is a radical change in the approach to life by the desert dwellers. Development is now a key word to the desert regions. Water is now available through canals and people have started using scientific approaches to conserve rainwater through improved systems. More information is also available on the state of natural resources, natural hazards and the impact scenario. GIS database is available through satellite remote sensing and thematic maps at different scale for assessing the resources at a given situation like drought and floods.

This publication "Environment, People and Development: Experiences from Desert Ecosystems" highlights some of the basic and technologically refined information from many parts of the Desert ecosystems of the world. Thanks to all the contributors for sharing their research experiences. This contribution would certainly open up and increase our understanding about the spatio-variablilty of desert ecosystems of the world.

Editor

List of Contributors

Abdelhamid, Ghassan
Training Advisor & System Analyst and Advisor at C4 Advanced Solutions, UAE Air Force, Space Reconnaissance Center, P. O. Box. 55260, Abu Dhabi, United Arab Emirates.
Email : trng@falconugs. ae.

Abd Elbasit, Mohamed A.M.
Arid Land Research Center, Hamasaka 1390, Tottori 680- 0001, Japan;
Email : m_abdelbasit@hotmail.com; mbasit@alrc .tottori -u.ac.jp
Desertification Research Institute, National Center for Research, Khartoum, Sudan.

Aina, Yusuf Adedoyin
Geomatics Technologies Department, Yanbu Industrial College, Yanbu. Saudi Arabia.

Ali, Rafat Ramadan
Soils and Water Use Dept., National Research Centre (NRC), Egypt

Ali, Abdelbagi M.
Arid Land Research Center, Tottori University, Tottori, Japan
Biotechnology and Biosafety Research Center, Agricultural Research Cooperation, Shambat, Khartoum, Sudan.

Alshehhi, Ali M.
Manager Falcon UGS, Space Reconnaissance Center, UAE Air Force & Air Defense, PO Box 51025, Abu Dhabi, UAE
Email : alishehhi@falconugs.ae

Basheer, Fadul
Um Jawaseer Desert Farm Project, ADRA-Sudan

Belem, M.
CNRST B.P.7047 Ouagadougou Burkina Faso

Chinto, Richard
Country Coordinator, Zambia

Dibloni, O.
CNRST B.P.7047 Ouagadougou Burkina Faso

El Gamri, Tarig
Water Resources Dept., Desertification Research Institute
The National Centre for Research, Sudan

et Frederic Dumay
Laboratoire GEGENA. EA3795 University of Reims
Champagne-Ardenne, Reims, France, 57, rue Pierre
Taittinger 51100 cedex.
Email : frederic.dumav@univreims.fr

Gaur, Mahesh Kumar
Associate Professor (Geography), Government Bangur PG College,
Pali - 306 401, India
Present Address : Senior Scientist (Geography),
Central Arid Zone Research Institute (CAZRI / ICAR),
Jodhpur – 342 003. India
Email : maheshjeegaur@yahoo.com

Hameli, Mahash S. Al
Director, Space Reconnaissance Centre, UAE Air Force &
Air Defense, PO Box 51025, Abu Dhabi, UAE.
Email : falcongs@emirates.net.ae

Iloweka, Ernest
President of FIEF NGO

Manganda
Executive Committee Member of TerrAfrica
Member of Thematic Group N° 13-Environment and Forests
Researcher and Scientist
Member of the National Coordination Team for Combat
Desertification
P.O.Box 16 239-Kinshasa-l
Democratic Republic of Congo
E-mail: mangadaernest@hotmail

Kar, Amal
Principal Scientist, CAZRI. Jodhpur. INDIA

Lioubimtseva, Elena
Professor, Geography and Planning Department, Grand Valley State University, Mackinac Hall B-4-202, 1 Campus Drive, Allendale, MI 49401, USA,
Email : lioubime@gvsu.edu

Magid, Isam M.A
Sudan University for Science and Technology-Faculty of Engineering

Mainguet, Monique
Directeur, Labortoire de Geographie Zonale pour le Developpement, Institut Universitaire de France 7, rue Pierre, Taittinses 51100 Reims, France

Mohamed EL-Bady
Geological Science Department, National Research Center

Mohamed Sabry
Giza, EL-Dokki, EL-Behoose St., Cairo, Egypt

Monique Mainguet
Laboratory of Zonal Geography for the Development,
University of Reims Champagne - Ardenne, Reims, France,
57, rue Pierre Taittinger 51100 cedex.
Email: monique.mainguet@univ-reims.fr

Mutamba, Emmanuel
Zambia,
Email : emutamba@yahoo.com

Niyogi, Shiladitya
Senior Marketing Director and Senior Project Director, with a Southern California firm, and has more than 14 years of experience in IT and Geospatial technologies.

Poda, Jean Noel
Director of Research
Assistant Director of IRSS/CNRST
Institut de Recherche en Science de la Sante (IRSS)
Centre Nationale pour la Recherche Scientifique et Technologique (CNRST)
IRSS / CNRST
03 B.P.7047
Ouagadougou 03, Burkina Faso
Email : podajnl@yahoo. fr

Raji, Bashiru Ademoal
Department of Soil Science, Faculty of Agriculture, Ahmadu Bello University, Zaria.

Shah, Jagruti
Gujarat Institute of Desert Ecology, Post Box No.83, Mundra Road. Bhuj-Kachchh, India.
Email : jhsphd@gmail.com

Shalaby, Adel
Land Use Dept., National Authority for Remote Sensing and Space Sciences (NARSS), Egypt

Tahir, M. A
Water Resources Dept., Desertification Research Institute
The National Centre for Research, Sudan

Wakponou Anselme
Department of Geography, University of Ngaoundere, P.O. Box: 454, Ngaoundere, Cameroon
Email : wakponouanselme@yahoo.fr

Yasuda, Hiroshi
Arid Land Research Center, Tottori University, Tottori, Japan

Contents

1

Desert, Desertification and Community Resilience

Mahesh Kumar Gaur

Introduction

Low and erratic precipitation, diurnal fluctuations in temperature and low humidity are the typical characteristics of deserts and their margins. Further, strong winds, sunny and very bright days and unpredictable periods of drought and famines are regular features. These lands are poor in fertility. Uninterrupted increase in human and animal pressures is violating carrying capacity of the natural resources, thus resultant is desertification.

Arid zones are described as a part of the drylands, and have the most severe edapho-climatic conditions. The major distinguishing feature for defining and for planning for the arid zone is the low rainfall (below 500 mm or Aridity Index <0.20) with more than 50% inter-annual variability. This makes a great difference in terms of the nature of the ecosystem, the socio-economic environment and the challenges for sustainability. High wind and solar regimes further increase the effect of rainfall variability and the whole complex makes a fragile ecosystem in which small disturbances may cause great loss to the sustainability, which are sometimes irreversible (Hou'erou, 1996). The hot arid zones of the world are economically and environmentally disadvantaged, with unique problems. The ecosystems of these zones are highly fragile with large risks that cause severe impediments to development programs. Of the total land area in the world, arid zones cover 18.8% and the hyper-arid zones account for 22.3% of the total arid zones (FAO, 1989).

These arid zones are diverse in terms of climate, soils, vegetation, animals, and the lifestyles and activities of the people. Little but variability in rainfall and the presence of distinctive periods of drought are the characteristics of the arid tropics. Often, the terms drought and aridity are used incorrectly. A drought is a departure from average or normal conditions in which shortage of water adversely impacts on the functioning of ecosystems, and the resident population of people, whereas aridity refers to the average conditions of limited rainfall and water supplies, not to the departures there from. In general, arid zones are characterized by pastoralism and little farming but there are always exceptions (Malhotra, 1984)

Desertification results from over exploitation of natural resources through over cultivation, over grazing, over irrigation practices, deforestation and poor natural resource management. This is especially true of desert and its margins where the climatic conditions particularly rainfall are highly variable. In a developing country like India, where population growth and poverty combine to place severe pressure on natural resources, the problem of desertification acquires serious magnitude. It has been estimated that about 50% of the total geographic area of India has been under various degradation hazards and about 2.1 million ha of land is getting degraded and deforested annually.

Characterized by climate variability, these lands sustain pastoralists and small-scale farmers, but are susceptible to desertification as a result of increasing human population, inappropriate government land-use policies, settlement, climate change, deforestation, expropriation of rangelands, land clearance, overgrazing, inappropriate irrigation practices, political instability and poverty. The livelihoods of more than 1.2 billion people inhabiting dryland areas in 110 countries are currently threatened by drought and desertification.

Recognizing the link between desertification and poverty, the United Nations Convention to Combat Desertification (UNCCD) stresses the importance of a "bottom-up participatory approach in identifying, implementing, monitoring, and evaluating projects that combat desertification and mitigate the effects of drought". The year 2006 was designated as the International Year of Deserts and Desertification to provide opportunities to highlight the difficult conditions faced by inhabitants of desertification affected areas.

Manifestations of Desertification

The manifestations of desertification are apparent in all categories of ecosystem services: provisioning, regulating, cultural, and supporting.

Some of these services are typically measured and quantified, such as food, forage, fibre, and fresh water; others may be inferred or implied through qualitative analysis. As indicated earlier, management approaches that prevent, reduce, or reverse these manifestations of desertification are available and practiced.

In desertified areas, people have responded to reduced land productivity and income by either increased use of other relatively marginal land (not yet degraded but having lower productivity) or by transforming more rangeland to cultivated land. Since policies to promote alternative livelihood opportunities are commonly not in place, migration to unaffected areas subsequently occurs. Initially it is from rural to urban areas, and then to locations of greater economic opportunity in other countries. These migrations sometimes exacerbate urban sprawl and can bring about internal and cross-boundary social, ethnic, and political strife.

Transformation of rangelands and silvo-pastoral dryland systems to croplands increases the risk of desertification due to increased pressure on the remaining rangelands or because of unsustainable cultivation practices. Although range lands are resilient under traditional mobile grazing practices—commonly called transhumance—in response to seasonal changes, reduced transhumance leads to overgrazing and rangeland degradation. Removal of the rangeland vegetation cover takes place both by overgrazing of forage and by transforming rangelands to cultivated systems worldwide. Removal of vegetation cover when combined with unsustainable soil and water management practices in the converted rangelands brings about soil erosion, soil structure change, and soil fertility decline. Between 1900 and 1950, approximately 15% of dryland rangelands were converted to cultivated systems to better capitalize on the food provisioning service; a somewhat faster conversion has taken place in the last five decades during the Green Revolution.

In many semiarid areas, there is a progressive shift occurring from grassland to shrubland that exacerbates soil erosion. During the second half of the nineteenth century, large-scale commercial stockbreeding quickly spread over the semiarid drylands of North and South America, South Africa, and Australia. Both the kind of imported herbivore and type of grazing management (including fire prevention) were not adjusted to the semiarid ecosystems. The resulting disturbance was therefore a "transition trigger" that, combined with drought events, led to a progressive dominance of shrubs over grass (sometimes called "bush encroachment"). The transition from land fully covered by grasses to one covered by scattered bushes creates greater bare soil surfaces, which encourages increased runoff velocity, resulting in higher soil erosion.

Desertification : Global Concern

The decade of the 1950's witnessed the first worldwide effort to call attention to the problems and potentials of arid regions. It started when the United Nations Educational, Scientific, and Cultural Organization (UNESCO) launched its Major Project on Scientific Research on Arid Lands in 1951. That project led to publication of a newsletter, the provision for funds for establishing and strengthening arid land research institutes, organization of conferences and symposia, and publication of a series of research reviews and special reports on a wide range of topics. The Major Project was terminated in 1962 and the arid land program was merged with the broader UNESCO natural resource program.

The impetus generated by the UNESCO project led to expanded interest in, and support of, arid lands studies throughout the world. By 1970, knowledgeable scientists were well aware of the magnitude of the land destruction that had taken place in the past, and that was becoming even more serious as population pressures increased.

At about the same time, one event served to focus world attention on desertification: the 1969 to 1973 drought in the African Sahel. Recognition of the severity of the drought affecting six countries on the southern border of the Sahara (Mauritania, Senegal, Mali, Upper Volta, Niger, and Chad) was slow to develop. Droughts, after all, were not unusual in the Sahelian countries; an equally bad or even worse one had struck the same region during the years from 1911 to 1914, and several other droughts had occurred before and after that time. It was only when the situation had become catastrophic that relief measures were undertaken on a large scale. Among the after-effects of the human toll and the millions of livestock that died due to the drought, was the call by the United Nations General Assembly for the convening of an International Conference on Desertification. The conference was held in Nairobi, Kenya, in August and September of 1977, attended by representatives of nearly 100 nations and many international organizations; governmental and nongovernmental. In addition to providing a forum by which the world's attention was drawn to the land destruction that was occurring, the conference addressed the problem of combating desertification of arid lands. In 1978, United Nations drew a plan of action to the land destruction that was developed and approved by the delegates. Responsibility for following up and coordinating the implementation of the Plan of Action to Combat Desertification (PACD) was entrusted to the United Nations Environment Program. (Dregne, 1986)

Involvement of United Nations : In 1992, the United Nations held the "Earth Summit" at Rio de Janeiro, when three conventions met - the Convention on Climate Change, the Convention on Biodiversity and the Convention to Combat Desertification. One of the goals of the Summit was to involve developing countries in combating desertification, particularly since they are the immediate victims of this process - a major cause of poverty to the extent that there are now "environmental refugees" who have lost their livelihood due to desertification. Although "dry lands" are less productive, they have the most intensive population growth rate. So the objective of the UNCCD was to reduce global poverty by combating desertification. It would be accomplished by participation and partnership between developed countries and developing countries. The *World Day to Combat Desertification and Drought*, is now celebrated every year the 17th of June.

The UN General Assembly (UNGA) designated 2011 as the International Year of Forests to bring attention to the value of forests and the social, economic and environmental costs of their loss. The theme of this year's World Day to Combat Desertification was "Forests keep drylands working." The General Assembly designated 2010-2020 as the *Decade to Combat Desertification* in 2007 to heighten public awareness of the threat posed by desertification, land degradation and drought to sustainable development. Along with the UNCCD, four other UN agencies; the UN Environment Programme (UNEP), the UN Development Programme (UNDP), the UN International Fund for Agricultural Development (IFAD) and the UN Department of Public Information (DPI) - have been mandated by the General Assembly to spearhead activities related to the Decade.

A Global Issue : According to Professor U.N Safriel of Hebrew University of Jerusalem, world could achieve "very little!" Although there has been a total agreement on the extent of the problem, but there were varied disagreements on definitions and on causes of desertification. Strategies for solutions have also been hotly debated. Behind much of the intellectual arguments, it becomes relatively simple to identify the motivation of various interest groups. Developed countries did not show any interest to change their lifestyles by adapting to different forms of energy and industrial constraints, or to finance massive programs in other developing countries. Developing countries were interested in portraying and retaining the worst case scenario in order to demand more money.

Global Solutions : Desertification should be defined by "lack of soil productivity" to enable us to focus on possible solutions for seemingly contradictory situations (Safriel, 2001). There is a need to focus not only on climatic conditions, but also on monitoring land usage without destroying its productivity. In fact, overuse of land resources often inspires ingenuity that promotes sustainable use of lands and biological activity.

Thus, desertification is a type of land degradation. The degradation converts useful but arid lands to less useful desert. It is characterized by:

1. Declining groundwater,
2. Increased erosion, and
3. Disappearance of native vegetation.

Areas Vulnerable to Desertification

As the world's population continues to grow, human induced stresses on biophysical resources increase proportionately. In the richer countries, environmental awareness, and the necessity to incorporate ecological considerations in land management and the ability to subscribe to tenets of sustainable agriculture have contributed to national strategies for a rational use of biophysical resources. The poorer third world countries, on the other hand, are unable to embrace these 'lofty' ideals and remained with reduced productivity and even an inability to feed them. It is evident that in the foreseeable future, unless the latter countries are assisted to better manage their resources and address food security through an important emphasis on soil and water conservation, the capacity of the planet as a whole to maintain an acceptable quality of life will be reduced.

According to Reynolds et al (2007), drylands cover about 51% of Earth's land surface and are home to more than 38% of the total global population of 6.5 billion. Some form of severe land degradation is present on 10 to 20% of these lands, the consequences of which are estimated to affect directly some 250 million people in the developing world, an estimate likely to expand substantially in the face of climate change and population growth (Table 1).

There are about 7.1 million sq. km. of land under low risk of human-induced desertification, 8.6 million sq. km. at moderate risk, 15.6 million sq. km. at high risk, and 11.9 million sq. km. under very high risk. Each of these classes represents a desertification tension zone. The major critical tension zone that requires immediate attention is the very high-risk class. There are 11.9 million sq. km. of land with about 1.4 billion inhabitants.

Table 1 : Extent of Drylands

Region	Arid	%	Semi Arid	%	Dry Sub-Humid	%	All Dry lands	%	Popu-lation
Asia (Incl. Russia)	6,164	13	7,649	16	4,588	9	18,401	39	1,444,906
Africa	5,204	17	5,073	17	2,808	9	12,933	43	267,563
Oceania	3,488	39	3,532	39	996	11	8,016	89	140,586
North America	379	2	3,436	16	2,081	10	5,896	28	86,990
South America	401	2	2,980	17	2,233	13	5,614	32	59,323
C. America & Caribbean	421	18	696	30	242	10	1,359	58	31,719
Europe	5	0	373	7	961	17	1,339	24	6,960
	15,910	**12**	**23,739**	**18**	**13,909**	**10**	**53,558**	**41**	**2,038,047**

Land area : millions of square kilometers. Population: thousands. (White and Nackoney, 2003)

Table 2 : Principal Deserts of the World

S. No.	Desert Name	1000 sq. miles	1000 sq.km.	Territory
1.	Sahara	3250	8420	Algeria, Chad, Libya, Mali, Mauritania, Niger, Sudan, Tunisia, Egypt, Morocco
2.	Australian	600	1550	Australia
3.	Arabian	500	1300	Saudi Arabia, Yemen
4.	Gobi	400	1040	Mongolia, China (Inner Mongolia)
5.	Kalahari	200	500	Botswana
6.	Takla Makan	125	324	Sinkiang, China
7.	Sonoran	120	311	Arizona, California, New Mexico
8.	Namibian	120	311	S.W. Africa (Namibia)
9.	Kara Kum	105	272	Turkmenistan, USSR
10.	Thar	100	259	N. W. India, W. Pakistan
11.	Somali	100	259	Somalia
12.	Atacama	70	181	Northern Chili
13	Kyzyl Kum	70	181	Uzbekistan-Kazakhstan
14.	Dasht-i-Lut	20	52	Eastern Iran
15.	Mojave	14	35	Southern California
16.	Desierto de Sechura	10	26	N.W. Peru
	Totals	5804	15041	Global

Source : Norris McWhirter, 1981.

In all, more than 110 countries have drylands that are potentially threatened by desertification. Africa, Asia and Latin America are the most threatened by desertification. (Fig. 1)

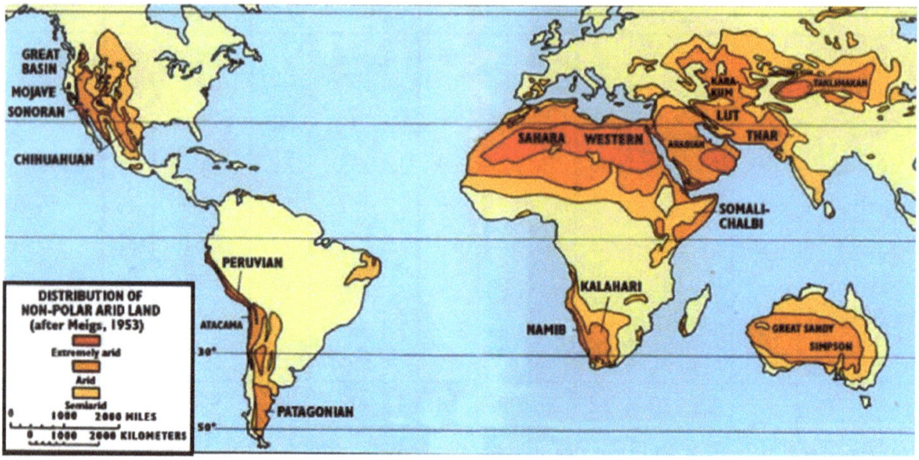

Source : http://pubs.usgs.gov/gip/deserts/what/world.html

Fig. 1 : Distribution of Aridlands

- 2/3 of the continent is desert or drylands in Africa. The region is afflicted by frequent and severe droughts. Many African countries depend heavily on natural resources for subsistence. Africa's desertification is strongly linked to poverty, migration, and food security;
- Asia contains some 1.7 billion hectares of arid, semi-arid, and dry sub-humid land reaching from the Mediterranean coast to the shores of the Pacific. In terms of the number of people affected by desertification and drought, it is the most severely affected continent;
- Although better known for their rain forests, Latin America and the Caribbean are actually about one-quarter desert and drylands. Poverty and pressure on land resources are causing land degradation in many of these dry areas;
- About 30% of the land in the United States is affected by desertification.

Land Quality

Land uses are primarily the result of human actions and decisions on land. In fact human activities arising from a multiplicity of social objectives are the immediate source of land cover change. To understand these social objectives, one needs to analyze the underlying driving forces that motivate or constrain the associated human activities. Biophysical driving forces and shocks (such as geomorphic processes, global and local climate change/ variability, etc.) are also responsible for changes in land cover and ultimately the land use.

Land quality is directly linked to quality of life; specifically, social and economic equity, and thus should be addressed in the socioeconomic context for sustainability. In the past, the focus of attention was on rehabilitation of degraded lands. Thrusts in the new Agenda are to evaluate the potential for land resource degradation, manage the resilience characteristics of the systems, and select technological options in the framework of the resilience properties (Eswaran, 1993). Consequently, the concepts of 'early warning indicators' and 'land resilience' are relevant (Greenland and Szabolcs, 1994). There are few methods to predict the onset of degradation, which is crucial to managing systems being stressed. The resilience capacity of systems is also less well established and this should be studied to implement remedial measures to rehabilitate degraded lands. Although soil resource information have been utilized for such purposes in developing countries, either the poor quality of information or its absence has prevented a more effective use for such purposes (Eswaran, 1992).

Table 3 : Estimates of land area belonging to vulnerability classes and corresponding number of impacted population

Vulnerability Class	Area Subject to Desertification		Population Affected	
	Area (million km^2)	Percent (Global land area)	Number (Millions)	Percent (Global Pop.)
Low	14.60	11.2	1,085	18.9
Moderate	13.61	10.5	915	15.9
High	7.12	5.5	393	6.8
Very High	7.91	6.1	255	4.4
Total	44.24	34.0	2,648	44.0

Note : The global population density map is limited to latitudes 72°N to 57°S.

Climate Change and Desertification

Climate has always been a dynamic entity, and it can affect natural systems through the consequence of climate variability and climate change. The accelerating and deepening impacts of climate change will touch everyone on Earth in some way, but those who stand to suffer most are the desert inhabitants and poor. Poor people already live on the margins; they have limited resources with which to adapt to changes in the climate. Most of the world's poor depend for their livelihoods directly upon highly climate sensitive resources: fields, forests, fisheries, and other natural assets are "the wealth of the poor." Climate change will make these resources less reliable in many places, lowering the prospects for many poor communities to escape from poverty.

Agriculture remains by far the most important sector in aridlands economy. However, the existences of climate variability influence agricultural sector on which majority of the population depend.

Almost 80 percent of aridlands population depends on rainfed agriculture, thus welfare and economic productivity are linked to the variable rains. This dependency on rainfed agriculture has thus made the region's economy extremely vulnerable to the effect of climate variability and climate change. Frequent droughts are an important aspect of climate variability and change that human being has been experiencing in current time. Consequently, area affected by drought is increasing in aridlands due to climate change, climate variability and other human induced factors (WMO, 1986; NMSA, 1996).

Currently, world climate in general has been changing due to mainly anthropogenic (human) factors. The major impacts of climate change are manifested through rainfall variability and recurrent drought. The risks associated with them are generally late onset and early cessations of rainy season that cause reduction in the growing season. The amount and distribution of rainfall in the growing season are also very erratic, indeed making crops vulnerable to moisture deficit.

Among others, agriculture is the most vulnerable and sensitive sector that is seriously affected by the impact of drought. Agricultural drought is currently more pronounced in the study area. Farmers inhabiting in the aridlands are exposed to the impact of agricultural drought that has resulted from the current climate change. This is due to the fact that most of farmer's cultivars (long duration crop variety) can no longer resist the exiting agricultural drought and intra-seasonal dry spell. As a result, most of the cultivars have gone obsolete, and yield reduction and crop failure becomes common. Thus, in order to adapt and/or mitigate the impact of agricultural drought through generating different improved drought resistant agricultural technologies, agricultural drought assessment through remote sensing is mandatory.

Climate change leads not only to higher temperatures, but also to changes in precipitation and evapotranspiration. In the India, summer droughts are expected to occur more often and last longer whereas rainfall is expected to be concentrated in sparse and scanty showers. These meteorological changes will not only alter the water balance of the soil-water-vegetation system, but will also affect habitat factors essential for plant growth such as soil alkalinity and the availability of moisture, oxygen and nutrients in the root zone of plants. Furthermore, vegetation is more likely feedback on climate change by adapting characteristics that determine actual evapotranspiration like rooting depth, vegetation cover and water use efficiency. All things considered, climate change will presumably result in large alterations to both scarce water balance components and vegetation composition. Changes in temperature and the spatial configuration of vegetation types will determine the chances of plant and animal species to migrate through the landscape and find new habitats to survive.

Climate change confronts nature management with a number of problems. The first is that climate change will change the physical nature of habitats. Aridlands of the world need to have a spatial strategy for protecting biodiversity in landscapes with high pressure on land use. Climate change affects biodiversity existence. Indirectly, climate change affects biodiversity through changing abiotic conditions, resulting in current habitat locations becoming unsuitable, while opportunities for habitat restoration arise elsewhere. Adaptive measures for other land use functions (e.g. water management, farming) may also change the spatial functioning of habitat networks.

This creates a second problem, on how to set climate-proof nature conservation targets. One may aim at conserving the originally targeted species and spatially follow the suitable environmental envelop by designating new nature reserves at the optimal locations. Another option is to redefine, for the existing nature reserves, the targeted species suited to the new environmental conditions. Additional complexity is introduced by feedbacks between abiotic and biotic ecosystem components or between functionally connected species (van der Putten *et al.*, 2004).

The third problem is the multitude of actors in charge of nature management, land use and water management in aridlands. They will have different strategies in dealing with climate change, also between specific geographic areas and individual nature reserves. Also their attitude towards cooperation with stakeholders in the rural areas might differ (farmers, recreation, etc.). Thus conflicts may arise from different strategies with respect to nature conservation areas.

To foster the process of climate-proofing nature conservation policy and targets a new approach is needed, aiming at:

- presenting the effects of climate change on biophysical site conditions inside, and outside, nature areas;
- guiding and facilitating the process of actors re-defining nature targets and biophysical management.

This has raised the question whether and how habitat networks should or can be adapted to facilitate climate-induced range shifts or mitigate effects of weather extremes in aridlands. Such adaptation is legitimated from the point of view of protecting species, but also to maintain adequate levels of functional biodiversity to maintain the adaptive capacity of ecosystems to provide ecosystem services (Hooper *et al.*, 2005).

As populations continue to grow and exert pressure on the resources, devastation caused by disasters will increase. In developing countries, disasters tend toward a higher rate of fatalities, in part due to inadequate infrastructure, lack of building codes, and poor land use. In the developed world, the cascading consequences of disasters increase as supply chains and critical infrastructure become more interdependent in a global economy. Combined decadal economic and insured losses to natural disasters have increased by a factor of nearly 7 since the 1980s.

Managing Climate Risk in Rural Kenya

Home to 10 million people, Kenya's arid and semi-arid lands have the lowest development indicators and highest incidence of poverty in the country. Over 60 percent of inhabitants currently live below the poverty line. Increasing population pressures, overgrazing, crop reduction in fallow fields, and recurring conflicts between pastoralists and farmers all pose serious development challenges. These will be exacerbated by climate change, which is expected to bring increased frequency and severity of both floods and droughts. (Government of Kenya, 2002)

In response to the devastating effects of past droughts—and the prospect of more as climate change intensifies—the Centre for Science and Technology Innovations has partnered with the Arid Lands Resource Management Project to work with communities in Makueni District in introducing a suite of measures to reduce vulnerability to climate change. Climate and weather forecasts are being downscaled and communicated to farmers to help them select appropriate planting times. Local production systems are being diversified through the use of drought-tolerant crop varieties and better systems for collecting and storing seeds. Farmers have been trained in soil and water conservation;

weather prediction and interpretation; selection of seeds to fit climatic and land conditions; and early land preparation and planting. As well, such technologies as sand dams and drip irrigation have been introduced to improve access to water. Credit systems are being strengthened, allowing community members to pursue diversified or alternative livelihood activities. (IISD, 2007)

Conclusion

Arid and semiarid landscapes comprise nearly a third of the Earth's total land surface. These areas are coming under increasing land use pressures. Despite their low productivity these lands are not barren. Rather, they consist of fragile ecosystems vulnerable to anthropogenic disturbance. At the same time, many deserts around the world are facing unprecedented land-use pressures.

The increasing pace of social change, innovation, and technological advances can combine to create additional vulnerabilities. Regional and global dependencies may make it difficult for individual operations or entire industries to tolerate disruptions that occur on the other side of the globe.

Recent national and international experience with natural and human-caused disasters highlights several realities. First, the planet on which we live—the planet on which we aspire to forge careers, establish marriages and families, grow economies, and seek peace and security—provides frequent and often unpredictable extreme events. Severe heat waves, cold snaps, and cycles of flood and drought determine what we call climate. Movement in the Earth's crust is manifested by earthquakes and volcanic eruptions. Environmental degradation, habitat loss, and reduction in biodiversity can occur incrementally but also through sudden devastation, such as through wildfire or an oil spill.

Second, extremes often trigger disruptions of communities that persist after the event that exceed a community's ability to recover on its own. These disasters are as much the result of human decisions as of nature. Land use, building codes, the engineering of critical infrastructure, distribution of wealth and poverty, and many other social decisions and actions shape the impacts of extremes and subsequent recovery.

Third, resilience to disasters is built at the community level. No community is immune to disasters, and no community is an island unto itself. The emerging role of critical infrastructure, just-in-time manufacturing, and the globalization of the economy means that all individuals and communities are interdependent.

Fourth, responsibility for building community resilience cannot rest with the public sector alone.

Communities are dynamic and respond to changes in population, political leadership, the economy, and environmental factors. Resilient communities can withstand hazards and disasters. They continue to operate under stress, adapt to adversity, and recover functionality after a crisis. Desert communities are one of the most resilient communities. They have always got up after every disaster.

The appearance and movement of sand blow-outs in arid environments are important as they signify the triggering of land degradation processes. Blowing sand has serious consequences for landscapes undergoing degradation. Aeolian removal and transport of dust, sand, and litter are the primary mechanisms of degradation, killing plants by burial and abrasion, interrupting natural processes of nutrient accumulation, and allowing the loss of soil resources by abiotic transport.

References

Biswas, D. K. 1984. Desertification and its control- Indian experience. In International Symposium on Integrated of Land Desertification proceedings MAB, National Committee of the people's of China for MAB, Sept. Ist-22nd 1984. Institute of Desert Research, Academia Sinica, Lanzhou, China.

Chadha, G.K. 1996. Wastelands in Rural India: Policy Initiatives and Programmes for their Development. Occasional Paper - 2, NABARD.

Dhir, R.P. 1983. Future Technological needs of Rajasthan Canal Command area with particular reference to stage II, Proceedings of seminar on irrigation of arid lands, Jaipur, pp. 47-60.

Dhir, R.P. 1997.Western Rajasthan Soils: their characteristics and properties. In: Desertification and its Control. ICAR, New Delhi. 102-115.

Dregne, H. E. 1986. Desertification of arid lands. In Physics of desertification, ed. F. El-Baz and M. H. A. Hassan. Dordrecht, The Netherlands: Martinus, Nijhoff.

Eswaran, H. 1993. Soil resilience and sustainable land management in the context of Agenda 21. In: Greenland, D.J. and I. Szabolcs (Eds.) Soil Resilience and Sustainable Land Use. CAB International. Wallingford, England. 21-32.

Eswaran, H., S.M. Virmani, and I.P. Abrol. 1995. Issues and challenges of dryland agriculture in Southern Asia. In: A.S.R. Juo and R.D. Freed (ed.) Agriculture and the Environment: Bridging Food Production and Environmental Protection in Developing Countries. ASA Spec. Pub. 60:161-180. Madison WI.

FAO, 1989. Arid Zone Agroforestry: A guide for field technicians. FAO, Rome. 147 p.

Geist H and Lambin E. 2004. Dynamic causal patterns of desertification. BioScience 54: 817–829.

GOI. 1995, Annual Report of the Department of Wastelands Development, Ministry of Rural Department.

GOI. 1995. Technologies in Wasteland Development, NWDB, Ministry of Environment and Forests.

Hooper, D.U., F.S. Chapin, J.J. Ewel, A. Hector & P. Inchausti, 2005. Effects of biodiversity on ecosystem functioning: a consensus of current knowledge. Ecol. Monogr. 75: 3–35.

Hou'erou, Henry N. Le. 1996. Climate Change, drough and desertification. Journal of Arid Environment. 34 : 133-185

Jodha, N.S. 1997. Trends in tree management in arid land use in western Rajasthan. In: Farms trees and farmers-response to agricultural intensification. (Eds J.E.M. Arnold and P.A. Dewees) EarthScan Publishing, London, 43-64.

Kaul, R.N. 2000. Success and sustainability of desertification control in the Aravalli hills of Haryana: An analysis In Desertification Control Bulletin, UNEP, Nairobi, Kenya 36.

Kothmann, M. M., *et al.* (editors), 1974. A glossary of terms used in range management. Second edition. Society of Range Management, Denver, Colorado. 36 p.

Malhotra, S.P. 1984. Traditional agroforestry practices on arid zone of Rajasthan. In: Agroforestry in arid and semi-arid zones. Eds. Shamkarnarayan *et.al.* CAZRI, Jodhpur, 263-266.

Millennium Ecosystem Assessment, 2005. Ecosystems and Human Well-being: Desertification Synthesis. World Resources Institute, Washington, DC.

Norris McWhirter, ed., 1981 Guinness Illustrated Encyclopedia of Facts, Bantam Books.

Rao, D.P., R.S.Dwivedi and A.T.Jayaseelan. 2003. Role of space technology in assessment and monitoring of drought and desertification In Human Impact on Desert Environment (Eds. Pratap Narain, S. Kathju, Amalkar M.P. Singh and Praveen Kumar), Arid Zone Research Association of India and Scientific Publishers, Jodhpur.

Reynolds, J. F., D. M. S. Smith, *et al.*, 2007. "Global Desertification: Building a Science for Dryland Development." Science 316(5826): 847-851.

Reynolds JF, Stafford Smith DM and Lambin E. 2003. ARIDnet: seeking novel approaches to desertification and land degradation. Global Change NewsLetter, 54, 5–9. International Geosphere-Biosphere Programme, Stockholm.

Reynolds JF and Stafford Smith M (Eds) 2002. Global Desertification: Do Humans Cause Deserts? Dahlem Workshop Report 88. Dahlem University Press, Berlin. Pp437.

Safriel, U.N. 2001. Ecological-oriented options for the sustainable development of drylands. In: Portnov, B.A. and A.P., Hare (Eds.). Desert Regions. Springer, Berlin, pp. 153-160.

Shankarnarayan, K.A. 1988. Monitoring of desertification - a conceptual frame In Desertification: Monitoring and Control (Ed A.K. Tewari), Scientific Publishers, Jodhpur.

Turner II BL, Skole II D, Sanderson S, Fischer G, Fresco L, Lemans R. 1995. Land-Use and Land-Cover Change (LUCC): Science/Research Plan. HDP Report No. 7. Stockholm: IGBP Secretariat.

Van der Putten, W.H., P.C. de Ruiter, T.M. Bezemer, J.A. Harvey, M.J. Wassen & V. Wolters, 2004. Trophic interactions in a changing world. Basic and Applied Ecology 5(6): 487-494.

White, Robin P, and Janet Nackoney 2003 Drylands, People, and Ecosystem Goods and Services: A Web-Based Geospatial Analysis. World Resources Institute Report.

World Meteorological Organization 2005. Climate and Land Degradation. Geneva, World Meteorological Organization. Report WMO-No. 989.

WRI. 2007. Weathering the Storm: Options for framing adaptation and development. P. see page 21.

A. CLIMATE CHANGE AND DESERTIFICATION

2

Desertification Paradigms and Measurements in Indian Context

Amal Kar

Introduction

Desertification, which is considered a major threat in the drylands of the world, is not about advancing boundaries of the existing deserts, but about land degradation in the arid, semi-arid and the dry sub-humid regions, which together form the drylands (Anon., 1995). Secular variation in climate that has resulted in periodic expansion and contraction of the desert boundaries and regional-scale changes in vegetation condition (Tucker *et al.*, 1991; Nemani *et al.*, 2003; Hickler *et al.*, 2005) is one of the drivers of desertification. Human activities constitute the other major driver. According to Geist and Lambin (2004) the prominent core variables that drive degradation are climatic, economic, institutional, national policies, population growth and remote influences.

Desertification has been defined differently at different periods of time. The term was first used by Lavauden (1927) to describe severely overgrazed rangelands in Tunisia (Dregne, 2000). Subsequently, Aubreville (1949) used the term to refer to excessive soil erosion due to deforestation in the French West Africa. Worldwide interest in the term desertification first surfaced in the context of the first UN Conference on Desertification at Nairobi in 1977 (UNCOD, 1977), which was held in the background of a long drought in the sub-Saharan Africa during the early 1970s. The term was then defined as diminution of the biological potential of land in any ecosystem. Since the definition failed to focus on the degradation problems of the drylands, especially in Africa, the 1992 UN Conference on Environment and Development (UNCED) negotiated a more acceptable definition as 'land degradation in arid, semi-arid, and dry sub-humid areas, resulting from various factors, including climatic variation as well as human activities' (Anon., 1992). The UN Convention to Combat Desertification (UNCCD) accepted this definition (Anon., 1995) for all practical purposes, but different national and regional needs and aspirations from UNCCD soon started to imply meaning to the definition and attempted to modify it. The most common usage so far has been to bracket desertification with land degradation, so that the environmental problems of drylands and non-dry lands are equally taken care of through international funding under the UNCCD. In Africa the emphasis is more on drought, and so 'drought and desertification' has become a familiar term. Incidentally, African drylands are the major beneficiaries of the UN effort on desertification.

Desertification usually takes place when the rates of natural geomorphic processes in the drylands are accelerated by human action or through extreme natural events like droughts, high-intensity rains, tectonic disturbances, etc. As human use of the drylands is increasing, especially through modern technologies and for higher returns without much concern for sustainability of the land resources, there is now increasing impact of human pressures on the landscape manifested as different types of degradation. Typically the degraded areas bear the signatures of wind and water erosion, soil salinity, waterlogging, industrial effluent discharges, mine wastes, denuded pastures and forests, etc. The scale of vulnerability of the landscape to accelerated degradation varies according to the landscape's inherent properties, including sediment characteristics, as well as plant cover, land use, etc., as well as on the climatic setting.

Globally it has been estimated that out of 5169 million hectare (mha) area of the drylands (constituting 40% of the global land surface), about 1035 mha (or 20% of the total area; 17% slight to moderately affected; 3% strongly affected) is affected by desertification (Thomas, 1995). While 467 mha was estimated to be susceptible to water erosion, 432 mha was suspected to be under wind erosion. Dregne and Chou (1992) and Dregne (2000) estimated that approximately 3% area of the drylands are irrigated croplands, 9% are rainfed croplands, and 88% rangelands. According to them about 30% of the irrigated croplands, 47% of the rain-fed croplands and 73% of the rangelands are degraded. They suggested that rehabilitation of the 18.60 million km^2 desertified area would cost at least \$213 thousand million to make them profitable at a cost-benefit ratio of 1:2.5 (Dregne and Chou, 1992; Dregne, 2000). Thomas (1995) suggested that only 20% of Asian drylands are light to moderately degraded, and 2% strongly. Based on the interpretation of global soil datasets Eswaran *et al.* (2001) suggested that ~7.1 million km^2 area was under low risk of human-induced desertification, 8.6 million km^2 at moderate risk, 15.6 million km^2 at high risk and 11.9 million km^2 at very high risk, the latter two zones representing potentially high tension zones.

Changing Paradigms and their Bearing on Measurement Needs

Defining desertification is becoming increasingly difficult with time. Since its first use, the term has undergone a number of etymological shifts and intended meaning, but a universally acceptable definition is still eluding the scientific community. Herrmann and Hutchinson (2005) discussed the changing contexts of desertification debate, while Baartman *et al.* (2007) provided a literature review of current research on desertification processes and solutions, with emphasis on the results from the European Union projects on desertification.

Over the last few decades the focus of desertification studies has gradually shifted from a dominantly biophysical-process-based enquiry to an amalgamated biophysical-cum-socio-economic analysis, the aim of the studies, especially those under public funding, changing from an exclusively land degradation control-oriented approach to a broader sustainability and livelihood oriented approach. This paradigm shift in the concept has encouraged a shift from a science-led top-down approach to a science-led participatory and societal approach (Kirkby, 2009), and scientists are increasingly encouraged to focus more on policy-relevant research (Vogt *et al.*, 2009 a,b).

One of the current desertification pathways, suggested by Safriel and Adeel (2005), considers that apart from drought, the non-sustainable management and land uses within the drylands often lead to over-exploitation of the natural resources and a decline in productivity, the consequent desertification, followed by poverty and migration. By contrast some researchers recognise that the typical land uses of the desert and dryland dwellers provide some of the best possible strategies to cope with the uncertainties, and that desertification does not automatically follow from those land uses, but is more contextual (Bharara, 1993; Mortimore, 1998; Warren, 2002).

Reynolds *et al.* (2007) suggested a different pathway of interaction between the demographic and socio-political drivers and the inherently low productivity of the drylands, and termed it as the dryland livelihood paradigm (or dryland development paradigm). According to them, it is necessary to identify and measure the biophysical and socio-economic factors (drivers) that put pressure on the land resources and cause significant degradation, leading to a longer-term fall in production from the land. To model the system properly it is necessary to perceive that the dryland human-environment systems are coupled, dynamic and co-adapting, so that their structure, function and inter-relationships change over time. It is also necessary to recognise that the systems, including the scales of processes, occur in nested, hierarchical planes, necessitating their operations over multiple scales, and that often the critical variables that determine the system dynamics are much slower than the processes that lead to geological 'disasters'. Finally, the thresholds of the variables are not to be considered as static but changing with time, which may lead to some unexpected outcomes like faster degradation or slower recovery (Reynolds and Stafford Smith, 2002; Reynolds *et al.*, 2007). Reynolds *et al.* (2011) suggested that desertification could be better understood through an integrated assessment model for the coupled human-environment dryland system, especially through integration of sub-models on environmental, social, economic and institutional variables, which may effectively link the scientific research with policy development.

A third paradigm, called the 'counter-paradigm' was proposed by Safriel and Adeel (2008) who suggested that the land users, to meet the challenges of inherently low productivity and other constraints, would apply innovative means to surmount the problems of desertification, poverty and other consequences. The efforts would be supported and sustained by the policies and infrastructural facilities, and thereby make the system more sustainable than before.

The debate reached a high point when a draft White Paper on monitoring and assessing desertification (for CCD's consideration) proposed that "desertification is best to be treated as an extreme case of land degradation, which is expressed in a persistent reduction or loss of biological and economic productivity of lands that are under uses by people whose livelihoods depend on this productivity, yet the reduction or loss of this productivity is driven by that use" (Vogt *et al.*, 2009a, p. 25). The proposal adds to the prevailing confusion, and attempts to shift the focus away from the Sahel and the non-drylands and by implication from a large number of poor nations. Possibly it also reflects the frustration of some alienated segments of the dryland dwellers for not being counted in the global effort. The second draft of the document has slightly modified the definition as "desertification is an end state of the process of land degradation; this process is expressed by a persistent reduction or loss of biological and economic productivity of lands that are under uses by people. The livelihood of many of these people at least partly depends on this productivity, yet the reduction or loss of the productivity is driven by its use" (Vogt *et al.*, 2009b, p. 25). The confusion still persists.

Despite the prevailing confusion on definition, the simplicity of the 1992 definition of desertification helped researchers to bring in sharp focus the special environmental and socio-economic problems of the drylands, and more specifically of the arid lands (including the hyper-arid lands) to the policy planners and the global financial institutions. It triggered gradual development of an assessment and monitoring system for land degradation in the drylands of many developing countries, as well as of the availability of funds for improvement in socio-economic conditions and the environment.

At UNCCD, desertification is not viewed now as a problem of land degradation alone and counter-measures for it, but as a larger societal problem that creates conditions favourable for such degradation, and hence needs to be addressed through overall improvement in the living condition and economic well being of the people. As part of this paradigm, all affected parties are to prepare a suitable National Action Programme and monitoring mechanism through appropriate framework for policy, legal, strategy, programme, institutional development and R&D. Themes like poverty alleviation, land reform, education, infrastructural development, etc., are therefore, now as important an element of the counter-measures in the affected countries, ncluding India, as soil and water conservation, wind erosion control, afforestation, etc. (Anon., 2001).

Indicators and Benchmarks for Measuring Desertification

It is recognised by researchers that at least the following biophysical processes need to be monitored for assessing desertification: wind erosion/ deposition, water erosion, salinisation and waterlogging, and degradation of vegetation. According to Dregne (1991) the processes are to be monitored in rainfed croplands, irrigated croplands and rangelands. Kharin *et al.* (1985) and Babaev *et al.* (1993) recognise that technogenic activities are important processes. There is, however, some confusion on what feature is to be measured to recognise a certain category of degradation (state indicator), as well as reasons for it (pressure) and responses of the society to control it. The basic reasons for the difference lie in the complexities or ambiguities in evaluation of the indicators, wide differences in the nature of problems in different regions, as well as large variations in the database collection system and availability of necessary infrastructures.

There is general agreement among researchers that a proper assessment and monitoring of desertification is possible only when these are based on key indicators and reliable benchmarks. Unfortunately, there is hardly any consensus on the appropriate indicators and benchmarks that can be applied at global, regional, sub-regional, national and local levels. The reliability of the methods is often questioned. Dregne (1983) and FAO (1984) described some of the most useful early methodologies for quantifying desertification. Grunblatt *et al.* (1992) showed a novel way of using quantified information in a GIS framework. A selection of the case studies, and the indicators on which these are based, is provided in UNEP's Atlas of Desertification (Middleton and Thomas, 1997).

Shifts in Measurement Domains of Desertification

With time, some shifts are taking place towards developing and adopting unbiased quantitative measurement techniques, which are helping to get spatially more reliable datasets without relying much on the very sparsely located individual observation sites. For example, the need for establishing monitorable 'benchmark' field sites has largely been dispensed with after the availability and use of moderate- to high-resolution satellite images that spread the benchmarks to all the pixels in an image. In fact, digital remote sensing is fast replacing the rigorous field methods in assessing the status of several biophysical indicators. Another major shift from the views held till the mid-1980s is on vegetation condition.

There is a growing feeling among researchers that desertification assessment should be based on a set of well-defined and measurable indicators at periodic intervals, and that measurement should include less of informed opinions, and more of the analysis of actual dimensions. While most of the biophysical variables are amenable to direct measurement, analyses of the socio-economic variables are largely based on informed opinions (Kar and Faroda, 1998; Kar and Takeuchi, 2003). This is one reason why socio-economic database does not usually figure much in the desertification modelling efforts, especially through GIS, save and except some weightage provided to the concerned factors.

The D-P-S-I-R Concept

Linking the causes and effects of desertification, as well as the societal response to desertification, is necessary for proper understanding of the facets of the problem and effective implementation of the counter-measures. It is now felt that apart from identifying the hot spots of desertification for control measures, there is a need to identify and measure the biophysical and socio-economic factors (drivers) that put pressure on the land resources and cause significant degradation, leading to a longer-term fall in production from the land. Towards this end, a conceptual model, consisting of Driving force (D), Pressure (P), State (S), Impact (I) and Response (R), which was first proposed in 1979 (Anon., 1979), has been useful in assessment of the problems, especially in the drylands of Europe (Enne and Zucca, 2000). While D and R represent the socio-economic factors, P, S and R represent the behaviour of the natural system. The P-S-R model was used earlier for the African situation. Although conceptually mature, implementation of the D-P-S-I-R model is not an easy task, given the complex relationships between the indicators involved. Improvement in the socio-economic indicators and linking those indicators with appropriate biophysical indicators is still a researchable matter and success is possible through modelling under a coupled GIS-remote sensing environment (Kar and Takeuchi, 2003; Kar, 2004).

Researches on developing realistic models to factor our understanding of the D-P-S-I-R variables properly are continuing at many learned centres, but such models demand a large database. Hopefully, once developed, the models will help in providing robust early warning of desertification (Kar and Takeuchi, 2003). One also feels that modelling the actual desertification is different from the modelling of risk of desertification. In many cases, the outputs speak of the risk alone, because the inputs to the models are the potential values based on theoretical calculations, rather than the actual values.

Indicators for Measuring the Implementation of Combating Activities

To effectively measure the progress of achievements of the implementation of strategic plans of the UNCCD at national to global scales, a number of impact indicators are now being evolved through scientific dialogue (Orr, 2011). Presently, the following 'General Indicators' are considered measurable: 1. Water availability per capita; 2. Change in land use; 3. Proportion of population above the relative poverty line; 4. Food consumption; 5. Capacity of soils to sustain agro-pastoral use; 6. Degree of land degradation; 7. Plant and animal biodiversity; 8. Drought index; 9. Land cover status; 10. Carbon stock above and below ground; 11. Land under sustainable land management.

These general indicators are expected to facilitate the measurement of following five 'Core Indicators': 1. Improvement in the livelihoods of people potentially impacted by the process of desertification/land degradation and drought (measured through general indicators 1, 3 and 4); 2. Reduction in the total area affected by desertification/land degradation and drought (2, 5, 6 and 8); 3. Maintenance of or increases in ecosystem function, including net primary productivity (7); 4. Increases in carbon stocks (10); 5. Areas of forest, agricultural and aquaculture ecosystems under sustainable management (5, 10 and 11).

These five core indicators will help to measure the effective implementation of the following three 'Strategic Objectives': 1. To improve the living conditions of affected populations; 2. To improve the condition of ecosystems; and 3. To generate global benefits through effective implementation of UNCCD.

If and when the above hierarchical and integrative scheme of indicator measurement is accepted, a series of agreed formulae will be required to provide the UNCCD a scientific basis of reasoning to the desertification conundrum and to guide and monitor properly the combat outcome through the principle of D-P-S-I-R. While on the drawing board, the measurement scheme is likely to raise several questions on methods of measurement, appropriateness, leveling issues in terms of D-P-S-I-R, national sensitivity, data availability, ambiguity of meaning, etc. For example, the degree of land degradation is a result of a number of causal factors that are complex and intertwined, taking into consideration many of the general indicators that have been put in the same plane with it. Determination of the poverty line in India has become a sensitive issue with widely differing perspectives of the planners and the commoners. Sustainable land management is in the realm of a concept that is still difficult to measure. Similarly, problems may arise in defining the capacity of a soil to 'sustain' agro-pastoral use, or in linking the status of biodiversity with desertification. Hopefully, these and other issues of concepts and measurements will be resolved soon and a universally acceptable and measurable system will be available for implementation. Nevertheless, this exercise is a huge step forward in implementation of the UNCCD's basic objectives through scientific reasoning.

Despite the awareness, the researches, and the funding for development for decades, drylands still continue to be the most handicapped among the major ecosystems, especially in Africa. Land productivity and human well-being in many of the global dryland countries remain much below the potentials available through technological interventions (Adeel *et al.*, 2005).

Indicators of Desertification : Examples from Arid Lands of India

State Indicators

Our studies in the arid lands of India suggest that some terrain features can be used as state indicators of desertification, provided we understand the spatial pattern of process-form interactions, and vulnerability of the landscape (Kar, 2009a). It is commonly known that the erosional and depositional processes leave their imprints on landforms. In the arid areas some of the effects of the processes stay for few days to few years. These include low sand streaks, nebkhas (also called shrub-coppice dunes) and small barchans, the areas of scalding, new flood deposits along ephemeral channels, etc. The occurrence of nebkhas in sandy plains is now recognised as a good indicator of recent sand deposition (Nickling and Wolfe, 1994; Tengberg, 1995; Kar, 2006). Some other features like stream bank erosion marks, gully networks, megabarchans, etc., may remain for tens to hundreds of years. The long-term processes leave their marks for thousands of years (e.g., old dunes, alluvial plains and fans, other planar surfaces, hills and uplands). Such elements of time and scale are inbuilt in the system, and create a hierarchical arrangement of the land units in a landscape system (Baker, 1986; Prince, 2002). Therefore, it is possible to look into the landscape system a nested arrangement of landforms as well as a nested pattern of processes (mainly the exogenic), which are working at different scales and periods (Kar, 1990, 1993a). It is the short-term processes and landforms which should be considered for indicators of desertification (Kar, 1996a), but inheritance of landscape properties from past processes need also to be considered (Kar, 1986, 1992; Singh *et al.*, 1994). Finding out which landform feature is a result of human activity and which one is due to natural factors is a difficult proposition under many situations, especially due to the fact that different driving forces lead to the same or identical state of landscape (e.g., degraded landscape in Kachchh Mainland that is primarily due to tectonic events; Kar, 1988, 1993b).

We tried to find out some critical field indicators, which could help in assessing desertification in the Indian part of the Thar Desert that occurs in the western part of Rajasthan state. The indicators were chosen with a view to recognise them primarily in the field, but also on aerial photographs and satellite imagery (to the extent possible). The climatic gradient in the desert, from wetter east to drier west, was kept in mind so that the variations in efficiency of different processes from east to west and spatial variability of their impact on landscape are understood before the indicators are used for scaling the severity of particular degradation process. The terrain features, indicators of assessment and suggested degree of severity for each of the three major biophysical degradation processes, i.e., wind erosion/ deposition, water erosion, and salinity/alkalinity are provided in Tables 1-3 (modified after Singh *et al.*, 1992). Criteria for measuring vegetation degradation are given in Tables 4 and 5.

Table 1 : Field indicators for assessing wind erosion/ deposition in the Thar Desert

Terrain	Average rainfall (mm)	Major indicators for assessment	Severity
Flat sandy plains with dominantly loamy sand to sandy loam soil	100-550	Fresh sand sheet up to 30 cm thick; few scattered new fence line hummocks and nebkhas up to 100 cm high	Slight
Moderately sandy undulating plains and sand dunes with loamy sand soils; thickly sand sheeted plains	Above 300	Presence of reactivated fresh sand of 50 to 150 cm thickness on stable dunes, sandy plains and fence line hummocks; many recently formed nebkhas	Moderate
Moderately sandy undulating plains and sand dunes with sand to loamy sand soils	Below 300	Reactivated and fresh sandy hummocks (nebkhas) and sand ridges of 90-300 cm height; sand sheets of 60-150 cm thickness between undulations; reactivated stable dunes with fresh sand deposits of 70 to 200 cm thickness; exposed plant roots to a depth of 40 to 100 cm in the sandy plains indicate erosion	Moderate
Moderate to strongly undulating sandy plains with closely spaced hummocks and high sand dunes with sand to loamy sand soils	100-550	Closely spaced sandy hummocks and sand ridges of 1 to 4 m height with fresh sand cover; sand deposits of 100-300 cm thickness usually present between undulations; highly reactivated sand dunes are covered by fresh sand and superimposed by crescentic bedforms of 2 to 4 m height	Severe
Barchan dunes and very thick sandy plains with loose sand throughout the profile	100-550	Areas of drift sand, especially as fields of barchans of 2 to 5 m height, which encroach upon roads, settlements and agricultural fields; also areas with very closely spaced nebkhas of 2-5 m height	Very severe

Table 2 : Field indicators for assessing water erosion in the Thar Desert

Terrain	Average rainfall (mm)	Major indicators for assessment	Severity
Plains with 30 to 90 cm deep medium to fine textured soils and good stand of tree vegetation	Above 300	Sheet erosion imperceptible, but rills of 0.3 to 0.5 m depth and 0.4 to 0.9 m width are usually present	Slight
Plains with 7 to 20 cm deep soils and poor stand of 300 vegetation	Below 300	Carbonate nodules exposed in pockets due to sheet erosion; widely spaced rills are present	Slight
Gently undulating plains with sand to loamy sand and sandy loam soils	Above 300	Sheet erosion as evidenced by few pebbles and carbonate nodules on the surface. Soil depth is generally less than 30 cm	Moderate
Hills, rocky/gravelly plains and sandy undulating plains with sand, sand and gravel and loamy sand soils	Below 300	Rills of 0.5 to 1.5 m depth and 1.0 to 2 m width; widely spaced gullies of 4-8 m depth and 6-15 m width	Moderate
Moderately sloping uplands, rocky/ gravelly surfaces and gentle to moderately sloping plains with sand, sand and gravel and loamy sand to sandy loam soils	Above 300	Exposed carbonate nodules and pebbles, suggesting complete stripping of soil through sheet wash; gullies of 6-10 m depth and 7-25 m width with narrow interfluves	Severe
Steeply sloping hill ranges, associated obstacle dunes and sandy undulating plains	Above 300	Closely spaced gullies of 10-15 m depth and 20-40 m width; almost no mappable interfluves	Very severe

There is an element of subjectivity involved in application of the above indicators for assessment, as many of the features cannot be rigorously measured over a large area, and interpreter's experience (and perhaps a personal bias) will be involved in discrimination of the units. Problems of equating rills and gullies everywhere with desertification are yet to be solved. Nevertheless the scheme attempts to remove some of the shortcomings in earlier efforts.

Table 3 : Field indicators for assessing salinity/alkalinity and waterlogging in the Thar Desert

Terrain	Average rainfall (mm)	Major indicators for assessment	Severity
Flat older alluvial plains and interdune plains with sandy to loamy and clay loam soils, underlain by salt-releasing formations (e.g. weathered granite)	100-500	Salt concentration in soil profile (EC) is 4-8 dSm^{-1}; irrigation with saline/ non-saline ground water; few salt encrustation visible during dry seasons	Slight
Flat older alluvial plains and interdune plains with sand to loamy sand soils, underlain by less permeable or impermeable strata	100-500	EC in soil profile is 4-8 dSm^{-1}; canal and tank irrigation; poor soil drainage; few salt encrustations visible during dry seasons	Slight
Flat older alluvial plains and interdune plains with sandy loam to clay loam soils underlain by nodular calcrete zone and salt-releasing formation, or by other less permeable and impermeable strata	100-500	EC in soil profile is 8-16 dSm^{-1}; ground water table at 2-3 m depth; especially along the canals where ground water shows rising tendency; poor crop performance; numerous patchy salt encrustations during dry seasons	Moderate
Low lying and depression areas in older alluvial plains and interdune plains with sandy loam to clay loam soils, underlain by impervious layer	100-500	EC in soil profile is >16 dSm^{-1}; ground water table within 2 m of the surface; very poor soil drainage; partial waterlogging; very poor crop performance	Severe
Waterlogging			
Low lying areas between sand dunes and buried courses of drainage channels with loamy sand to sandy loam soils, underlain by impermeable strata.		Water logging in pockets, more area is likely to be waterlogged due to rapidly rising ground water table to the critical limit.	Severe
-Do-		The deep standing waters in the cultivated fields which have been completely desertified in terms of agricultural productivity	Severe

Table 4 : Field criteria for assessing desertification due to vegetation degradation in rangelands

Terrain characteristics	Parameters for assessment of vegetation degradation	Degradation score from rating of parameters	Severity
Dominantly dune covered, or sandy undulating, or rocky and/or shallow soils areas	Botanical composition of climax species (%), total basal cover, existing biomass as percentage of potential biomass; basal cover (%) of climax species	5-8	Slight
Dominantly dune covered, or sandy undulating, or rocky and/or shallow soils areas	Botanical composition of climax species (%), total basal cover, existing biomass as percentage of potential biomass; basal cover (%) of climax species	9-12	Moderate
Dominantly dune covered, or sandy undulating, or rocky and/or shallow soils areas	Botanical composition of climax species (%), total basal cover, existing biomass as percentage of potential biomass; basal cover (%) of climax species	13-16	Severe
Dominantly dune covered, or sandy undulating, or rocky and/or shallow soils areas	Botanical composition of climax species (%), total basal cover, existing biomass as percentage of potential biomass; basal cover (%) of climax species	17-20	Very severe

Table 5 : Rating of parameters for assessing vegetation degradation

Parameters	Rating				
	1	2	3	4	5
	Parameter values				
Botanical composition of climax species (%)	40-25	25-10	10-1	1	0
Total basal cover (%)	8-4	4-3	3-2	2-1	<1
Existing biomass as percentage of potential biomass	>60	60-40	40-25	25-10	<10
Basal cover (%) of climax species	>50	50-25	25-20	20-5	<5

Vegetation degradation is now assessed from NDVI and other vegetation indices derived from satellite imagery. Ustin *et al.* (2005) provide a summary of major remote sensing based indicators developed so far. Although vegetation is a major casualty in desertification, caution is required in assessing vegetation degradation. This is especially in view of the long debate on whether the changes in vegetation cover in the Sahel represented natural spatio-temporal changes, related to short-term climatic variability, or long-term land degradation, as well as other related debates. There is now a growing belief that some of the earlier conclusions on desertification, based on an analysis of changes in vegetation status (through NDVI, etc.), especially in the African continent, were misleading. It is felt that those changes reflected some temporary adjustments of the vegetation communities to short-term spatio-temporal variability of climate, and did not suggest any long-term land degradation (Ahlcrona, 1988; Warren and Khogali, 1992; Olsson, 1993; Thomas, 1993).

In the context of common grazing lands in western Rajasthan, which are severely degraded due to long continued overuse, the collection of required data is more difficult as benchmarks for pre-degradation state are difficult to establish. The resilience of the system needs also to be given proper weightage, because numerous studies confirm that a slight protection to the landscape and some reseeding dramatically improves the natural vegetation cover and composition (Saxena, 1977; Kumar *et al.*, 2009).

There are other kinds of land degradation in Thar Desert, like waterlogging which is a major problem in canal command areas, and which can be easily identified from the marshy lands and pools of water, or from the information on trends of rising ground water. Fortunately, its expansion is now under some control. Mine wastes and industrial wastes are becoming significant with growing industrialization and urbanization (Kar *et al.*, 2009). Sodicity and soil compaction have developed in some soils due to use of ground water with high residual sodium carbonate (RSC). Irrigation with water containing >5 me L^{-1} RSC in the rainfall zone of 200 to 250 mm average annual rainfall usually leads to sodicity, but the sodium absorption ratio (SAR) of the soils is also to be considered (Joshi and Dhir, 1991).

Pressure Indicators

As we have argued in earlier paragraphs, acceleration to the processes due to natural and human activities lead to the formation of some terrain features that can be taken as state indicators of desertification. In many cases the pressure indicators for the features noted can be suggested from knowledge of the natural geomorphic processes and kind of human pressure. There is, however, no thumb rule in this field. Factors are to be evaluated critically. For example, a sudden increase in bare sand thickness may not always be a result of faulty land use; the feature can form even independent of any land use change, especially during severe sandstorms. In the semi-arid areas rill and gully erosion usually take place in pulses of fast advancement, separated by years with almost no advancement. The processes are activated during periods of high-intensity above-normal rainfall. A good vegetation cover can minimise the rate of advancement of rill and gully heads, but under a constantly degrading environment it is difficult to say if land use during a particular period can be held responsible for the measured advancement. It is most likely that rainfall amount and rainfall intensity will determine the rate of rill and gully formation, while the sites of their occurrences may be linked to fractures and other weaker zones within the sediments.

In the tectonically mobile areas like Kachchh Mainland the high density of rills and gullies and the high rates of soil erosion are associated with a slow uparching of the land between the Arabian Sea in the south and the Great Rann of Kachchh in the north (Kar, 1993b). Interestingly, the vast Banni natural grassland of the relict mudflat adjoining the Kachchh Mainland in its north, has a typical east-to-west disposition of belts of high and low salinity that have come up due to the slow rise of the near-shore deposits of the erstwhile remnants of the receded Arabian Sea along defined east-west lineaments. Digital terrain modelling has shown at least two E-W oriented higher surfaces across the Banni, that are separated by lower surfaces. On the higher surfaces the slope is steeper towards north and gentler to the south, producing mini-cuestas and mimicking similar slopes in the Kachchh Mainland and the Pachham Island. These are the results of repeated earth movements along the E-W lineaments during the late Quaternary period. Since the movements are also controlled by a 'hinge zone' running through a N-S belt across Mandvi-Mundra in the south to Pachham Island in the north, the Banni also gets tilted down westward with each earth movement and gets gradually drowned in that direction. At the same time, its eastern part gets elevated, the maximum being along the hinge zone through Bherandiala and Bhojardo. This complex and dynamic landscape history has developed three clearly defined zones of high-level, mid-level and low-level mudflats with successively higher salinity and lesser suitability for grassland down the slope (Kar, 1993c). Even though invasion of *Prosopis juliflora* bushes into the system is supposed to have degraded the grassland through aggressive colonization, the space-time narratives of the colonization vis-à-vis salinity level and grassland potentials need to be understood in terms of Banni's terrain dynamics.

Considering that many biophysical indicators can develop from a variety of causes it is difficult to prejudge the features and link them to a particular pressure indicator. It is more realistic to search for reasons after an indicator is marked in an area.

Socio-economic Indicators

So far the major contribution of research has been towards finding out the biophysical indicators of desertification. Socio-economic indicators have received very little attention, save and except some studies to gather opinions of village folk regarding traditional knowledge on drought and its mitigation, or the impact of desertification on the socio-economic fabric. Dregne (1983) described some socio-economic indicators like measuring changes in land use and herd composition, child health, utilization of fuelwood, local perceptions, etc., which need to be elaborated.

An attempt was made to identify few socio-economic indicators of desertification in a village called Khabra Kalan in the central Thar. From opinion surveys and interpretation of secondary data it transpired that the use of tractors for dryland agriculture in a sandy terrain was a major reason for wind erosion, but hiring a tractor for few hours was cheaper than to keep draught animals for a year. In other words, economic considerations override the environmental considerations. Number of people using tractor in a sandy terrain within high wind erosivity zone may be considered as a socio-economic indicator, while tractor ploughing becomes a major responsible process (pressure indicator).

Villagers considered that encroachment of grazing lands and other marginal lands for cultivation have led to poor production and accelerated erosion. The number of adult cattle units (ACUs) per hectare of grazing land increased manifold in the village, and so did the pressure of fuelwood and timber collection. At the same time range care was absolutely nil, thus leading to lesser biomass production and increased vegetation degradation. In other words, the valued judgment of the villagers suggested that decreasing size of grazing lands, and collection of fuelwood and timber from local sources could be used as socio-economic indicators of the state of degradation. Since these are related to human and animal pressures, increase in the number of human and animal population may be considered as pressure indicators.

To keep the land fallow and to practice rotational cropping are now almost impossible, especially due to increased population and consequent fragmentation of land holdings. Opinion survey revealed that the present average per family agricultural land in the village was 10.1 ha and that too in many fragmented segments; in the previous generation the average holding was 17.8 ha (few fragmentation), and before that 44.0 ha (almost no fragmentation). These three generations together accounted for a period of about 50 years. As the land holdings became smaller and more fragmented the farmers had little choice to select land parcels for cultivation. Fallowing, therefore, had to be abandoned, and whatever land holdings one was left with had to be ploughed for subsistence. In the absence of adequate farmyard manure and fertilizers the nutrient status of the land became precarious, and so the production started showing a declining trend. The villagers opined that the decline in yield from shallow soils was more dramatic.

In order to find out the reasons we carried out some tests on the soils, which indicated that available P in the shallow soils declined sharply with continuous cultivation, and could not be replenished as quickly as available N. Because of the depth limitation, the negative rate of replenishment meant a gradual impoverishment of the nutrient budget and consequent decline in yield. The deep sandy soils had a higher reserve of available P and could withstand the pressure of continuous cultivation for a longer time (Tsunekawa *et al.*, 1997). The opinions of the farmers were thus validated. Once the views of the villagers were evaluated scientifically, it was possible to suggest continuous cultivation of shallow soil as a pressure indicator and decline in crop yield as a state indicator. Similarly, the decreasing length of period under fallow and fragmentation of land holdings were the immediate causes for possible decline in crop yield. Once the statement is verified scientifically, these factors, as well as the population increase that lead to such exigencies, can be termed as pressure indicators, while the decline in crop yield that speaks for the productivity status of the land, can be designated as state indicator. In other words, a number of pressure indicators can lead to the same state indicator.

Farmers in the village rated continuous cultivation as a major cause of land degradation, followed by tractor ploughing and indiscriminate cutting of trees and bushes. There are valid reasons to consider these as indicators of desertification in a similar terrain, provided there is no counterbalancing by other factors. Many such opinions can be tested and then used as socio-economic indicators. Our numerous interactions with villagers across the desert suggest that most of the opinions of rural folk have sound scientific basis. Some of these can be tested with modern scientific tools and then used to improve the data obtained from independent measurement of biophysical indicators.

Improvement in the socio-economic indicators and linking those indicators with the biophysical indicators are researchable matters, which need attention. A mechanism has to be developed for providing a closer link between the two kinds of indicators, with a feedback mechanism. Despite this awareness among researchers, however, fewer studies have been carried out on the socio-economic aspects of desertification, including a critical examination of the conceived ideas on impacts of human use of the land. Another problem is to attach a scale of severity to the social indicators, like one does in the case of biophysical indicators. Some of the notable studies include, e.g., Tiffen *et al.*, 1994; Swift, 1996; Mortimore, 1998; Amissah-Arthur *et al.*, 2000; Lambin *et al.*, 2001; and Warren, 2002.

Desertification Mapping in India

The first mapping of desertification in India was carried out by Central Arid Zone Research Institute (CAZRI) in 1977 for the UN Conference on Desertification (UNCOD) at Nairobi. The processes of desertification in arid western Rajasthan were mapped at 1: 2 million scale as per the guidelines and nomenclature supplied by the UN. Simultaneously an assessment of desertification in the region was carried out, which revealed that due to climate change the desert boundary had fluctuated more than once during the Quaternary period from its present position along the Aravalli Ranges. This was attested by numerous semi-stabilized sand dunes and relict fluvial landforms both within and outside the present desert boundary, bearing testimony to the past aridity and landscape stabilization. Signatures of a slight amelioration of climate in the desert during the Twentieth Century were also found along the eastern fringe in the forms of higher fluvial dissection of the sandy landscape. At the same time, the study identified numerous landscape signatures of increasing human and livestock pressures within and outside the desert, especially since the mid-twentieth century (Ghose *et al.*, 1977). Through aerial photo interpretation and field mapping, a detailed case study on changes in landscape properties due to desertification was carried out in a small representative administrative block of Jodhpur district, the Luni Development Block. It was estimated that between 1958 and 1977 approximately 1990 ha of land was affected by wind erosion/deposition, and that another 18,000 ha land, mostly in the sandy undulating terrain, could be affected in 'near future' if the deep ploughing of dune slopes, overgrazing of sandy terrain and indiscriminate removal of trees and shrubs continued unabated (Singh *et al.*, 1978).

Almost from that time a number of attempts were made to assess the extent of land degradation at the country level, irrespective of whether they existed in the drylands or not. The assessments have been made by different organisations (e.g., scientific institutes, state line departments, ministry officials, etc.), using different sets of criteria (perceived degree of erosion/ deposition, etc. from terrain features; reliance on secondary data, etc.). Most of the early assessments were based on compilation of the different kinds of low-yielding or waste/ barren lands at state level, which are mentioned in the land revenue records of the villages. Visual remote sensing of land use/wasteland categories formed part of some the assessments. Estimates of land degradation in India, therefore, vary from 53 mha to 187 mha (Samra and Sharma, 2005; Table 6).

Table 6 : Estimates of land degradation/wastelands in India

Agency (with year)	Extent (mha)	Reported method
National Commission of Agriculture (1976)	148.09	Secondary information
Ministry of Agriculture (1978)	175.00	Secondary information
Society for Promotion of Wastelands Development (1984)	129.58	Secondary information
National Remote Sensing Agency (1985)	53.28	Visual remote sensing of wastelands
Ministry of Agriculture (1985)	173.64	State-level statistics
Ministry of Agriculture (1994)	107.43	State-level statistics (revised)
NBSS&LUP (1994)	187.70	GLASSOD

Harmonization of Heritage Data at National Level

Harmonization of the different datasets from different agencies has been arrived at recently by a team of scientists from Indian Council of Agricultural Research (ICAR), namely CAZRI (Jodhpur), CSSRI (Karnal), CSWCRTI (Dehradun), NBSS&LUP (Nagpur), as well as from National Remote Sensing Centre (NRSC), which suggests that 120.72 mha of the country is degraded (Anon., 2010; Table 7).

Table 7 : Harmonized estimates of degraded and wastelands in India

Degradation category	Arable land (mha)	Open forest (mha)	Total (mha)
Water erosion (>10 t/ha/y)	73.27	9.30	82.57
Wind erosion	12.40	-	12.40
Exclusively salt-affected	5.44	-	5.44
Salt-affected & water eroded	1.20	0.10	1.30
Water logging	0.88	-	0.88
Exclusively acidic soils	5.09		5.09
Acidic and water eroded soils	5.72	7.13	12.85
Mining/industrial wastes	0.19		0.19
Total affected	104.19	16.53	120.72

Systematic Mapping of Desertification based on Remote Sensing

In 1991 CAZRI prepared a 1:1 million scale map of desertification that was based on visual interpretation of 1:250,000-scale false colour composites of Landsat-TM images of western Rajasthan for 1989-1991, which was backed up by extensive field information. A set of quantified parameters was developed as indicators of the biophysical processes as well as interpretation keys for identification of the degradation in field (Tables 1-5). The degree of severity of degradation was assumed to reflect qualitatively the state of biological production as follows: i) slight - almost no decline in biological production under normal weather situation; ii) moderate - loss of production up to 25 percent of the expected; iii) severe – loss of production between 25 and 50 percent of the expected, and iv) very severe - loss of production more than 50 percent of the expected (UNEP, 1977).

Based on the parameters, image patterns on FCCs and available field knowledge, a two-stage mapping scheme was followed. The first stage involved mapping of the dominant land uses at 1:1 million scale, showing irrigated croplands, un-irrigated croplands, rangelands and degraded forests, with superimposed average annual isohyets to suggest the control of rainfall on spatial distribution of major land uses. The second stage involved mapping of desertification processes at 1:250,000 scale, which was finalised at 1:1 million scale. According to the above mapping, out of 20.875 mha area mapped, 19.175 mha (92% of the total mapped area) was degraded, and the rest was free from any degradation. While 33% area was slightly affected, 35% area was moderately affected and 24% severely. Wind erosion constituted the largest area (~76% of the total mapped; 25% slightly affected, 35% moderately and 15% severely). Water erosion was mapped in 15% area, while water logging and salinity was mapped in 6% area (Singh *et al.*, 1992, 1994). The 1:1 million scale desertification map of western Rajasthan was updated by CAZRI in 2005 with minor modifications, and was published in Hindi and English (CAZRI, 2005).

Systematic Land Degradation Mapping at Country Level

In view of the non-availability of a reliable country-wide land degradation status map, a task of desertification/ land degradation status mapping (DSM) was taken up in 2004 by the Space Applications Centre (SAC), in association with several national institutes, including Central Arid Zone Research Institution. To develop a methodology for country-wide mapping, pilot studies were first carried out in 16 test sites spread over the drylands of the country, and including both hot and cold arid areas. Taking guidance from earlier mapping criteria, and using the Indian Remote Sensing satellite (IRS-1D) Linear Imaging Self-scanning Sensor (LISS-III) images, the maps were prepared at 1:50,000 scale (SAC, 2007a).

This was followed by framing of the methodology for 1:500000 scale mapping and its finalization at 1:1 million scale map, keeping in view the sensor limitations of the Advanced Wide Field Sensor (AWiFS) on board the Indian Remote Sensing Satellite, IRS-P6-Resourcesat, which was to be used for mapping land degradation in the major land uses. Efforts were made to minimize the classification units so that the map becomes easy to read. A three-level, hierarchical classification system was prepared that included land use/land cover (level 1), processes of degradation (level 2), and severity of degradation (level 3). Altogether 12 land use/ land cover categories, 13 degradation processes and 2 severity classes (high and low) were identified for mapping.

Conceptually it followed the earlier mapping procedure of CAZRI. It was recognized that due to sensor limitations, distinction between slight and moderate degradation would be difficult through visual interpretation. Therefore, only two categories of severity were identified and mapped: low (corresponding to slight and moderate degradation) and high (corresponding to severe and very severe degradation). Cloud-free satellite data for Kharif, Rabi and summer seasons of 2003, 2004 and 2005 were used. Ground truth was collected to verify the mapped features. Tables 8-10 show the classification system used for the purpose, while Tables 11-15 provide the indicators/criteria for the mapping (Anon., 2003).

Table 8 : Level-1 mapping of land use/ land cover for nation-wide mapping of desertification

Land use category	Mapping code
Agriculture - Unirrigated	D
Agriculture - Irrigated	I
Forest	F *
Grassland/Grazing land	G
Land with scrub	S **
Barren / Rocky area	B/R #
Dune / Sandy area	E
Water body / Drainage	W
Glacial / Peri-glacial (in cold region)	C/L
Others (Urban, Man-made etc.)	T

*Rocky areas within forest can be annotated as only FV3-R in the map.

**Vegetal degradation in Land with Scrub around notified forests can be delineated as SV.

Encroachment in forest area especially agricultural practices, is FV3

#Barren and Rocky areas to be delineated separately as B or R and shown in others category of the legend. B(Sc) to be used for indicating scree areas in cold deserts.

Table 9 : Level-2 mapping of processes of degradation for nation-wide mapping of desertification

Degradation category	Mapping code
Water Erosion	w *
Wind Erosion	E
Waterlogging	I
Salinization / alkalinization	s/a **
Vegetation degradation	V
Mass Movement (in cold areas)	G
Frost heaving (in cold areas)	H
Frost shattering (in cold areas)	f
Man-made (Mining/Quarrying, Brick Kiln, Industrial Effluents, City Waste, Urban Agg., etc.)	m

*Gully/ravines should be shown as Xw3, where X is the Land use / land cover class of surrounding area.

**Salinization or alkalinization should be shown as 's or 'a' separately. Where both occur, they should be shown together i.e. $s_x a_y$, where x and y are respective degree of severity.

Table 10 : Level-3 mapping of severity of degradation for nation-wide mapping of desertification

Slight	1
Moderate	2
Severe	3

Table 11 : Criteria for nation-wide mapping of the severity of water erosion

Status	**Desertification classes**		
	Slight	**Moderate**	**Severe**
	Non-arable land		
Type of erosion	SHEET EROSION AND/OR single rills (depth = 0.5 m and width = 0.4-0.9 m)	Rill erosion, and/ or formation of gullies (depth= 0.6-3.0 m and width= 1.0-3.5 m)	Network of gullies / ravines (In Gullies: depth 3 m =10 m and width= 3.5-20.0 m; in Ravines: depth= >10 m, width= 20-40 m)
Density of channels, linear km per sq. km	<0.5	0.5-1.5	1.5-3.0
Removal of top soil horizon, %	<25	25-50	>50
	Arable land		
Removal of top soil horizon	<25	25-50	>50
Loss of yield of main crop, %	<25	25-50	>50

Table 12 : Criteria for nation-wide mapping of the severity of wind erosion

Status	**Desertification classes**		
	Slight	**Moderate**	**Severe**
	Non-arable land		
Sand sheet in cm	< 30 cm per hummock up to 100 cm over plains	< 50-150 cm /stable dune and sandy hummock (east of 300 mm isohyet);< 90-300 cm/reactivated sand/plant roots up to 40-100 cm (west of 300 mm Isohyet)	< 1-4 m dunes/100-300 interdunal sand/ Barchans 2-4 cm (mostly west of 300 mm isohyet);2-5 m active/drifting dunes: very severe

Contd.

Percent area covered with sand dunes	<30	30-70	>70
Percent area covered with sod forming plants	50-30	30-10	<10
	Arable land		
Removal of top horizon, %	<25	25-50	>50
Blow-outs, percentage of area	<5	5-10	>10
Loss of yield of main crops, %	<25	25-50	>50

Table 13 : Criteria for nation-wide mapping of the severity of salinity / alkalinity

Status	Desertification classes		
	Slight	**Moderate**	**Severe**
Soil salinity/ alkalinity	4-8 dS/m; <15	8-30 dS/m; 15-40	>30 dS/m; >40
Soil salinization, solid residue, %	0.20-0.40	0.40-0.60	>0.60
Salinity of ground water, g/litre	3-6	6-10	10-30
Salinity of irrigation water, g/litre	0.5-1.0	1.0-1.5	>1.5
Seasonal salt accumulation, t/ha	16-30	30-45	45-90
Loss of yield of main crop, %	<15	15-40	>40

Table 14 : Criteria for nation-wide mapping of the severity of water logging

Status	Desertification classes		
	Slight	**Moderate**	**Severe**
Water logging	Seasonal (affecting one crop); 4-6 months	Affecting two crops; >6 months of submergence	Inland marshes

Table 15 : Criteria for nation-wide mapping of the severity of vegetation degradation

Status	Desertification classes		
	Slight	**Moderate**	**Severe**
Plant community	Climax or slightly changed	Long lasting secondary	Ephemeral secondary
Percentage of climax species	> 75	75-25	< 25
Decrease of total plant cover, %	< 25	25-75	> 75
Loss of forage, %	< 25	25-75	> 75
Loss of current increment of wood, %	< 25	25-75	> 75

The mapping revealed that leaving aside the islands, India had 105.48 mha (32.07% of country's area, excluding the islands) under different land degradation categories. Notable among these, water erosion covered 33.56 mha, followed by vegetation degradation (31.66 mha) and wind erosion (17.56 mha). In the Himalayas, frost shattering covered 10.21 mha. Among the arid and semi-arid states, Rajasthan had the maximum degraded area (21.8% of the country), followed by Jammu and Kashmir (12.8%), Maharashtra (12.7%), and Gujarat (12.7%). The total area affected by desertification within arid region was 34.89 mha, while that in the semi-arid region was 31.99 mha (SAC, 2007b). Details of the mapped desertification units in arid western Rajasthan and arid Gujarat are given in Tables 16 and 17 (Kar *et al.*, 2007a, 2009).

Realising the need for mapping land degradation in the country at 1:50000 scale, the National Remote Sensing Centre, Hyderabad, initiated in 2007, another mapping programme in association with CAZRI and several other national institutions. The maps will form a part of the National Natural Resources Census.

All the above classifications were based on visual interpretation of images and eye estimation of degradation in the field. An element of subjectivity was, therefore, involved, and interpreter's experience played a critical role in discrimination of the units. This, however, is a potential source of confusion in future monitoring, as the perception of degradation may not necessarily be the same as that of the earlier set of interpreters.

Table 16 : Desertification in different land uses in arid western Rajasthan (based on visual interpretation of 1:500,000 scale AWiFS satellite imagery (2003-04)

Land use	Type of degradation	Area (km²)	% of total mapped area
Cropland (rainfed)	Wind erosion (slight-moderate)	62868	29.88
	Wind erosion (severe)	12584	5.98
	Water erosion (slight-moderate)	1143	0.54
	Water erosion (severe)	39	0.02
	Salinity (slight-moderate)	1928	0.92
	Salinity (severe)	118	0.06
Cropland (irrigated)	Wind erosion (slight-moderate)	27906	13.26
	Wind erosion (severe)	380	0.18
	Water erosion (slight-moderate)	1350	0.64
	Salinity (slight-moderate)	527	0.25
	Salinity (severe)	2	0.00
	Water logging (slight-moderate)	177	0.08
	Water logging (severe)	6	0.00
Scrubland	Vegetation (slight-moderate)	4655	2.21
	Vegetation (severe)	61	0.03
	Wind erosion (slight-moderate)	11640	5.53
	Water erosion (slight-moderate)	926	0.44
	Water erosion (severe)	64	0.03
	Salinity (slight-moderate)	135	0.06
	Salinity (severe)	530	0.25
Sand dune/ sandy area	Wind erosion (slight-moderate)	17462	8.30
	Wind erosion (severe)	22046	10.48
	Water erosion (slight-moderate)	416	0.20
Forest	Vegetation (slight-moderate)	1066	0.51
	Vegetation (severe)	72	0.03
Barren land	Wind erosion (slight-moderate)	4675	2.22
	Salinity (severe)	1311	0.62
Mining	Mining (slight-moderate)	111	0.05
	Mining (severe)	91	0.04
Rocky		2129	1.01
Total degraded area		176418	83.85
Settlement		274	0.13
Not affected		33692	16.01
Water bodies /drainage		15	0.01
Total mapped area		210399	100.00

Table 17 : Desertification in different land uses in arid Gujarat (based on visual interpretation of 1:500,000 scale AWiFS satellite imagery (2003-04)

Land use	Type of degradation	Area (km^2)	% of total mapped area
Cropland (rainfed)	Wind erosion (slight-moderate)	1875.42	2.50
	Water erosion (slight-moderate)	9693.33	12.93
	Water erosion (severe)	373.29	0.50
	Salinity (slight-moderate)	3066.33	4.09
	Salinity (severe)	278.05	0.37
Cropland (irrigated)	Wind erosion (slight-moderate)	2072.99	2.76
	Water erosion (slight-moderate)	13128.49	17.51
	Water erosion (severe)	1618.08	2.16
	Salinity (slight-moderate)	46.66	0.06
Scrubland	Vegetation (slight-moderate)	3851.77	5.14
	Water erosion (slight-moderate)	4741.31	6.32
	Water erosion (severe)	2561.91	3.42
	Salinity (slight-moderate)	462.03	0.62
Forest	Vegetation (slight-moderate)	239.90	0.32
	Vegetation (severe)	2328.93	3.11
Grassland	Vegetation (severe)	1188.35	1.58
Barren land	Salinity (slight-moderate)	272.07	0.36
	Salinity (severe)	24460.60	32.62
Rocky		24.89	0.03
Total degraded area		72287.40	96.40
Settlement		50.73	0.07
Not affected		2464.25	3.29
Water bodies/ drainage		181.20	0.24
Total mapped area		74983.58	100.00

Looking Beyond Visual Interpretation

In order to minimize the operator bias and to maintain certain degree of precision while mapping large areas and integrating information from different variables, it is necessary to quantify the parameters systematically and process them through pixel-based modeling in GIS. It is now increasingly possible to get the quantifiable data over large area and in time series from digital remote sensing products. However, it is necessary to address several sensor-related issues before embarking upon actual modeling work, like changes in the time of observation by the satellite, characteristics of the scanners (including bandwidth and spectral resolution), atmospheric corrections, etc.

In many cases the biophysical data sets are taken as surrogates for human impact. For biophysical measurement digital remote sensing has proved to be a reliable tool capable of reasonably accurate assessment of condition over a large area in a reasonably small time period, especially for measuring the spatio-temporal patterns of a large number of land surface variables. Increased spatial resolution of the sensors, choice of object-specific spectral bands and better revisit facilities of the satellites have now helped researchers to quantify the patterns of land surface variability and their changes more confidently (e.g., Pickup and Chewings, 1988; Hill *et al.*, 1995; Hill and Schutt, 2000), including the analysis of wind and water erosion, salinization, changes in net primary productivity, changes in albedo, etc. However, not all the biophysical variables are yet equally amenable to remote sensing. (e.g., most of the subsurface features, soil characteristics, vegetation species, and quantity of sediment lost, etc.). New results from advanced remote sensing satellites and Ground Penetrating Radars (GPRs) are fast changing our views on this aspect. For example, groundwater changes from GRACE satellites measured micro-gravity changes (Rodell *et al.*, 2009), while precipitation measurements from TRMM, dust and cloud measurements from MODIS, CALIPSO and MISR, and the results from AMSRE and other upcoming satellites on soil moisture changes are vastly changing our perspectives on the capabilities of remote sensing to measure environmental parameters.

At the local scale, where much of the information is to be gathered from field level, high-resolution geo-referenced data (including hyperspectral data) will help in understanding the patterns of change, and in matching the interpreted results with landscape and socio-economic trends observed in the field. Up-scaling from the local level to district and state levels and then to national and regional levels, rather than a downscaling approach will help to eliminate many of the ambiguities and doubts concerning unit boundaries.

Test Cases of Desertification Monitoring in a Digital Environment

A. Measuring wind erosion

To test the workability of the D-P-S-I-R concept in a digital environment for monitoring of wind erosion and its likely impacts, a study was carried out in a cluster of villages in the sandy terrain of the southern part of Churu district (Rajasthan) in eastern part of Thar Desert. The study involved time-series analysis (1971-2001) of the following variables in 339 contiguous village polygons of the region: human and livestock population densities as driving forces, quantification of cultivation and grazing pressures, digital analysis of ortho-rectified, geo-referenced and atmosphere-corrected Landsat data for a measure of sand reactivation (a state variable), and analysis of NDVI condition and dust load as impact variables (Kar, 2011a).

Signatures of sand reactivation were analysed from Landsat-derived Principal Component Analysis (PCI), Soil Brightness Index (SBI) and Colour Brightness Index (CBI) classifications and evaluated with supervised classification (maximum likelihood). CBI provided the most acceptable outputs after the brighter signature of gypsiferous and salt-affected areas were filtered out.

Cultivation pressure was measured through intensity of cropland use between 1970-71 and 2000-01. Grazing pressure was calculated from optimum carrying capacity of grazing lands in the area vis-a-vis adult cattle units. Using village as a mapping unit for driving force and pressure variables, and satellite pixel for state variable, a spatio-temporal analysis was carried out under GIS for 339 villages of the region, which revealed good agreement between the moderate and high cultivation pressures with moderate and high sand reactivation, but the agreement was poor for grazing pressure. Multi-criteria evaluation of the driving force, pressure and state variables using Ordered Weighted Average technique and fuzzy set membership function revealed that 3% area had severe risk of desertification, 20% moderate and 62% slight. The at-satellite radiance from land surfaces in red band revealed values of 11mW cm^{-2} Sr^{-1} mm^{-1} and higher in western part, and lower in the eastern part, reflecting an identical spatial pattern of albedo. Since this is considered a measure of land degradation, the pattern was compared with results from multi-criteria evaluation, but agreement was poor. Analysis of sequential rabi season AVHRR-NDVI images for 1983-2001 showed how greenness and land productivity shifted from E to W with groundwater irrigation, and hinted that the sandy landscape here needs

2-3 decades to stabilize after irrigation starts, but can fast return to unstable state if cultivation pressure continues without adequate water application.

For the Thar Desert as a whole, the changing winter NDVI pattern was evaluated with atmospheric dust concentration pattern (June AOD) measured by the Total Ozone Mapping Spectrophotometer, annual rainfall, wind speed and tractor use, which suggested that a decline in rabi-irrigated area and increased tractor-ploughing might be the major responsible factors for recent increase in dust concentration. In other words, it was possible to use surrogates of biophysical processes to quantify land degradation in the sandy terrain, and to better understand the driving forces (Kar, 2011 a).

In a different context, Kar (1993a) developed a climatic wind erosion index (WEI) for Thar Desert, and then linked the different index values with the spatial pattern of the degree of reactivation of the sandy landscape, especially the dune-covered areas. The study shows that the areas having index values between 120 and 480 are very highly vulnerable to wind erosion and feature numerous fields of barchans under natural settings (e.g., roughly to the west of a line joining Chohtan, Barmer, Shergarh, Bikaner in India and Khangarh in Pakistan). Areas having >480 values are extremely vulnerable due to higher wind speed and lower precipitation effectiveness, where fields of megabarchans occur (e.g., the Shahgarh Bulge area to the southwest of Jaisalmer, and the adjoining Chhor area of Pakistan). Using the gridded climatic datasets for the region for 1951-2000 and the simulated future climate data for 2001-2100, Kar (2009b, 2011c) then showed that during the second half of last century the maximum sand reactivation potentials existed during 1961-65, followed by 1986-90, while the lowest potentials were reached during 1991-95 when the rainfall was higher and wind strength very weak. If the simulation of future climates is correct, there is high possibility that by 2016-20 the WEI will increase sharply beyond the previous highs of the 1960s and the 1980s, and the trend may persist till the end of the century, save and except some dips when rainfall might increase, like in the 2030s and 2070s. The implications are that with changing climate the stabilization of sandy landscape will be a foremost requirement to minimize large-scale sand reactivation, dune movement and dust emission. The task is already becoming difficult with the large-scale deep ploughing of the sandy soil through tractors, denudation of the natural shrub and tree covers for fuel, fodder and cropping, and fewer efforts for restoration of the degraded pastures and other denuded lands, and a general rise in wind speed.

B. Measuring water erosion

To find out the possibilities of using remote sensing data for quantification of water erosion, a case study was undertaken in Sirohi and Revdar tehsils in the south-eastern part of Thar Desert, receiving an average annual rainfall of >550 mm in ~26 days. The topography is rugged with barren or partly forested hills, interspersed with large pediplains having shallow sandy soils and wide ephemeral channels that have occasional narrow strips of alluvium, except in some pockets where vestiges of an earlier dry climate with strong wind are noticed in the form of stable and gullied echo dunes. The plains, despite their shallow soils, are largely under double cropping through irrigation wells, and provide an impression of a stable landscape with negligible erosional hazard. Field campaigns on the other hand reveal examples of soil stripping, scalding and gully formation, but quantitative information on erosion is lacking. To correctly assess the problem for large-area mapping, we tested a physically-based soil erosion model that has been successfully applied in the Mediterranean drylands of Europe and North Africa. Called the Morgan-Morgan-Finney model (Morgan, 2001, 2005), it was last refined by Morgan and Duzant (2008), but was still being validated for different components. Both the models demand large amount of field and remote sensing data (e.g., rainfall amount and intensity, soil bulk density, WHC, effective hydrologic depth, soil resistance, cohesion, actual and potential ET, land cover, average plant height and canopy cover, fraction ground cover, slope steepness, etc. in grid format). We relied largely on the 2008 model, but also used the 2001 model for certain calculations for which the 2008 model had data limitations. We call this amalgamated version as Revised Morgan-Morgan-Finney Model (re-MMF model). The ortho-rectified SPOT Vegetation data on percent cover at 1 km grids was taken as the base grid framework, over which the analyzed data from GPCC, TRMM, SRTM, TM and ETM+ grids were put. Using re-MMF we calculated the kinetic energy of rainfall (J m^{-2}), runoff (mm), particle detachment by raindrop and runoff (kg m^{-2}), transport capacity of runoff (Mg ha^{-1}) and soil erosion rates (Mg ha^{-1}) for a representative high-rainfall year (monsoon, 2006). The simulation results were encouraging and matched the expected pattern, though needed some more refinement. The expected runoff during the monsoon period was >500mm in 31% area in Sirohi tehsil, and 52% area in Revdar tehjsil. About 25% area of both the tehsils might have experienced <200mm runoff during the monsoon. The dominant transport capacity during the period was simulated as 500-2500 t ha^{-1} (~24% in Sirohi and ~31% in Revdar). About 24% area of

Sirohi tehsil was expected to have very severe soil erosion rate of >80 t ha^{-1} under such situation, while Revdar tehsil was expected to have very high soil erosion rate in 31% area. The simulation results thus showed a higher potential of water erosion in Revdar tehsil.

Flash floods also lead to localized high erosion and sedimentation across the desert, and thus changes in land productivity. For example, in July 1979, high intensity monsoon rainfall for 5 days at a stretch in the southern part of the desert affected nearly 108799 ha of agricultural land with either rill and gully formation or coarse sand deposition due to shifting channels. Crops were damaged over 87186 ha area. Similarly, two spells of high intensity monsoon rains in July and August, 1990, led to flooding and soil erosion in the same region (Sharma and Vangani, 1992; Vangani, 1997). In August, 2006 high-intensity rainfall for three days in Barmer-Jaisalmer tract not only resulted in year-long flooding of some interdune plains, but also led to erosion of topsoil and nutrients from the upper catchment and their deposition in the lower catchment. The stream network, revived after more than half a century, however, did not deviate from the simulated channel courses for the area (Kar *et al.*, 2007b; Kar, 2011b). Part of the human misery in the wake of the flood lay in unscientific land use practices. It has been noticed that the shallow sandy channels of the desert are hardly able to accommodate much of the run-off generated during the long spells of high intensity monsoon rainfall, and leads to flooding. The floodwater often gushes out along some old and abandoned channels. It is only when such old courses are under human use systems that obstruct the free passage of floodwater that the suffering becomes glaring (Kar, 1995).

C. Other questions involved

Some forms of degradation, like subsurface water logging, soil salinization, soil fertility changes or soil acidification are best assessed and mapped through field sampling, as remote sensing consistently yield poor results. With the present level of technological development, monitoring of salinity-alkalinity will require accurate point-based field information at certain time intervals. Indiscriminate use of groundwater for irrigation, leading to pumping of poor quality water and depletion of water level below the rechargeable limit, has become a source of land degradation in many parts of the drylands. Suitable methods of integrating groundwater irrigation with land quality degradation, therefore, need to be developed. Remote sensing will play a major role in accurate and fast mapping of other forms of land degradation, and in understanding the complex inter-relationship between the biophysical processes involved, especially through several digital techniques. Once the parameters are quantified and scaled properly, it is possible to integrate the results from different driving forces and pressures in a GIS framework to model the degradation scenario over space and time (Grunblatt, 1992; Hellden and Tottrup, 2008; Rasmy *et al.*, 2010; Santini *et al.*, 2010; Kar, 2011a).

In our opinion, much of the socio-economic investigations will continue to be based on secondary database and opinion surveys in near future. Impacts of development work can be measured through the same indicators that are used for desertification status. The indicators being developed by UNCCD may be a great help in this respect. As we have noted earlier, a set of reliable indicators has to be chosen first, and then tested. During the period when biophysical indicators are tested in sample areas, it may be worthwhile to carry out opinion survey on the development programmes being run by Government, NGOs and others to counter desertification. Direct benefit of many programmes like plantation, etc. takes time to reach the local people.

Conclusion

Concepts of desertification and approaches to understand its facets are changing with time. Quantification of variables and their modeling are becoming more in demand for scenario building of the past, present and future states, as well as for finding out the impacts of investments in controlling desertification. As digital technology and IT revolutions are spreading, it is also becoming possible to remotely sense several land surface parameters in near-real time, and to integrate the variables in GIS. A thorough understanding of the process of desertification and its causes and effects is, however, a pre-requisite for taking up appropriate digital techniques for assessment and monitoring. The quantity, rates, direction and the spatial dimension of the processes and the responses, have to be analysed, co-evaluated with the atmospheric and socio-economic variables in a GIS environment, and suitably modeled to find out the trends in desertification and the factors responsible. Simulation modeling, using remote sensing and GIS will play critical roles in the analysis. However, simulation models have to be used cautiously, and should be based on sufficient local information.

References

Adeel, Z., Safriel, U., Niemeijer, D., White, R., Kalbermatten, G., Glantz, M., Salem, B., Scholes, B., Niamir-Fuller, M., Ehui, S. and Yapi-Gnaore, V. 2005. *Esosystem and Human Well-being: Desertification Synthesis.* Millennium Ecosystem Assessment, World Resources Institute, Washington.

Ahlcrona, E. 1988. *The Impact of Climate and Man on Land Transformation in Central Sudan*. Meddelanden Fran Lunds Universitets Geografiska Institutioner, No. 103, 140p.

Amissah-Arthur, A., Mougenot, B. and Loireau, M. 2000. Assessing farmland dynamics and land degradation on Sahelian landscapes using remotely sensed and socio-economic data. *International Journal of Geographical Information Science* 14: 583-599.

Anonymous 1979. *Mastering the Probable and Managing the Unpredictable*. Inter-Futures Study Team, Organisation for Economic Cooperation and Development, and International Atomic Agency, Paris.

Anonymous 1992. Managing fragile ecosystems: Combating desertification and drought. In, *Chapter 12, Agenda 21, United Nations Conference on Environment and Development, Rio de Janeiro*. A/CONF.151/4 (Part II), United Nations, New York, pp. 46-66.

Anonymous 1995. *United Nations Convention to Combat Desertification*. Interim Secretariat for the Convention to Combat Desertification, Geneva, 71 p.

Anonymous. 2001. *India: National Action Programme to Combat Desertification in the Context of United Nations Convention to Combat Desertification (UNCCD). Volume I: Status of Desertification*. Ministry of Environment & Forests, Government of India, New Delhi, 294 p.

Anonymous. 2010. *Degraded and Wastelands of India: Status and Spatial Distribution*. Indian Council of Agricultural Research, and National Academy of Agricultural Sciences, New Delhi, 133 p.

Anonymous. 2003. Proposed benchmark and indicator system for desertification monitoring and assessment in Asian region. *Appendix II, TPN1 Meeting on Benchmark and Indicators Development and Mapping for Desertification Monitoring and Assessment, jointly with The International Symposium on Space Technology Applications for Sustainable Dryland Development and Desertification Monitoring*, 22-24 October, 2003, Beijing.

Aubreville, A. 1949. *Climats, Forets et Desertification de l'Afrique Tropicale*. Societe d' Editions Geographiques, Maritimes et Coloniales. Paris, 255 p.

Baartman, J.E.M., Godert, W.J., van Lynden, W.J., Reed, M.S., Ritsema, C.J. and Hessel, R. 2007. *Desertification and Land Degradation: Origins, Processes and Solutions – A Literature Review*. Report 4, Scientific Reports Series, ISRIC, The Netherlands, 71 p.

Babaev, A.G., Kharin, N.G. and Orlovsky, N.S. 1993. *Assessment and Mapping of Desertification Processes: A Methodological Guide*. Desert Research Institute, Academy of Sciences of Turkmenistan, Ashkabad.

Baker, V.R. 1986. Introduction: Regional landform analysis. In, *Geomorphology from Space: A Global Overview of Regional Landforms* (Eds., N.M. Short and R. Blair), pp. 1-26. Special Publication 486, NASA, Washington DC.

Bharara, L.P. 1993. Livestock migration and pastoralism in Rajasthan Desert. In, *Desertification and its Control in the Thar, Sahara and Sahel Regions*, (Eds., A.K. Sen and Amal Kar), pp. 219-247. Scientific Publishers, Jodhpur.

CAZRI. 2005. *Western Rajasthan: Desertification Map*.

Dregne, H.E. 1983. *Desertification of Arid Lands*. Harwood Academic Publishers, Chur, 242 p.

Dregne, H.E. 1991. Global status of desertification. *Annals of Arid Zone* 30: 179-185.

Dregne, H.E. 2000. Desertification: Problems and challenges. *Annals of Arid Zone*, 39: 363-371.

Dregne, H.E. and Chou, N-T. 1992. Global desertification: Dimensions and costs. In, *Degradation and Restoration of Arid Lands* (Ed., H.E. Dregne), pp. 249-281. Texas Tech. University, Lubbock.

Enne, G. and Zucca, C. 2000. *Desertification Indicators for the European Mediterranean Region: State of the Art and Possible Methodological Approaches*. ANPA - National Environmental Protection Integrated Strategies, Promotion, Communication Department, Rome, and NRD - Nucleo di Ricerca sulla Desertficazione, Universita degli Studi di Sassari, Sassari. 260 p.

Eswaran, H., Lal, R. and Reich, P.F. 2001. Land degradation - an overview. In, *Responses to Land Degradation; Proc. Second International Conference on Land Degradation and Desertification*. Khon Kaen, Thailand (Eds., E.M. Bridges, I.M. Hannam, L.R. Oldeman, F.W.T. Pening de Vries, S.J. Scherr and S. Somepatpanit). Oxford, New Delhi.

FAO. 1984. *Provisional Methodology for Assessment and Mapping of Desertification*. Food and Agriculture Organisation of the United Nations, Rome.

Geist, H.J. and Lambin, E.F. 2004. Dynamic causal patterns of desertification. *BioScience* 54: 817-829.

Ghose, B., Singh, S. and Kar, A. 1977. Desertification around the Thar: A geomorphological interpretation. *Annals of Arid Zone* 16: 290-301.

Grunblatt, J., Ottichilo, W.K. and Sinange, R.K. 1992. A GIS approach to desertification assessment and mapping. *Journal of Arid Environments* 23: 81-102.

Hellden, U. and Tottrup, C. 2008. Regional desertification: A global synthesis. *Global and Planetary Change*, 64: 169-176.

Herrmann, S.M. and Hutchinson, C.F. 2005. The changing contexts of the desertification debate. *Journal of Arid Environments* 63: 538-555.

Hickler, T., Eklundh, L., Seaquist, J.W., Smith, B., Ardo, J., Olsson, L., Sykes, M. and Sjostrom, M. 2005. Precipitation controls Sahel greening trend. *Geophysical Research Letters* 32L: 21415.

Hill, J. and Schutt, B. 2000. Mapping complex patterns of erosion and stability in dry Mediterranean ecosystems. *Remote Sensing of Environment* 74: 557-569.

Hill, J., Megier, J. and Mehl, W. 1995. Land degradation, soil erosion and desertification monitoring in Mediterranean ecosystems. *Remote Sensing Reviews* 12: 107-130.

Joshi, D.C. and Dhir, R.P. 1991. Rehabilitation of degraded sodic soils in an arid environment by using residual Na-carbonate water for irrigation. *Arid Soil Research and Rehabilitation* 5: 175-185.

Kar, A. 1986. Physical environment, human influences and desertification in Pushkar-Budha Pushkar lake region, India. *The Environmentalist* 6: 227-232.

Kar, A. 1988. Evidence for neotectonism from the Indian desert. In *Geomorphology and Environment* (Eds., S. Singh, and R.C. Tiwari), pp. 300 310. Allahabad Geographical Society, Allahabad.

Kar, A. 1990. The megabarchanoids of the Thar: Their environment, morphology and relationship with longitudinal dunes. *The Geographical Journal* 156: 51 61.

Kar, A. 1992. Morphological processes, human influences and land degradation in the Indian desert. In, *New Dimensions in Agricultural Geography*, Volume 2: The Ecology of Agricultural System (Ed., N. Mohammad), pp. 123 150. Concept, New Delhi.

Kar, A. 1993a. Aeolian processes and bedforms in the Thar Desert. *Journal of Arid Environments* 25: 83-96.

Kar, A. 1993b. Neotectonic influences on morphological variations along the coastline of Kachchh, India. *Geomorphology* 8: 199-219.

Kar, A. 1993. Digital terrain analysis, granulometry and remote sensing of the Banni land of Kachchh for development related morphological classification. In, *Applied Geomorphology in the Tropics* (Ed., P.R. Sharma), pp. 87 104. Rishi Publications, Varanasi.

Kar, A. 1995. Geomorphology of arid western India. In, *Quaternary Environment and Geoarchaeology in India* (eds., S. Wadia, R. Korisettar and V.S. Kale), pp. 168-190. Memoir 32, Geological Society of India, Bangalore.

Kar, A. 1996a. Desertification processes in arid western India. In, *Towards Solving the Global Desertification Problem*, vol. 4, (Eds., T. Miyazaki and A. Tsunekawa), pp. 20-29. National Institute for Environmental Studies, Tsukuba.

Kar, A. 2004. Advanced planning for desertification: Need for a monitoring and warning system. In, *Coping with Natural Disaster: Indian Context* (Ed., K.S. Valdiya), pp. 229-249. Universities Press, Hyderabad.

Kar, A. 2006. Formation and morphology of shrub-coppice dunes in the Thar Desert. *Deccan Geographer* 44: 51-64.

Kar, A. 2009a. An analysis of climate change through quantification of forms and processes in the Thar Desert. In, *Earth System Sciences* (Eds., A. Kumar, R.A. Kushwaha and B. Thakur), Vol. 2, pp. 363-381. Concept, New Delhi.

Kar, A. 2009b. Wind erosion potentials in the Thar-Megathar region from AD 1951 to 2100: Patterns and implications. *Abstracts, International Conference on Nurturing Arid Zones for People and the Environment: Isues and Agenda for the 21st Century*, pp. 24-25. Central Arid Zone Research Institute, Jodhpur.

Kar, A. 2011a. Geoinformatics in spatial and temporal analyses of wind erosion in Thar Desert. In *Geoinformatics in Applied Geomorphology* (Eds., S. Anbazhagan, S.K. Subramanian and X. Yang), pp. 39-62. CRC Press, USA.

Kar, A. 2011b. Flood havoc in Thar Desert: Is nature to blame? In, *Water Crisis in the Indian Subcontinent* (Eds., Zahid Husain and L. Cajee), pp. 3-17. Bookwell, Delhi.

Kar, A. 2011c. Caring for the Thar. *Geography and You* 11:64-69.

Kar, A. and Faroda, A.S. 1998. *Indicators and Benchmarks for Measuring Desertification: A View from the Thar Desert.* Consultant's Report to UN Secretariat on Convention to Combat Desertification, Geneva, 24 p.

Kar, A. and Takeuchi, K. 2003. Towards an early warning system for desertification. In *Early Warning Systems*, pp. 37-72. UNCCD Ad Hoc Panel, Committee on Science and Technology. UN Convention to Combat Desertification, Bonn, Germany.

Kar, A., Moharana, P.C. and Singh, S.K. 2007a. Desertification in arid western India. In *Dryland Ecosystem: Indian Perspective* (Eds., K.P.R. Vittal, R.L. Srivastava, N.L. Joshi, A. Kar, V.P. Tewari, and S. Kathju), pp.1-22. Central Arid Zone Research Institute, Jodhpur, and Arid Forest Research Institute, Jodhpur.

Kar, A., Moharana, P.C., Singh, S.K., Goyal, R.K. and Rao, A.S. 2007b. Barmer flood, 2006: Causes and consequences. In, *Flood of August 2006 in Arid Rajasthan: Causes, Magnitude and Strategies* (eds., A.S. Faroda and D.C. Joshi), pp. 26-39. State of Art Report, Scientific Contribution No. INCOH/SAR-29/2007; INCOH, Roorkee.

Kar, A., Moharana, P.C., Raina, P., Kumar, M., Soni, M.L., Santra, P., Ajai, Arya, A.S. and Dhinwa, P.S. 2009. Desertification and its control measures. In, *Trends in Arid Zone Research in India* (Eds., Amal Kar, B.K. Garg, S. Kathju and M.P. Singh), pp. 1-47. Central Arid Zone Research Institute, Jodhpur.

Kharin, N.G., Nechaeva, N.T., Nikolaev, V.N., Babaeva, T., Dobrin, L.G., Babaev, A., Orlovsky, N.S., Redzhepbaev, K., Kirsta, B.T., Nurgeldyev, O.N., Batyrov, A. and Svintsov, I.P. 1985. *Methodological Principles of Desertification Processes Assessment and Mapping (Arid lands of Turkmenistan taken as example).* Desert Research Institute, Turkmen SSR Academy of Sciences, Askhabad.

Kirkby, M.J. 2009. Desertification: The broader context. *Proceedings, International Conference on Desertification*. Universidad de Murcia. Web: http://congresos.um.es/iccd/2009/

Kumar, S., Singh, G., Parmar, P.J., Pandey, R.P., Sharma, S.C., Sundarmoorthy, S., Singh, J.P., Kasera, P., Purohit, C.S., Sen, D.N. 2009. Plant diversity conservation for sustainability. In, *Trends in Arid Zone Research in India* (Eds., Amal Kar, B.K. Garg, S. Kathju and M.P. Singh), pp. 151-179. Central Arid Zone Research Institute, Jodhpur.

Lavauden, L. 1927. Les forets du Sahara. *Revue des Eaux et Forets* 6: 265-277; 7: 329-341.

Middleton, N. and Thomas, D. (Eds.). 1997. *World Atlas of Desertification* (Second Edition). UNEP and Arnold, London.

Morgan, R.P.C. 2001. A simple approach to soil loss prediction: A revised Morgan-Morgan-Finney model. *Catena* 44: 305-322.

Morgan, R.P.C. 2005. *Soil Erosion and Conservation*. Third Edition, Blackwell, USA.

Morgan, R.P.C. and Duzant, J.H. 2007. Modified MMF (Morgan-Morgan-Finney) model for evaluating effects of crops and vegetation cover on soil erosion. *Earth Surface Processes and Landforms* 32: 90-106.

Mortimore, M.J. 1998. *Roots in the African Dust*. Cambridge University Press, Cambridge.

Nemani, R.R., Keeling, C.D., Hashimoto, H., Jolly, W.M., Piper, S.C., Tucker, C.J., Myneni, R.B. and Running, S.W. 2003. Climate-driven increases in global terrestrial net primary production from 1982 to 1999. *Science* 300: 1560-1563.

Nickling, W.G. and Wolfe, S.A. 1994. The morphology and origin of nebkhas, region of Mopti, Mali, West Africa. *Journal of Arid Environments* 28: 13-30.

Olsson, L. 1993. Desertification in Africa - A critique and an alternative approach. *GeoJournal* 31: 23-31.

Orr, B.J. 2011. *Scientific review of the UNCCD provisionally accepted set of impact indicators to measure the implementation of strategic objectives 1, 2 and 3*. White Paper, version 1, 04 February 2011. UNCCD, Bonn.

Pickup, G. and Chewings, V.H. 1988. Forecasting patterns of soil erosion in arid lands from Landsat MSS data. *International Journal of Remote Sensing* 9: 69-84.

Prince, S.D. 2002. Spatial and temporal scales for detection of desertification. In, *Global Desertification: Do Humans Cause Deserts*? (Eds., J.F. Reynolds and D.M. Stafford Smith), pp. 23-40. Dahlem University Press, Berlin.

Rasmy, M., Gad, A., Abdelsalam, H. and Siwailam, M. 2010. A dynamic simulation model of desertification in Egypt. *The Egyptian Journal of Remote Sensing and Space Sciences* 13: 101-111.

Reynolds, J.F. and Stafford Smith, D.M. (Eds.). 2002. *Global Desertification: Do Humans Cause Desertification?* Dahlem Workshop Report 88, Dahlem University Press, Berlin.

Reynolds, J.F., Stafford Smith, D.M., Lambin, E.F., Turner, B.L. II, Mortimore, M., Batterbury, S.P.J., Downing, T.E., Dowlatabadi, H., Fernandez, R.J., Herrick, J.E., Huber-Sannwald, E., Jiang, H., Leemans, R., Lynam, T., Maestre, F.T., Ayarza, M. and Walker, B. 2007. Global desertification: Building a science for dryland development. *Science* 316:847-851.

Reynolds, J.F., Grainger, A., Stafford Smith, D.M., Bastin, G., Garcia-Barrios, L., Fernandez, R.J., Janssen, M.A., Jurgens, N., Scholes, R.J., Veldkamp, A., Verstraete, M.M., Von Matitz, G. and Zdruli, P. 2011. Scientific concepts for an integrated analysis of desertification. *Land Degradation and Development* 22: 166-183.

Rodell, M., Velicogna, I. and Famiglietti, J.S. 2009. Satellite-based estimates of groundwater depletion in India. *Nature* 460: 999-1002.

SAC 2007a. *Desertification Monitoring and Assessment using Remote Sensing and GIS: A Pilot Project under TPN-1 UNCCD*. Space Applications Centre, ISRO, Ahmedabad.

SAC 2007b. *Desertification and Land Degradation Atlas of India.* Space Applications Centre, ISRO, Ahmedabad.

Safriel, U. and Adeel, Z. 2005. Dryland systems. In, *Ecosystems and Human Well-being: Current State and Trends*, vol. 1 (Eds., R. Hassan, R. Scholes and N. Ash), pp. 625-658. Island Press, Washington.

Safriel, U. and Adeel, Z. 2008. Development paths of drylands: Thresholds and sustainability. *Sustainability Science* 3: 117-123.

Samra, J.S. and Sharma, P.D. 2005. Quality of soil resources in India. *Souvenir, International Conference on Soil, Water and Environmental Quality – Issues and Strategies*, pp. 1-23. Indian Society of Soil Science, New Delhi.

Santini, M., Caccamo, G., Laurenti, A., Noce, S. and Valentini, R. 2010. A multi-component GIS framework for desertification risk assessment by an integrated index. *Applied Geography* 30: 394-415.

Saxena, S.K. 1977. Vegetation and its succession in the Indian Desert. In, *Desertification and its Control*, pp. 176-192. ICAR, New Delhi.

Sharma, K.D. and Vangani, N.S. 1992. Surface water resources in the Thar desert and adjoining arid areas of India. In, *Perspectives on the Thar and the Karakum* (eds., A. Kar, R.K. Abichandani, K. Anantharam, and D.C. Joshi), pp. 130-140. Department of Science and Technology, Govt. of India, New Delhi.

Singh, S., Ghose, B. and Kar, A. 1978. Geomorphic changes as evidence of palaeoclimate and desertification in Rajasthan desert, India (Luni Development Block: A case study). *Man & Environment* 2: 1 13.

Singh, S., Kar, A., Joshi, D.C., Kumar, S. and Sharma, K.D. 1994. Desertification problem in western Rajasthan. *Annals of Arid Zone* 33: 191 202.

Singh, S., Kar, A., Joshi, D.C., Ram, B., Kumar, S., Vats, P.C., Singh, N., Raina, P., Kolarkar, A.S. and Dhir, R.P. 1992. Desertification mapping in western Rajasthan. *Annals of Arid Zone* 31: 237-246.

Stocking, M. and Murnaghan, N. 2000. *Land Degradation: Guidelines for Field Assessment.* Norwich.

Swift, J. 1996. Desertification: Narratives, winners and losers. In, *The Lie of the Land: Challenging Received Wisdom in African Environmental Change and Policy* (Eds., M. Leach and R. Mearns), pp. 73-90. James Currey, Oxford.

Tengberg, A. 1995. Nebkha dunes as indicators of wind erosion and land degradation in the Sahel zone of Burkina Faso. *Journal of Arid Environments* 30: 265-282.

Thomas, D.S.G. 1993. Sandstorm over a tea cup? Understanding desertification. *The Geographical Journal* 159: 318-331.

Thomas, D.S.G. 1995. Desertification: causes and processes. In, *Encyclopedia of Environmental Biology* (Ed., W.A. Nierenberg), Vol. 1, pp. 463-473. Academic Press, San Diego.

Tiffen, M., Mortimore, M. and Gichuki, F. 1994. *More People Less Erosion: Environmental Recovery in Kenya.* John Wiley & Sons, Chichester.

Tsunekawa, A., Kar, A., Yanai, J., Tanaka, U. and Miyazaki, T. 1997. Influence of continuous cultivation on the soil properties affecting crop productivity in the Thar Desert, India. *Journal of Arid Environments* 36: 367-384.

Tucker, C.J., Dregne, H.E. and Newcombe, W.W. 1991. Expansion and contraction of the Sahara Desert from 1980-1990. *Science* 253: 299-301.

UNCOD 1977. *Round-up, Plan of Action and Resolutions.* United Nations Conference on Desertification, Nairobi.

UNEP. 1977. *UNEP and Fight Against Desertification*. Nairobi, Kenya.

Ustin, S.L., Jacquemoud, S., Palacios-Orueta, A., Li, L. and Whiting, M.L. 2005. Remote sensing based assessment of biophysical indicators for land degradation and desertification. *Proceedings, Workshop on Remote Sensing and Geoinformation Processing in the Assessment and Monitoring of Land Degradation and Desertification*, pp. 2-21. Trier.

Vangani, N.S. 1997. water erosion and flash flood hazards in Indian arid ecosystem. In, *Desertification Control in the Arid Ecosystem of India for Sustainable Development* (eds., S. Singh and Amal Kar), pp. 131-138. Agro Botanical Publishers (India), Bikaner.

Vogt, J., Hill, J., Safriel, U., Sokona, Y. and Bied-Charreton, M. 2009a. *Integrated Methods for Monitoring and Assessing Desertification/Land Degradation Processes and Drivers (Land Quality)*. White Paper of DSD Working Group 1, Version 1, 22 May 2009. Dryland Science for Development Consortium (DSD), European Commission Joint Research Centre.

Vogt, J., Hill, J., Bied-Charreton, M., Rohrig, J., Safriel, U., Sokona, Y., von Maltitz, G., and Zougmore, R. 2009b. *Integrated Methods for Monitoring and Assessing Desertification/Land Degradation Processes and Drivers (Land Quality)*. White Paper of DSD Working Group 1, Version 2, 19 August 2009. Dryland Science for Development Consortium (DSD), European Commission Joint Research Centre.

Warren, A. 2002. Land degradation is contextual. *Land Degradation and Development* 13: 449-459.

Warren, A. and Khogali, M. 1992. *Assessment of Desertification and Drought in the Sudano-Sahelian Region (1985-1991)*. United Nations Sudano-Sahelian Office, New York.

3

Desertification in the Eastern Margin of Arid Western Rajasthan – A Geomorphological Assessment

P.C. Moharana

Introduction

Desertification is now accepted as a major environmental issue because of the fact that 69.5 percent area of the world's dry lands and more than 1.2 billion people in over 110 countries living in this habitat are affected by various land degradation processes. During last 20 years, researchers have focused their discussion either on a condition or a cause or a set of processes to assess different paradigms of desertification. Several studies carried out in arid western part of Rajasthan (Singh *et al.*, 1992, Kar, A. 1993b, Kar, 1995, Dregne, H.E. 2000, Kar *et al.*2007) show that wind erosion is still a major problem. Though, there are many instances of location specific problems like water erosion, salinity/ alkalinity and water logging, about 75% area in arid western Rajasthan is affected by wind erosion hazard of different intensities (SAC, 2007,Kar *et al.* 2009). Decreasing rainfall gradient from east to west and increasing wind strength in that direction are responsible for the spatial variability in sand reactivation pattern. Water erosion in the form of sheet, rill and gully erosion is a major problem in areas, where average annual rainfall exceeds 350 mm, the rainfall intensity is high and the topographic and soil characteristics favour high run-off. Destruction of vegetation cover and excessive ploughing of shallow soils, especially on the sloping lands, accelerate the problem.

The western part of Rajasthan has about 62 % of India's arid zone and the Great Indian Desert or the Thar Desert. The arid western Rajasthan has been divided into 4 agro-climatic zones (Fig.1): the Arid western plain (IA), the Rajasthan irrigated plain (IB), the Transitional plains of inland drainage (IIA) and the Transitional plains of Luni basin (IIB). The region under discussion here is covered under zone IIa and IIb which covers about 6.7 mha area. This is roughly the eastern most extension of the Thar Desert which is roughly co-terminus with the Aravalli hills. This mountain system has a NE to SW alignment in the state of Rajasthan. Jhunjhunu, Sikar, Nagaur, Pali and Sirohi are the districts falling within these two zones. These 5 districts cover 46905 sq km area which is about 22.4% of the total geographical area of western Rajasthan. The average annual rainfall in this area varies from 450 to 600 mm, therefore, comparatively, it has a wetter environment. Again, the presence of Aravallis makes the fluvial process, comparatively more active in the eastern margin of western Rajasthan. Therefore, landforms in this region show varied signatures of water erosion. We discuss the role of fluvial process in the formation of landforms and the state of desertification in this region with a case study from western Rajasthan.

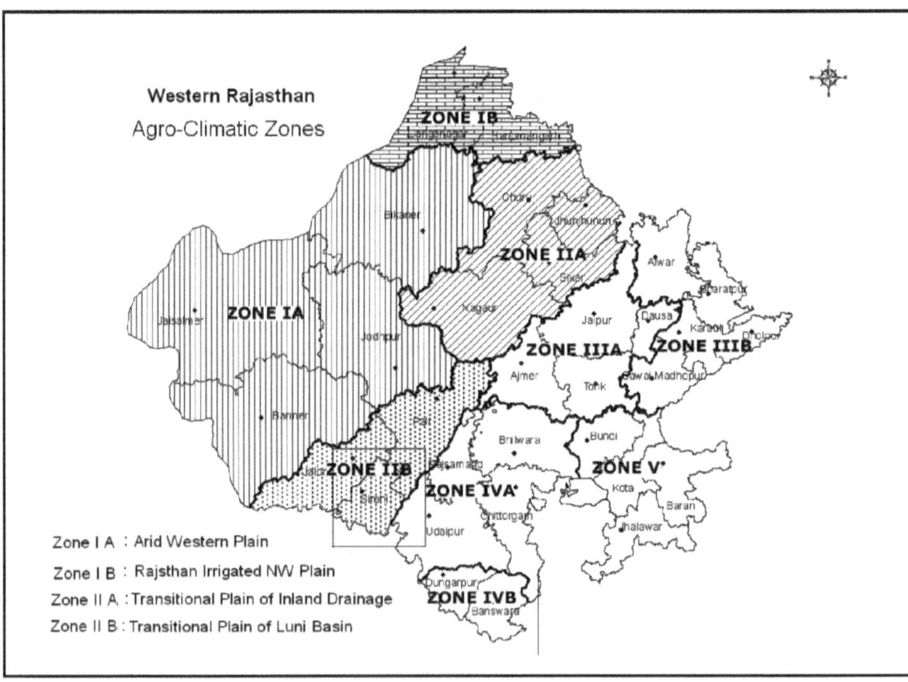

Fig. 1 : Agro-climatic zones of Western Rajasthan

Climate and Environment

The present study has been made in the arid part of Sirohi District. Sirohi, Shivganj and Revdar are the tehsils or the administrative units which will be covered under the present discussion (Fig 2). Environmentally, it is relatively a wetter zone; Sirohi and Sheoganj experience a mean annual rainfall of 577 mm occurring in 26 rainy days. Revdar is relatively a higher rainfall region, with a mean of 615 mm in 27 rainy days. As per 2007 meteorological data, southwest monsoon rainfall had contributed for 93% to the annual. Normal Agro-meteorological records show that excessive rainfall conditions has led to floods in this region occurring in 4 out of 10 years during 1941-50 and 1971-80. The long-term trend in annual rainfall showed an increase in rainfall at the rate of 78 mm/100 years at Sirohi and 66 mm/100 years at Sheoganj. Comparing the average rainfall during the periods 1901-90 and 1991-2007, it is observed that the rainfall has decreased from 598 mm to 575 mm at Sirohi whereas, it has increased from 520 mm to 544 mm at Sheoganj. The trend analysis of rainfall shows a decrease @3.7 mm/yr at Sirohi and an increase @0.5 mm/yr at Revdar. The mean minimum temperature in Sirohi district vary between 10.5 and 14.3^{0}C, occasionally falling as low as 6.0^{0}C during winter month due to passing western disturbances. Summer air temperatures vary between 38.5^{0}C and 41.0^{0}C with peak values as high as 44.5^{0}C in summer (CAZRI 2009).

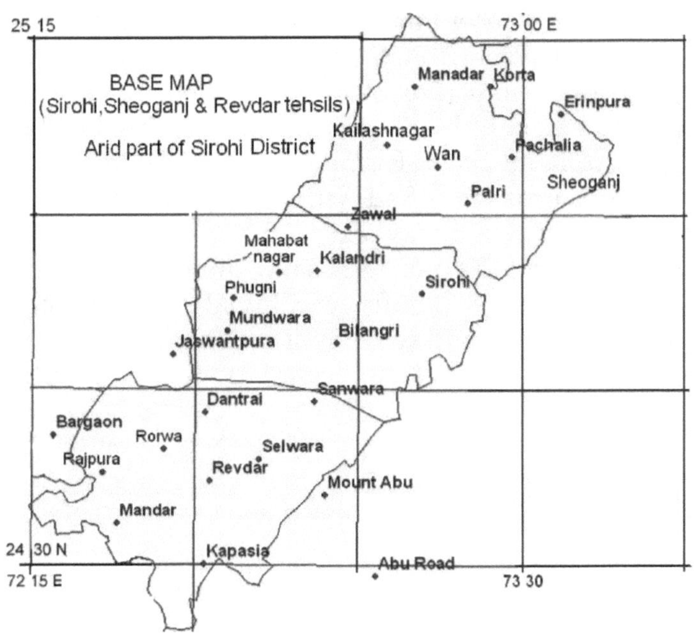

Fig. 2 : Location Map

Physiography and Drainage

The region has rocky/gravelly uplands in the central part mainly in the south of Sheoganj and north of Sirohi tehsil. Hills in the form of Aravalli offshoots form the margin of fringe area of the arid boundary of the district. Sandy uplands occur in patches mainly in Revdar tehsil and rest of the area is plains. About 18% area in the study area is covered by rocky/gravel waste, being confined to 240 - 340 m contours and are best described as rocky/gravelly pediments because of well developed slopes. About 90 % areas under these surfaces occur in Sirohi tehsil and 10% area falls in the northern part of Revdar tehsil. Highest relief of these surfaces occurs in Belangri, Balda, Phugni, Churiya in Sirohi tehsil with maximum elevation of 437 m at Angor and Goyali near Sirohi town. The region is drained by few major river systems; the SE-NW flowing Krishnawati and Kalpaganga in Sirohi tehil which joins Bandi river and NE-SW flowing Sukal and Sipu river in Revdar tehsil which are tributaries to Banas River. All these streams with higher order stream network are responsible for carrying lot of sediments during floods and forming extensive alluvial plains. Riverbeds are narrow and deep in the upland and shallow downslope with little vegetation.

Terrain and Landform Characteristics

Physiographical subdivision suggests that the region can be divided into two first order land units; uplands and flat plains. Based on interpretation of IRS -L3 FCC images of 2005-06 and limited field checks, a landform map (Fig 3) of the study area was prepared. Distribution and extent of landform units is presented in Table. Based on this map, the uplands can be further categorised into rocky and sandy types. Major landforms identified within rocky uplands include; Hills with intermontane valleys, rocky pediments, residual hills and colluvial plains. The sandy terrain comprises sand dunes, hummocks and the flat plains include older and younger alluvial plains. (Table 1). Based on the origin, topography and analysis of sediment characteristics, these landforms could be linked to tectonic and fluvial and Aeolian origin.

Table 1 : Landform units

Landform units	Area (sq km)	Area (%)
Landforms of tectonic and fluvial origin		
Hills	542.93	17.29
Rocky / gravely pediment	397.95	12.67
Colluvial plain	287.18	9.15
Intermontanne valleys	4.21	0.13
Younger alluvial Plain	172.03	5.48
Older Alluvial plain	1599.95	50.96
Riverbed & waterbody	42.15	1.34
Landforms of Aeolian origin		
Sand dunes and interdunes	93.38	2.97
Total area	**3139.78**	**100.00**

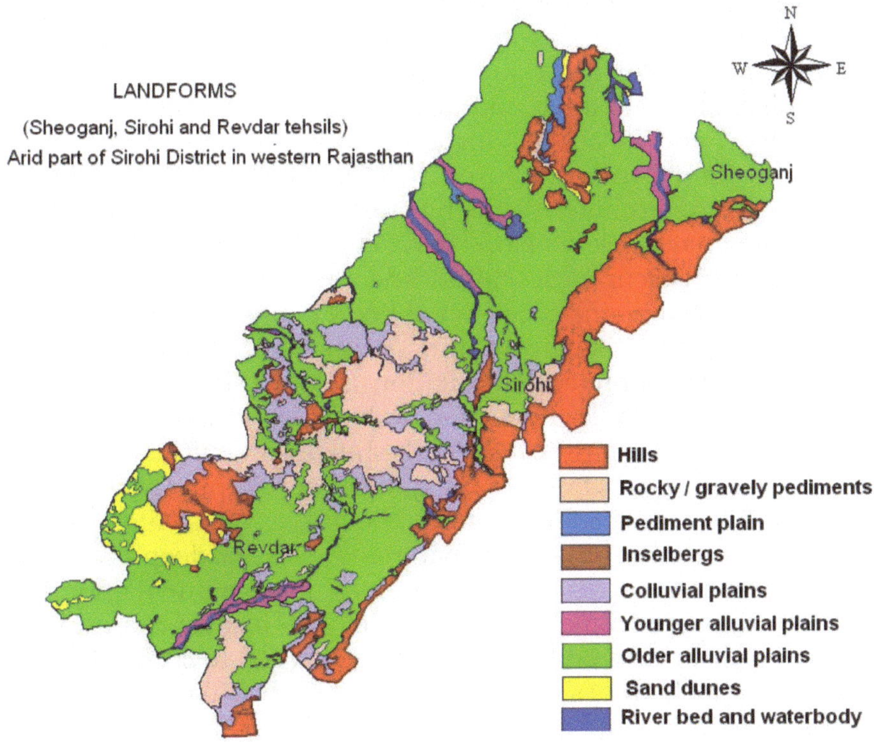

Fig. 3 :

Morphological Characteristics of Landforms

Hills and rocky pediments are prominent features of rocky terrain covering about 30 % area mainly occupying the eastern fringes of the Aravallis. Majority of hills are granites namely Ernipura granites in Sirohi tehsil at Phugni, Kakendra, Sildar, Akuna and Mohabatnagar and few patches in Revdar tehsil at Rorwa and Jirawal, Jalor granites in Jirwal and Nivaj in Revdar tehsil and Mirpur granites near Krishangarh in Sirohi tehsil. Rhyolite hills occur near Khanbal and Krishnaganj in Sirohi tehsil. Hills of mica-schist and phyllites of Sirohi group occur in a majority of places in Revdar tehsil. Isolated hills or inselbergs occur in Maria-Albari-Sanpur-Sirori area in Sirohi tehsil and in Dhan-Malawa-Rohwa-Manpura-Raipur in Revdar tehsil. At micro-level, the hill slopes exhibit cavernous and spheroidal weathering mostly on granite hills and to some extent on the rhylite hills near Phugni (Plate 1). Descending from the isolated hill slopes, the most common landform is the rocky pediment (Plate 2) which covers nearly 12.67% area. These pediments are dotted with granite/rhyolite wastes, rocky outcrops, boulders and domes sculptured with eroded topography. Extensive formations of these rocky plains on an average elevation of 300-320 m above msl occur in the west of Sirohi town and near Wan, Griti, Kalandri, Chauriwal, Nun, Albari and Belangri. Morphologically, these features are criss-crossed by numerous fractures at Jaila-Phalwadi-Kakendra in the western part of Sirohi tehsil. Such features can be observed on the IRS-L3 satellite images. Down slope to rocky pediments and hills, colluvial plains (Plate 3) occur with shallow and deep sand cover in 9.15% area. Younger alluvial plains occur in the flood plains of major rivers of the region; Sipu and Krishnawati. Most of their interfluves show eroded plains with dissections and undulations. Older alluvial plains cover the maximum (50.9%) area. These plains show network of rills and gullies mainly in the foothills of Aravalli hills showing manifestations of fluvial process in this region. Some of the gullies near Raipur (Plate 5) are very deep (8m deep, and 6-8 m wide) while gullies on an oran near Dhan (Plate 4) are shallow (1-1.40m wide and 1-1.25m deep). Sand dunes mainly occur in the west of Revdar town near Raipur-Rajpura-Rohua and Dantrai where dunes are of obstacle types formed against the hills. While relief of peaks on the hills varies from 500 to 800m, the average relief of obstacle dunes varies between 200 and 300 m. These are big dunes (25 to 40 m high) and their slopes exhibit gully dissections (1.2m-6m wide and 20-30m deep) formed due to numerous streams emanating from the above hills. A patch of low sand dunes (8-15 m high) also occur in the west of Revdar near Bant. Location specific description of terrain features is described in the following table (Table 2)

Plate 1 : Cavrenous weathering on granite rocks and formation of tafonis near Phugni

Plate 2 : Rocky pediment in Sirohi tehsil

Plate 3 : Colluvial plains in Sheoganj

Plate 4 : Shallow and network gullys near Dhan

Plate 5 : Deep gullys near Raipur

Table 2 : Terrain characteristics and land use of arid part of Sirohi district

S. No	Place	Latitude	Longitude	Land use	Terrain characteristics
1	Bithura	25 13.069	72 52.288	Open scrub land	Hill foot slope, pediment, 5-10 m wide, 5 m deep gullies dissecting sand dunes
2	Bithura	25 13.069	72 52.288	Cropland	Sandy plain, 150 m down slope to hills
3	Bithura	25 13.068	72 52.267	Cropland	Sandy plains, deep, cultivated
4	Jharoli	25 08.131	72 49.242	Cropland	Flat buried plain
5	Jaroli	25 04.980	72 49.769	Open scrub land	Gullies on the foothills of barren hills

Contd.

6	Wan	25 05.10	72 52.60	Hills	Barren rocks on hills with no soil cover, boulders on slope
7	Ora-Sabli -Madani	25 02.30	72 48.05	Cropland	Alluvial plains, undulating interfluves, channel cut area
8	Goyali	24 54.71	72 50. 04	Open scrub land	Stony waste near River Kameri
9	Kameri river	24 50.450	72 45.850	Riverbed	Gravel bed, 500-600 m wide and shallow
10	Bank of Krishna -vati R	24 55.454	72 45.600	Cropland	Younger alluvial plain, very deep soil
11	Padiv	24 55.568	72 45.527	Stony waste	Eroded rocky plains, rock exposures, barren land, concrete mining
12	Kalandri	24 55.360	72 43.759	Cropland	Moderately deep older alluvial plains, gravels at 1m depth
13	Choural	24 54.939	72 39.116	Rocky waste	barren rocky plain, juliflora invaded, granite exposures
14	Mohabat nagar	24 57.89	72 38.10	Riverbed	Gravel bed river
15	Mohabat nagar	24 57	72 36.09	Hills	Granite, grooves/hollows on granite, spheroidal weathering
16	Phugni	24 54.5	72 35.32	Rocky waste	Weathered granite out crops
17	Maria	24 52.25	72 33.02	Cropland	Alluvial plains, very deep clay soil, brick mining
18	Maria	24 52.254	72 32.241	Cropland	Flat sandy plain on foothills of eroded hills
19	Ablari	24 46.20	72 33.5	Hills / rocky upland	Surface dotted with granite hillocks/domes
20	Sanpur	24 47.5	72 36 .33	Rocky waste	Pediment on the eroded hill slopes
21	Siakra	24 46.585	72 33.788	Rocky waste	Degraded oran, scattered granite mounds
22	Balda	24 47.45	72 40.34	Forest land	Balda reserved forest with dots of granite
23	Belangir	24 48.3	72 43.50	Cropland	Cropland, stony surface, shallow, red soil
24	Mama -wali	24 51.573	72 44.337	Open scrubland	Rocky/stony plains, patches of cropland, very shallow soil,
25	Sanwara	24 44.142	72 41.934	Rocky waste	Rocky plains in the north of Sirori

Contd.

26	Dantrai	24 42.888	72 35.587	Open scrubland	Scattered patches of granite rocks
27	Dhan	24 42.792	72 33.827	Hills	granite outcrops, boulders with weathering
28	Dantrai	24 42.843	72 33.832	Open scrubland	Gullies : 200 m long, 1-1.40 m wide, 1-1.25 m deep
29	Bhamra	24 36.115	72 29.046	Open scrubland	Gullied area, granite outcrops in patches, river side
30	Barman	24 36.067	72 27.673	Cropland	Water eroded area, gullies
31	Malpura	24 37.857	72 27.337	Cropland	River bank, Alluvial plains
32	Rorwa	24 37.722	72 26.109	Sand dunes	Sand dunes dissected by gullies, width: 1-1.20 m to 5-6m,
33	Rorwa	24 37.72	72 25.125	Sand dune	dune top & dune valley
34	Rajpura	24 37.3	72 26.512	Hills	Barren isolated hills
35	Har -matiya	24 40.766	72 19.276	Sand dune	30m high dune, sample 23 (top), 24 (windward side bottom)
36	Raipur	24 41.256	72 22.141	Open scrubland	Gullied land, 8 m wide, 6-8 m deep, vertical ridges
37	Uthaman	25 01.823	72 58.391	Hills	Leveled land down slope, very deep soil, brick mining
38	Morchad	25 04.795	72 53.088	Cropland	Flat deep sandy plain
39	Bamara	25.10.346	72.00.831	Hills	Highly weathered ridges on top of denuded hills

Present Land Use

Based on interpretation of IRS-LIII data and field checks, a broad classification of present land use map was prepared for the region. Details of distribution of these units are presented in table (Table 3). Analysis of data showed that the first order units; croplands (irrigated double crop and rainfed mono-crop) pasture/grazing lands and forest/plantation lands occupy 43.57%, 3.95% and 13.18% area respectively while wastelands (33.55%) and gullied lands (0.63%) also cover substantial area.

Table 3 : Present land use and its statistics in study area (Based on interpretation of IRS-L3 images, 2005-06 of kharif, winter and zaid period)

Land use	Sheoganj		Sirohi		Revdar	
Area >	Sq. km	%	Sq.km	%	Sq.km	%
Irrigated double crop	104.75	11.7	322.16	27.4	410.55	37.71
Rainfed mono-crop	362.59	40.5	65.85	5.6	110.25	10.13
Saline waste	5.82	0.65	0	0	0	0
Gullied land (0.3%)	3.13	0.35	0	0	16.77	1.54
Sandy waste	1.79	0.2	0	0	42.93	3.94
Rocky/stony/gravely	98.48	11	482.1	41.03	428.88	39.39
Forest	181.74	20.3	215.93	18.37	0	0
Pasture/grazing	59.98	6.7	27.66	2.35	37.01	3.4
Plantation	18.80	2.1	0	0	0	0
Others (settlement/ water body)	58.19	6.5	61.5	5.25	42.41	3.89
	895.28	**100**	**1175.2**	**100**	**1088.8**	**100**

Types, Distribution and Morphology of Degraded Landforms

Digital interpretation and mapping of satellite images showed that these surfaces can be described under three major landscapes; a) Sandy uplands b) Colluvial/rocky-gravelly plains and c) Alluvial plains.

Erosion of Sandy Uplands

Sand dunes form the major landform units under sandy upland region which cover about 3 % area of the study area. Maximum area under this unit was identified and mapped in Bithura-Manadar-Wan-Mochhal in Sheoganj tehsil and Nivaj-Raipur-Rohua-Rorwa-Manpura-Rajpura sector in Revdar tehsil. Based on their morphology, these sand dunes can be divided into; Obstacle and Parabolic types. Their morphology in very large scale GE (Google Earth) and IRS LISS III images shows signs of fluvial dissections. In the following paragraphs, we present details of the state of degradation from two of the case studies from the district.

Bithura - Wan sector: This region lies in the central part of Sheoganj tehsil in Sirohi District. The hills representing the southern extension of Aravalli hills show homoclinal pattern, where several first order channels emerge from its eastern and western flanks and run in all possible directions. The eastern part is drained by a major ephemeral river: Khari nala which forms a natural boundary between Pali and Sirohi district. At some places near Bithura, Wan and Mochal, the hill slopes are covered with sand dunes which are otherwise known as obstacle dunes (Plate 6). The altitude of hills varies from 640 m to 835 m but the ridgeline of obstacle dunes lies between 260 and 280 m contours. The height of dunes varies from 30-45 m and their slopes extend by 200-500 m on the pediment slope of the hills. Due to action of surface run-offs, these slopes show fluvial dissections mostly by first order channels. The highest stream order in this sector, however is restricted to 2nd order streams which run for 5-7 km on both sides of the hills. On satellite images, such eroded areas show a definite erosion pattern, dull grey to red tone indicating growth of sparse vegetations or shrubs along the shallow valleys. Downslope, the plains have gathered more sediment deposits that supports kharif crops.

Plate 6 : Low obstacle dunes in the Aravalli foothills near Bithura in Sheoganj tehsil

Nivaj-Raipur-Rohua-Rorwa-Manpura-Rajpura sector: This represents a major sandy upland in Revdar tehsil, comprising obstacle and low parabolic dunes. While obstacle dunes are formed against a hill in the south of Nivaj, the parabolic dunes and sandy plains occur in the western part of the tehsil. The altitude of hills varies from 330 to 432m while obstacle dunes formed against these hills are 18-55 m high. Unlike in Sheogang tehsil, here, the sand dunes extend up to 4-5 km^2 area. The area under parabolic sand dunes and plains are dissected by channels (Plate 7) which join to form major streams like Dior, Thandawara which finally meet the Sipu River as tributaries. The interfluves of these streams are either sandy uplands or sandy plains and show fluvial dissections. Bank cut erosion can also be seen on both sides of Sukali nadi at Dantrai,

Bhamra, Malawa, Bikanwas, along the Sipu River at Marol, Nimbora, Sapol, Selwara, Pipliya and along Devangan nadi in the south of Anadra and Sarliya.

Plate 7 : Dissected parabolic dunes in Revdar tehsil

Erosion of Rocky Uplands

Hills and rock out crops which occupy about 30% area are in different state of degradation either due to natural or due to human actions. The morphology of such landforms indicates various stages of weathering and erosion. The rocks particularly, granites show grooves, cavernous and spheroidal weathering, domes and sheet rocks. The Between Kalandri and Mohabat nagar in the north to Dantrai and Sirori in Sirohi tehsil an extensive area is dotted with stony-rocky-gravely waste. Places like Dantrai, Dhann, Nagoni, Pamera, Sanwarain Revdar and Ablari, Sanpur, Jaila, Akuna, Mohabbhat nagar, Kalandri, Belangari and Balda in Sirohi tehsil are dotted with granite hillocks, domes and mounds. The topography and the morphology of landforms would suggest the role of water action for the denudation of the rocky uplands. The terrain is low rocky upland with a mean elevation of 221 m. Major landforms include isolate residual hills or inselbergs, rocky pediments, granitic mounds or boulders. Some of the boulders exhibit tafoni formations which are large rounded pits that form by a physical process called cavernous weathering. Outer surface layer or the tafoni is harder than the interior. Isolated granite exposures or boulders occur frequently between Nun and Jaila. Spectacular granitic rocks with horizontal and vertical cracks occur near Phugni. These are low formations where the surface shows signs of spheroidal weathering actions. Similarly, granitic outcrops are also near Dhan, Wan and Rorwa where such landforms are in different stages of denudation. The weathered layers could be detached scale by scale which is otherwise, also known as onion-peeling or exfoliation process. Similar descriptions can be seen from the literatures which suggest that exfoliation joints are associated with landforms which have never been deeply buried. The presence of sheet jointed fractures systems make the rock more vulnerable to erosion in the form of concentric slabs. The uplands comprise inselbergs, rocky flats, rock exposures and barren rocky lands. Isolated boulder shaped formations also occur frequently as in Wan. There are number of inselbergs, morphologically, can be of rugged and domed types (Plate 8). There are a number of such instances at Kalandri, Mohabat Nagar and near Kameri riverbed (Table 2) where the rocky-gravelly pediments show different stages of erosion. The morphology of such degraded surface varies from well-jointed cracks to gravel formations.

Plate 8 : Rugged and domed granitic rock outcrops in Sirohi tehsil

Drainage Network and Water Erosion

Both Sheoganj and Sirohi tehsil area has a well network of streams which could be divided into three major 5th order stream catchments: the N flowing Sukri River with a catchment area of 1379 km^2 and length of main stream as 131 km and the west flowing Ora river (upto Ora Dam) with a catchment area of 108 km^2 and length of main stream as 24 km. Because of steep slopes and high rainfall the average density in the three catchments varies from 1013 m / km^2 (in the Sukri cathment) to 1099 m/km^2 (in the Ora catchment). The Krishnawati catchment has an average drainage density of 1048 m/km^2.

Discussion and Conclusion

Landforms are the natural environments for the growth of soils, vegetation and drainage networks. Geomorphic processes helped by climate are the key factors to the formation of a variety of landform units. Climate is a major determining factor which accentuates the role of natural processes and lands are degraded.

The studied region has three different situations; dominant rocky plain, dissected sandy dune upland and a vast alluvial plain drained by a well developed drainage network. Water erosion is major agent of erosion because of a hilly terrain, moderate to steep slopes, and shallow soils and the region receives higher than average annual rainfall. Besides these factors, expansion of cultivated land to marginal sandy areas and denudation of forest lands for fuel, food and fodder are enhancing the erosion potentials of water. Water erosion has sculptured all these landforms as evident from their morphology and sediment characteristics. In most of the places, hills are barren and show different state of denudation in the form of spheroidal weathered rocks, columnar cracks and joints, dome shaped inselbergs, erosion on the rocky pediments, residual hills. Similarly, alluvial plains show rills and gully formation and sand dunes show dissected topography. In Sheoganj tehsil, at Bammara, the granite hills have highly weathered columnar ridges with vertical cracks. With time to come, the cracks are liable to expand to disintegrate. Another situation is that of water erosion activities in a sandy terrain in Revdar tehsil. The obstacle sand dunes as well as low parabolic dunes show well developed dissected surfaces in the form of gullies of various dimensions which has impacts on the soil loss and sedimentation in the lower plains during major rainfall events. Some examples of soil erosion of higher magnitude has happened while during flood 1973. Another problem is the shortage of water despite the fact that this region experiences higher rainfall. Runoff flow has reduced after 1973. Since then there has not been any major rainfall event. Therefore, many of the streams have remained dry or without any major flow for many years. While big dams can arrest the sudden runoffs, topography of the terrain may not be appropriate to construct the type of embankments required. However studies based on DEM and simulation assessment of stream networks (Moharana and Kar 2002) can help better understanding of fluvial process in sandy terrain. The reason is related to terrain geomorphology. The hills have almost no pediment slopes, runoff length is shorter, and therefore, ground water recharge is very less. At many cases, streams emanating from hills do not reach the main stream under normal rainfall situation. It is important to assess the streams which are capable of receiving runoffs during rains and also identify those streams which are catchments to some effective dams like at Silwara. However, degradation owing to runoff dissections and gully formations can be effectively tackled through soil and water conservation practices through small watersheds, as well as increasing the vegetation cover in the affected areas.

References

Dregne, H.E. 2000. Desertification: Problems and challenges. *Annals of Arid Zone,* 39: 363-371.

Geomorphological investigations of Rajasthan Desert, Monograph No7, Central Arid Zone Research Institute, Jodhpur.

Ghose, B., Pandey, S and Lal, G. 1967. Quantitative geomorphology of drainage basins I central Luni basin, western Rajasthan. *Zeitschrift fur Geomorphologie,* 146-160.

Ghose, B.,Kar, A., and Hussain, Z. 1979: The lost courses of Sraswati river in the Great Indian Desert : New Evidences from Landsat imagery. *The Geographical Journal,* 145(3): 446-451

Kar, 1995, Geomorphology of arid western India, Memoirs of Geological Society of India, 32, 168-190.

Kar, A., Moharana, PC, Singh, S.K., Goyal, R.K. and Rao, A.S. 2007: Barmer flood, 2006, causes and consequences. IN, Flood of August 2006 in arid Rajasthan, Magnitude and Strategies (Eds. A.S.Faroda and D.C.Joshi), pp.26-39. State of Art Report,Scientific Contribution No. INCOH/SAR-29/2007;INCOH, Roorkee.

Kar, A., Moharana, P.C., Raina, P., Kumar, M., Soni, M.L., Santra, P., Ajai, Arya, A.S. and Dhinwa, P.S. 2009. Desertification and its control measures. In, *Trends in Arid Zone Research in India* (Eds., Amal Kar, B.K. Garg, S. Kathju and M.P. Singh), pp. 1-47. Central Arid Zone Research Institute, Jodhpur.

Kar, 1994. Lineament control on channel behavior during 1990 flood in the southeastern Thar Desert. *International Journal of Remote Sensing,* 15:2521-2530.

Kar, A. 1993b. Present day geomorphic processes as key to the reconstruction of Quaternary landform history in the Thar Desert, *Journal of Geological Society of India,* 41(6): 513-517.

Moharana, P.C and Kar, A. 2002. Watershed simulation in a sandy terrain of Thar Desert using GIS. *Journal of Arid Environments,* 51: 487-498.

4

Climate and Environmental Changes in the Drylands of Central Asia

Elena Lioubimtseva

Introduction

Drylands account for over 40% of the global land surface, or approximately 5.2 billion hectares (de Sherbinin, 2002). They provide home for more than 2 billion people or nearly 40% of the population of the world (White and Nackoney, 2003). Asia alone accounts for about 32% of the World's drylands. Broadly defined by their modest water supply drylands have many regional names, such as semideserts, grasslands, savannas, steppes, and pampas. Dryland ecosystems are a vital part of the earth's human and physical environments and play a major role in global biophysical processes. They provide much of the world's grain and livestock, forming the habitat that supports many vegetable species, fruit trees and micro-organisms. With an exception of Australia and the United States, almost all drylands are situated in developing countries with populations living in extreme poverty. Populations of drylands are frequently some of the poorest in the world. Food security and water availability of these areas has always been unstable due to high variability in both rainfall amounts and intensities and the occurrence of prolonged periods of drought. The increasing pressures caused by the global climate change and massive land-use changes further increase human vulnerability in dry regions. The impacts of climate change in drylands are likely to lead to still larger populations being affected by water scarcity and the risk of declining food security and increase the risk of environmental migrations and violent conflicts caused by the decline of resources, which are important to sustain livelihoods (Barnett and Adger, 2007).

Among other drylands of the world, transitional economies of Central Asia are particularly vulnerable to impacts of climate change because of the combination of biophysical, geopolitical, social, and economic stressors already present in this region (Lioubimtseva and Henebry, 2009). In additional to impacts of the global climate change and regional land-use changes, the recent degradation of the Aral Sea has severely affected the region's climate, ecosystems, economy and livelihoods (Micklin, 2007).

The usage of the term "Central Asia" needs to be clarified. In the Russian-speaking literature this region is typically defined as "Middle Asia" but excludes Kazakhstan, while the term "Central Asia" includes Kazakhstan, as well as parts of China and Mongolia (Cowan, 2007). In the English-language publications (as well as some Russian publications of the past decade) the term "Central Asia" sometimes refers to the five Asian republics of the former Soviet Union but also is used to describe the larger area, including Dzhungaria, the Takla-Makan, and Gobi (Merzlyakova, 2002; Lioubimtseva, 2002, 2004; Lioubimtseva *et al.*, 2005; Cowan, 2007). A detailed discussion of geographic usage of these terms can be found in Cowan (2007). Further in this paper I use the term "Central Asia" referring to five republics of the former Soviet-Union after collapse of the USSR.

This paper focuses on impacts of global and regional climatic and environmental changes on post-Soviet Central Asia and discusses human vulnerability to climate change. It consists of five sections. This introductory section is followed by section two, which defines the study area and its present-day climatic and environmental characteristics. Section three discusses the regional climatic changes caused by the Aral Sea degradation. Section four examines climate change scenarios and their potential impacts on the region. The final section draws some conclusions on the current and future vulnerability of the society and environment of Central Asian to climate change.

Study Area and Historical Trends of Climate Change

Based on the UNEP classification (UNEP, 1992) roughly 50% of Kazakhstan, Uzbekistan and Turkmenistan located west from 68-70°E are classified as arid, with aridity index (the ratio of precipitation and potential evapotranspiration) ranging between 0.05 and 0.2 and rainfall between 25 and 200 mm. The rest of the region is semi-arid, with aridity index varying between 0.2 and 0.5 and the annual rainfall within 200-

500 mm range (Lioubimtseva, 2002). Pockets of sub-humid zone can be found only on the slopes of the mountains in Tajikistan and Kyrgyzstan due to high elevation. Drylands of Central Asia feature continental climate with hot and dry summers and moist, relatively warm winters in Turkmenistan and south-eastern Uzbekistan but cold winters with severe frosts in the rest of the region. Temporal variability of precipitation is very high and precipitation has a distinctive spring maximum throughout most of the region. Very high daily temperature variance is recorded with frequent sand storms and intense sunshine. As in many other arid and semi-arid regions, climate of Central Asian deserts and semi-deserts is highly variable. The major controls on precipitation change in Central Asia include latitudinal shifts of the westerly cyclonic circulation and position of the Siberian high (Lioubimtseva, 2002). The North Atlantic Oscillation (NAO) exerts an important control over the pattern of winter time atmospheric circulation variability over arid and semi-arid zones of Central Asia. Over the past four decades, the pattern captured in the NAO index has altered gradually from the most extreme and persistent negative phase in the 1960s to the most extreme positive phase during the late 1980s and early 1990s. The level of the Caspian and Aral Seas have a great impact on the balance of moisture and heat in the lower atmosphere. The level of the Aral Sea, entirely depends on the run-off of the Syr Darya and Amu Darya Rivers and ultimately dependent on the rhythm of glaciations in the Pamir and Tian Shan mountains. On the other hand, the level of the Caspian Sea primarily correlates with climatic variability over European Russia because it entirely depends entirely on climatic conditions and run-off in the basins of the Volga and Ural Rivers. The detailed description of the region's physical geography and climate can be found in Lioubimtseva (2002) and Lioubimtseva *et al.* (2005).

Palaeoclimatic and archaeological data indicate that climate of arid and semi-arid Central Asia has experienced many natural fluctuations in the past. The landforms of Central Asia carry relict features both of relatively short humid intervals with runoff higher than now a days, and long arid periods (Boomer *et al.*, 2000; Lioubimtseva, 2002; Lioubimtseva *et al.*, 2005; Boroffka *et al.*, 2006). Pollen and archaeological data from the Aral Sea basin suggest that climate change was followed by significant ecosystem changes. Marine fossils, relict shore terraces, archaeological sites, and historical records point to repeated major recessions and advances of the Aral Sea (Varuschenko *et al.*, 1987; Kes *et al.*, 1993). Although the structure of the Aral Basin can be traced back to the late Neogene, the Aral Sea itself has only existed in its present form for the past 10,000 years and its Holocene history has been shaped by regional climatic variations, the development of the associated drainage system, and anthropogenic forces (Boomer *et al.*, 2000). Significant cyclical variations of regional climate and sea level during this period resulted from major changes in river discharge into it caused by climatic variability and several natural diversions of the Amudarya River away from the Aral Sea (Micklin, 1988; Vinogradov and Mamedov, 1991; Kes *et al.*, 1993; Boomer *et al.*, 2000). Early studies about the Aral Sea (Shnitnikov, 1969) demonstrate that the level of the Aral Sea could be linked to the global climatic variations. Environmental reconstructions based on pollen and archaeological data suggest that that the Aral Sea experienced a period of almost complete desiccation during the Late Glacial Maximum (around 20-18,000 years before present) and again in the Younger Dryas (12,800 and 11,500 years before present), when climate of Central Asia was featured by colder winter temperatures, cooler summers and greater aridity (Boomer *et al.*, 2000, Lioubimtseva, 2004; Tarasov *et al.*, 2007). The climate transition from the Last Glacial Maximum conditions to the Holocene was most likely triggered by changes in seasonal sunlight distribution caused by oscillations in the Earth's orbit and the tilt of the Earth's axis (Lioubimtseva *et al.*, 2005). The Djanak arid phase of the Younger Dryas was followed by an increase in temperatures and precipitation during the Early and Mid-Holocene. A trend towards greater humidity during the Holocene culminated around 6,000 years ago, a phase known in Uzbekistan and Turkmenistan as the Liavliakan pluvial (Vinogradov and Mamedov, 1991; Lioubimtseva at al., 1998). According to pollen reconstructions, mean annual precipitation in the deserts of Central Asia could be three times higher than at present in the middle of Holocene, whent desert landscapes were probably entirely replaced by mesophytic steppes, with well-developed forest vegetation along the river valleys (Sorrel *et al.*, 2007). A general trend of aridization that started approximately 5,000 years ago was interrupted by multiple minor climatic fluctuations in this region at a much finer temporal scale (Lioubimtseva *et al.*, 2005).

Historical records available from the weather stations show a steady increase of annual and winter temperatures in this region since the middle of the 20th century. On the other hand, the precipitation trends are highly variable across the region, both spatially and temporally, reflecting the great natural rainfall variability and landform diversity. Precipitation records available in this region since the end of the 19th century show a slight decrease in the western part of the region, little or no changes throughout most of the region, and relatively significant increase in precipitation recorded by the stations surrounded by irrigated lands during the past 50-60 years. This precipitation decrease in the area between the Caspian and Aral Sea mainly occurred since 1960 and coincides with the Aral Sea desiccation. Both the degradation of the Aral Sea and the dramatic fluctuations of the Kara-Bogaz-Gol Bay caused by the construction in 1980 of the Caspian-Kara-Bogaz-Gol Dam (followed by its demolition in 1992) have caused significant changes in albedo, hydrological cycle, and mesoclimatic changes throughout western parts of Kazakhstan, Uzbekistan, and Turkmenistan (Lioubimtseva and Cole, 2006).

Regional Climate Changes Caused by the Aral Sea Degradation

Covering a surface area of 67,500 km^2 in 1960, the Aral Sea was the world's fourth largest inland water body in area, behind the Caspian Sea in Asia, Lake Superior in North America and Lake Victoria in Africa (Micklin, 2010). The extensive deltas of the Syr Darya and Amu Darya sustained a diverse flora and fauna. Since 1960 the surface of the Aral Sea has shrunk to less than 17,000 km^2 (Small *et al.*, 2000; Shibuo *et al.*, 2007) and, according to the satellite imagery, it separated into four small lakes by 2009 (Micklin, 2010). During the same time the lake volume decreased by 90%, lake level dropped 23 m, area shrunk 75% and salinity has increased from 10 to more than 100 g/l (Micklin, 2007). Desiccation of the Sea has also caused an extensive massive modification of land cover in its immediate vicinity. The Aral Sea degradation has resulted in significant changes in surface albedo and temperature, soil moisture, evapotranspiration, cloud cover, precipitation patterns, wind speed and direction, atmospheric transparency, and many other mesoclimatic changes in the area adjacent to the Sea (Middleton, 2002; Small *et al.*, 2001; Shibuo *et al.*, 2007).

The thermal capacity of the remaining lake has substantially decreased due to reduction of its surface, volume and depth. Significant changes in the interactions between the land and boundary layer of the atmosphere have resulted in dramatic air temperature changes in this region. Several studies (Chub, 2000; Small *et al.*, 2001; Micklin, 2007) identified significant climatic changes resulting from the desiccation of the Aral Sea as opposed to the impacts of the global changes. The climate records from around the sea show dramatic temperature and precipitation changes since 1960. Mean, maximum, and minimum temperature near the Aral Sea have changed by up to 8°C, increasing both seasonal and diurnal amplitudes (Small *et al.*, 2001), as the lake effect has diminished. The magnitude of such changes decreases with distance from the coast, with changes extending about 200 km from the original 1960 shoreline. Precipitation records also show a shift in amount and seasonality.

The Aral Sea desiccation caused significant climate change not only in the coastal area but affected the entire system of atmospheric circulation in its basin. Summer and winter air temperatures at the stations near the sea shore increased by 1.5-2.5°C and diurnal temperatures increased by 0.5-3.3°C (Chub, 2000; Micklin, 2007). Near the coast the mean annual relative humidity decreased by 23% and recurrence of drought days increased by 300% (Middleton, 2002). The annual cycle of temperature and precipitation has also changed. A seven-fold rise in the albedo of the area previously occupied by the Aral Sea caused a three-fold increase in reflected solar radiation and increased overall continentality of the climate. Some regional modeling scenarios suggest that the rise of the air temperature in Central Asia is likely to cause further increase in evaporation by 8-15% (Chub, 2000; Small *et al.*, 2001).

According to Small *et al.* (2001), an increase in diurnal temperature range of 2-3°C is observed in all months throughout the Aral Sea area. Small and colleagues (2001) have examines the Aral Sea surface temperature (SST) trends between 1960 and 1996 using *in situ* buoy and boat SST measurements from the State Oceanographic Institute in Russia, combined with Multi-Channel SST derived from the Advanced Very High Resolution Radiometer satellite imagery. They found that the highest change in the Aral SST occurred in spring – a 4 to 5°C increase in April and in May. SST increase in summer was about 3°C, and there were no changes between August and October. During the same period of time SSTs decreased by 4-5 °C in November and December (Small *et al.*, 2001).

The basin-scale water diversion and irrigation, along with the desiccation of the Aral Sea have considerably increased evapotranspiration and thereby decreased net water flux from the atmosphere to the surface. Increased evaporation cools the irrigated areas, and the decrease of net atmosphere water influx is likely to affect the entire basin. Modeling experiments by Shibuo *et al.* (2007) indicate important effects of water diversion on the regional climate at the scale of the entire Aral Sea basin. The excessive irrigation in the southeastern part of the Aral Sea basin appears to have significantly increased evapotranspiration and cooled this area in the process. By contrast, temperature increases are considerable in non-irrigated areas, where hydro-climatic changes reflect local effects of the Aral Sea shrinkage itself in addition to the regional manifestation of global climate change. The main reported precipitation increase in the Aral Sea basin is also localized to the southeastern part of the basin is likely to be an effect of the local evapotranspiration increase due to irrigation. By contrast all stations in immediate vicinity of the Aral Sea have experienced precipitation decrease during the same period of time (Lioubimtseva and Cole, 2006). Such influence of the local land-cover changes have been reported in many studies in other arid and semi-arid regions (Pielke, 2007).

In addition, the exposure of the former lake-bed areas, especially on the eastern side of the Aral Sea, represents an enormous source of highly saline wind-blown material (up to 1.5% salt in the total mass of hard particles transported by the wind). The dry bed of the Aral Sea has become one of the biggest sources of dust aerosols in the world. Salty dust blown into the atmosphere is another important factor that needs to be considered in model simulations of both global and regional climates. Dust tends to cool the earth by reflecting sunlight back into space, and it decreases rainfall by suppressing atmospheric convection (Lioubimtseva *et al.*, 2005). This effect is not taken into account in the regional climate change scenarios and the future increase of aridity in the western part of Central Asia might be underestimated by the models.

As Aral Sea states, Kazakhstan and Uzbekistan are immediately affected by the regional climate change caused by the reduction of water volume in the Aral Sea and impacts of salt and dust storms carrying particles from the former sea-bed for hundreds kilometers around. The salinity of the Aral Sea has already exceeded 100-150 g/l (Small *et al.*, 2001; Aladin *et al.*, 2005; Micklin, 2007). The same processes that contributed to the Aral Sea degradation—excessive irrigation and mismanagement of water—have also resulted in the rise of groundwater table, which then became contaminated with high levels of salts and other minerals (Small *et al.*, 2001). Groundwater quality ranges in the region from a minimum of 1.5 g/l total dissolved solids to 6 g/l, which is 20 times higher than in the US (about 300 mg/l) and drinking water reaches levels of up to 3.5 g/l total dissolved solids (Lioubimtseva and Cole, 2006). In Karakalpakstan, an autonomous republic of Uzbekistan adjacent to the Aral Sea, about 65% of drinking water samples tested did not meet national standards of 1 g/l (AQUASTAT, 2011). At the same time water consumption in the Central Asian countries of the former USSR continues to increase and is projected to reach 122 km^3 per year by 2025 (Shiklomanov and Rodda, 2001). There is a growing concern that water stress in Central Asia may lead to open water conflicts between the states and also weaken them to such extend that they lose their capacity to address other threats to stability and development (Sievers, 2002; Glantz, 2005; Lioubimtseva and Henebry, 2009).

Restoration and conservation efforts in Kazakhstan have recently resulted in improvement of microclimatic conditions in vicinity of the

Small Aral (northern part of the Aral Sea). Separated by a dyke from the Large Aral, the Small Aral Sea is now showing steady signs of water level increase and decline of water salinity (Miklin, 2008). In the early 1990s Kazakhstan constructed an earthen dyke to block outflow to the south but in April 1999 the dyke collapsed. The second 13-km dyke with a gated concrete dam for water discharge was completed in November 2005. Heavy runoff from the Syr Darya in the ensuing winter jump-started the Small Aral's recovery. The water rose from 40 to 42 meters—the intended design height—in only eight months. Area increased by 18 percent, and salinity has dropped steadily, from roughly 20 to about 10 g/l today (Micklin, 2007). Fishers are once again catching several species in substantial numbers—most important, the highly prized pike perch and carp (Aladin *et al.*, 2005).

Climate Change Impacts and Human Vulnerability

Warming of the global climate system is unequivocal, as is now evident from observations of increases in global average air and ocean temperatures, widespread melting of snow and ice and rising global average sea level. Changes in atmospheric concentrations of greenhouse gases (GHGs) and aerosols, and land cover alter the energy balance of the climate system. Global GHG emissions due to human activities have grown since pre-industrial times, with an increase of 70% between 1970 and 2004 (IPCC AR4, WGI, 2007).

This author has conducted several assessments of the annual, seasonal and monthly climate change scenarios for Central Asia and their probabilities computed by the Atmosphere Ocean Global Climate Models (AOGCMs) used in the Third and Fourth Assessment Reports by the Intergovernmental Panel on Climate Change (IPCC TAR, 2001; IPCC AR4 WGI, 2007). Detailed discussion of the climate change scenarios for Central Asia can be found in Lioubimtseva (2004), Lioubimtseva *et al.* (2005), Lioubimtseva (2005), and Lioubimtseva and Henebry (2009). AOGCM scenarios from twenty different models under different IPCC SRES policy scenarios (Nakicenovic *et al.*, 2000) indicate a generally good agreement that the current trend of temperature increase in arid Central Asia is likely to continue. The ranges of precipitation projections produced by AOGCMs are still uncertain for Central Asia and drylands elsewhere. The majority of climate models project a slight decrease in precipitation rate over most of the region with a stronger decrease in the west and southwest and a very slight increase in the north and east of this region (Lioubimtseva, 2005). However, given the low absolute amounts of precipitation and high inter-annual, seasonal, and spatial patterns of precipitation across the region, the changes in precipitation rate projected by climate models should be addressed with caution. It is the change in the temporal and spatial variability of precipitation and its seasonal distribution—rather than absolute precipitation values—that are more important for the assessment of human vulnerability of this arid region, but they are also more difficult to assess and project.

Temperature changes are predicted to increase by 3-5°C around 2080 and all the AOGCMs used in the IPCC AR4 agree that the warming will be accompanied by a further increase of aridity (Lioubimtseva and Henebry, 2009). The rates of the projected changes significantly differ across seasons, with much higher temperature changes generally expected during the winter months. Despite significant differences in the ranges of change among the scenarios, the majority of the recent AOGCM experiments tend to agree that precipitation is likely to increase both northward (European Russia and Central Siberia) and southward (Northern India, Iran, Pakistan) from Central Asia. For the Central Asian plains, however, the expectation is for increasingly dry conditions with a slight increase in winter rainfall, but decreases particularly in spring and summer. This trend towards higher aridity is projected to be more significant west from 70-72 °E. The AR4 supports these findings, pointing out that Central Asia, particularly its western parts, is very likely to become drier during the coming decades (IPCC AR4 WGI, 2007). The AOGCM scenarios appear to be consistent with the observed temperature and precipitation trend over the past decades in most of the region. However, it is uncertain the extent to which the observed and projected trends result primarily from the global restructuring of atmospheric circulation and changes in the teleconnections controlling macroclimatic conditions over Central Asia versus mesoclimate changes induced by regional land use change.

There are several sources of uncertainties associated with climate change scenarios predicted by the computer models. The resolution of AOGCMs is quite coarse (a horizontal resolution of between 250 and 600 km, 10 to 20 vertical layers in the atmosphere and sometimes as many as 30 layers in the oceans). Many physical processes, such as those related to clouds, occur at more detailed scales and cannot be adequately modeled; instead, their known properties are averaged over the larger scale in a technique known as parameterization (IPCC TAR, 2001). Other uncertainties relate to the simulation of various feedback mechanisms in models concerning, for example, water vapor and warming, clouds and radiation, ocean circulation and ice and snow albedo (Arnell, 2004; IPCC, WGI, 2007).

Human Vulnerability to Climate Change

Today drylands of Central Asia represent an area with multiple environmental, social and economic stresses. The well-being and security of this vast region depend on interplay of several groups of factors, such as institutional changes (agricultural reforms, management practices, policies, legislation, and international agreements) and the subsequent regional land-use changes, climate variability and change, and internationalization of the global economy (Leichenko and O'Brien, 2002).

Central Asia is projected to become warmer and probably drier during the coming decades. Aridity is expected to increase across the entire region, but especially in the western part of Turkmenistan, Uzbekistan, and Kazakhstan, where the Aral Sea degradation has already significantly modified the region's climate and economy. These seasonal climatic shifts are likely to have profound implications for agriculture when some parts of the region can be winners (cereal production in northern and eastern Kazakhstan can benefit from the longer growing season, warmer winters and slight increase in winter precipitation), while others can be losers (particularly western Turkmenistan and Uzbekistan, where frequent droughts will negatively affect cotton production, increase already extremely high water demands for irrigation, and exacerbate the already existing water crisis and human-induced desertification). The ongoing severe drought, particularly during past decade, has already resulted in multiple water disputes and increased tensions among the states of the Aral Sea basin. Knowing that the aridity and water stress are likely to increase, new political and economic mechanisms are necessary to ease such tensions in future.

The ability of Central Asia to adapt to hotter climate is already severely limited by the existing water stress and the regional land degradation and poor irrigation practices. Central Asia inherited many environmental problems from the Soviet Union but now, more than twenty years after independence, the key land and water-use related problems remain severe. Decline of agriculture after independence, documented by agricultural statistics, was significant enough to produce a signal in the temporal series of remote sensing data (Chuluun and Ojima; 2002; Lioubimtseva, 2005; Lioubimtseva and Henebry, 2009; Kariyeva and van Leewuven, 2011). Agricultural transformation had extremely high social cost but to date agricultural reforms and transition to market remain problematic in most of the region. Increasing rural poverty and unemployment, growing economic and social inequality, and shortage of adequate living conditions, healthcare and water-management infrastructure have significantly increased human vulnerability of the majority of population in the region. Any assessment of climate change impacts should be put in the context of human factors, such as political and institutional changes, economic development and globalization, changes in the land-use practices and livelihoods. There is compelling evidence from many parts of the world that there is a strong inverse relationship between human vulnerability to climate change and sustainable development (Parry, 2007; Polsky *et al.*, 2007).

Non-climatic stresses are likely to increase vulnerability to climate change and reduce its adaptive capacity because of resource deployment to competing needs. Increases in surface temperature and frequency of droughts in the Aral Sea basin, decline of precipitation and the length of the growing season due to the microclimatic changes caused by the Aral Sea degradation, soil salinization and degradation, degradation of vegetation cover, water loss due to inadequate irrigation practices, chemical runoff from agriculture, combined with degradation of living standards, malnutrition and limited access to drinking water and sanitation, health care collapse and outbreaks of many chronicle and infectious diseases, and many other regional stresses would require unprecedented amount of resources to alleviate just some of them. In the context of dry climates of Central Asia short-term, unplanned reactive coping strategies aiming to address separately some of these stresses usually provide only an immediate solution for a limited areas or groups of the population, but in the long-term they only exacerbates the problem. Focusing on effects but not on the causes of the problem they can only further aggravate the ongoing adverse environmental changes in the long term. For example, there is a continuous migration of the population from Karakalpakstan, an autonomous republic within Uzbekistan, adjacent to the Aral Sea, to eastern Uzbekistan and Kazakhstan. During the first years after collapse of the USSR the estimated number of environmental migrants from the Aral Sea area was more than 100,000 people and in the recent decade the net emigration from the areas adjacent to the Aral Sea has doubled from over 3,000 to over 6,000 persons per year (Akiner, 2000). About 10% of working-age population of this region is leaving home to work every year (Elpiner, 2003, Elpiner, personal communication 2011). Some studies in other parts of the world suggest that migration is a feasible climate adaptation strategy (Patz *et al.*, 2005; Adger, 2006). Considering, however, that these environmental refugees are usually individuals who had the skills, opportunity, and psychological aptitude to migrate and adjust to different lifestyles in other regions or countries, it is likely that the population left behind would have even lower capacity, skills and potential to adapt to the regional environmental crisis (Lioubimtseva and Henebry, 2009).

To cope with the multiple regional stresses in the context of multiple increasing stresses, both related and non-related to climate change, it is important to consider such adaptive strategies that could place equal importance on environmental, social, and economic considerations. Development of such adaptation strategy involves inevitable trade-offs between environmental, economic, and socio-cultural and political considerations and priorities. There is evidence from around the World that development and implementation of adaptation strategies and policies are successful only when they are driven by the interests of stakeholders - groups of individuals and communities vulnerable to the risks of climate change (Schröter *et al.*, 2005; Adger, 2006; Polsky *et al.*, 2007). At the national and regional scale adaptations are usually undertaken by the governments on behalf of the entire society or particular groups but regardless the geographic scale, these decisions, policies and projects must be driven by the "place-based" initiatives and integrate the needs of various communities at multiple scales. Communities rarely face only one effect or risk of climate change at a time and the interaction of multiple vulnerabilities often can lead to amplification of risks (Schröter *et al.*, 2005; Liechenko and O'Brien, 2002). Climate change impacts are interconnected with land-use changes, socio-economic changes, and many other processes that interact in the human-environmental system. Therefore, adaptations can be sustainable only if they target multiple processes and risks in the integrated manner, reaching across various aspects of human life (food security, water resources, health, quality of life, etc.) at multiple geographic and temporal scales. For example, reduction of cotton monoculture, diversification of crops, and application of no-tillage techniques in agriculture could not only help to increase food security but also would decrease the use of water, improve soils through the nitrogen fixation in soil. These measures would be useful as carbon sequestration measures contributing to climate change mitigation. The renovation of the existing irrigation network, and introduction of more advanced irrigation techniques, such as drip irrigation could significantly reduce the loss of water resources, but also would improve crop productivity, reduce the soil losses due to salinization, and help to reduce the risks of water contamination and transmission of many vector-borne and water-borne diseases.

In conclusion, comprehensive systematic regional and national policies targeting sustainable economic, social, and environmental aspects of development are imperative to mitigate adverse impacts of climate and land-cover changes on the drylands of Central Asia.

References

Aladin, N., J.-F. Crétaux, I. S. Plotnikol, A. V. Kouraev, A. O. Smurov, A. Cazenave, A. N. Egorov, F. Papa, 2005. Modern hydro-biological state of the Small Aral Sea. *Environmetrics*, Vol. 16 (18), pp. 375-392.

Akiner, S., 2000. *Central Asia: A Survey of the Region and the Five Republics.* UNHCR Centre for Documentation and Research, WRITENET Paper No.22/1999, United Nations High Commissioner for Refugees, Geneva, 50 p.

Adger, W.N., 2006. Vulnerability. Global Environmental Change, Vol. 16 (3), pp. 268.

Arnell, N.W. 2004. Climate change and global water resources: SRES emissions and socio-economic scenarios. *Global Environmental Change*, Vol. 14(1), pp. 31-52

AQUASTAT. 2011. Food and Agriculture Organization, Rome, http://www.fao.org/nr/water/aquastat/main/index.stm, last access December 2011.

Barnet, Jon and W. N. Adger. 2007. Climate change, human society and violent conflict. *Political Geography*, Vol. 26 (6), pp. 639-655.

Boomer, I., N. Aladin, I. Plotnikov, R.Whatley. 2000. The palaeolimnology of the Aral Sea: a review. Quaternary Science Review, Vol. 19 (13), pp. 1259-1278.

Boroffka, N., H. Oberhänsli1, Ph. Sorrel, F. Demory, Ch. Reinhardt, B. Wünnemann, K. Alimov, S. Baratov, K. Rakhimov, N. Saparov, T. Shirinov, S. K. Krivonogov, U. Röhl. 2006. Archaeology and climate: Settlement and lake-level changes at the Aral Sea. *Geoarcheology*, Vol. 21 (7), pp. 721-734.

Chub, V.E. 2000. *Climate change and its impact on the natural resources potential of the Republic of Uzbekistan.* (Izmenenija klimata I ego vlijanije na prirodno-resursnij potensial respubliki Uzbekistan), in Russian, Gimet, Tashkent, 253 p.

Chuluun, T. and D. Ojima. 2002. Land use change and carbon cycle in arid and semi-arid lands of East and Central Asia. *Science and China (Series C)*, Vol. 45, pp. 44-54.

Cowan, P.J. 2007. Geographic usage of the terms Middle Asia and Central Asia. *Journal of Arid Environments*, Vol. 62 (2), pp. 359-363.

de Sherbinin, A. 2002. A Guide to Land-Use and Land-Cover Change (LUCC). Land-Use and Land-Cover Change, A CIESIN Thematic Guide, Palisades, NY: Center for International Earth Science Information Network of Columbia University. Available on-line at http://sedac.ciesin.columbia.edu/tg/guide_main.jsp.

Elpiner L.I. 2003. A scenario of possible effect of changes in the hydrological conditions on the medical and environmental situation: on the problem of global hydroclimatic changes. *Water Resources*, Vol. 30 (4), pp. 434-444.

Glantz, M.H. 2005. Water, climate, and development issues in the Amu Darya basin. *Mitigation and Adaptation Strategies for Global Change*, Vol. 10(1), pp. 1381-2386.

IPCC 2001. *Climate Change: The Scientific Basis Contribution of Working Group I to the Third Assessment Report of the Intergovernmental Panel on Climate Change (IPCC)* J. T. Houghton, Y. Ding, D.J. Griggs, M. Noguer, P. J. van der Linden and D. Xiaosu (Eds.) Cambridge University Press, UK, pp 944.

IPCC, WGI, 2007. *Climate Change 2007: The Physical Science Basis. Contribution of Working Group I to the Fourth Assessment Report of the Intergovernmental Panel on Climate Change (IPCC)* Solomon, S., Qin, D., Manning, M., Chen, Z., Marquis, M., Averyt, K.B., Tignor, M., Miller, H.L.(eds.), Cambridge University Press, Cambridge, United Kingdom and New York, NY, USA, 996 pp.

IPCC, WGII, 2007. *Climate Change 2007: Impacts, Adaptation and Vulnerability. Contribution of Working Group II to the Fourth Assessment Report of the Intergovernmental Panel on Climate Change*, Parry, M.L., Canziani, O.F., Palutikof, J.P., van der Linden, P.J., Hanson, C.E., (eds.), Cambridge University Press, Cambridge, UK, and New York, NY, USA, 976pp.

Kariyeva J. and W. van Leewuven. 2011. Environmental Drivers of NDVI-Based Vegetation Phenology in Central Asia. *Remote Sensing*, Vol. 3(2), pp. 203-246; doi:10.3390/rs3020203.

Kes, A. S., E. D. Mamedov, S. O. Khondkaryan, G. N. Trofimov, and K.V. Kremenetsky. 1993. Plains of northern central Asia during the Late Pleistocene and Holocene: Stratigraphy and palaeogeography (Stratigrafija i paleografija ravninnih oblastej Srednej Azii v pozdnem pleistocene i golocene) (in Russian), pp. 82–87. In: *Evolution of Landscapes and Climates of the Northern Eurasia* (Velichko, A. A., Ed.). Nauka Publishers, Moscow.

Leichenko R.M. and K.L. O'Brien. 2002. The dynamics of rural vulnerability to global change: the case of southern Africa. *Mitigation and Adaptation Strategies for Global Climate Change*, Vol. 7, pp. 1-18.

Lioubimtseva, E., B. Simon, H. Faure, L. Faure-Denard, and J. M. Adams. 1998. Impacts of climatic change on carbon storage in the Sahara-Gobi desert belt since the Late Glacial Maximum. *Global Planetary Change*, Vol. 16–17, pp. 95–105.

Lioubimtseva, E., 2002. Arid Environments, In: M. Shahgedanova (editor) *Physical Geography of Northern Eurasia*, pp.267-283, Oxford University Press, Oxford.

Lioubimtseva, E. 2004. Climate change in arid environments: revisiting the past to understand the future. *Progress in Physical Geography*, Vol. 28 (4), pp. 502-530.

Lioubimtseva, E. 2005. Environmental changes in arid Central Asia inferred from remote sensing data and ground observations. *Arid Ecosystems*, Vol. 11 (26-27), pp. 64-72.

Lioubimtseva, E., Cole, R; Adams, JM; Kapustin, G. 2005. Impacts of climate and land-cover changes in arid lands of Central Asia. *Journal of Arid Environments*, Vol. 62, (2), pp. 285-308.

Lioubimtseva E. and R. Cole. 2006. Uncertainties of Climate Change in Arid Environments of Central Asia. *Reviews in Fisheries Science*. Vol. 14 (1,2), pp. 29 - 49.

Lioubimtseva E. and Henebry G.M. 2009. Climate and environmental change in arid Central Asia: Impacts, vulnerability, and adaptations. *Journal of Arid Environments*. Vol. 73, pp. 963-977.

Merzlyakova, I. 2000. The mountains of Central Asia and Kazakhstan. In: M. Shahgedanova, Editor, *The Physical Geography of Northern Eurasia*, Oxford University Press, Oxford, pp. 377–402.

Micklin, P. 1988. Dessication of the Aral Sea: A water management disaster in the Soviet Union. *Science*, Vol. 241, pp. 1170–1176.

Micklin, P. 2007. The Aral Sea disaster. *Annual Review of Earth and Planetary Sciences*, Vol. 35, pp.47-72. DOI:10.1146/annurev.earth.35.031306.140120.

Micklin, P. 2010. The past, present, and future Aral Sea. *Lakes & Reservoirs: Research and Management*. Vol.15, pp. 193–213.

Middleton, N.J. 2002. The Aral Sea. In: M. Shahgedanova, Editor. *Physical Geography of Northern Eurasia*, Oxford University Press, Oxford, pp.497-510.

Nakicenovic N., J. Alcamo, G. Davis, B. de Vries, J. Fenhann, S. Gaffin, K. Gregory, A. Grübler, T. Y. Jung, T. Kram, E. L. La Rovere, L. Michaelis, S. Mori, T.Morita, W. Pepper, H. Pitcher, L. Price, K. Riahi, A. Roehrl, H.-H.Rogner, A. Sankovski, M.Schlesinger, P. Shukla, S. Smith, R. Swart, S. van Rooijen, N. Victor, Z. Dadi. 2000. *IPCC Special Report on Emissions Scenarios*, IPCC Special Reports, Cambridge University Press, Cambridge, 599 p.

Parry M.L., O.F. Canziani, J.P. Palutikof, P.J. Van der Linden, and C.E. Hansen. Eds., 2007. *Climate Change 2007: Impacts, Adaptations and Vulnerability. Contribution of Working Group II to the Fourth Assessment Report of the IPCC*, Cambridge University Press, Cambridge, 1000 pages.

Patz, J.A., D.Campbell-Lendrum, T.Holloway and J. A. Foley. 2005. Impact of regional climate change on human health. *Nature*, Vol. 438 (17): 310-317.

Pielke Sr., R.A., J. Adegoke, A. Beltran-Przeukurat, C. Hiemstra, J. Lin, U.S. Nair, D. Niyogi, and T.E. Nobis. 2007. An overview of regional land-use and land-cover impacts on rainfall. *Tellus*, DOI: 10.1111/j.1600-0889.2007.00251.x.

Polsky, C., R. Neff, and B. Yarnal. 2007. Building comparable global change assessments: the vulnerability scoping diagram. *Global Environmental Change*, Vol. 17(3-4) pp. 472-485.

Shibuo, Y., J. Jarsjo,and G. Destouni. 2007. Hydrological responses to climate change and irrigation in the Aral Sea drainage basin. *Geophysical Research Letters*, Vol. 34, L21406, doi:10.1029/2007GL031465.

Shiklomanov, I.A., Rodda, J.C. (Eds.), 2001. *World Water Resources at the Beginning of the Twenty-First Century*, Cambridge University Press, Cambridge, UK, 449 p.

Schröter, D., C. Polsky and A. G. Patt. 2005. Assessing vulnerability to the effects of global climate change: an eight step approach. *Mitigation and Adaptation Strategies for Global Change*, Vol.10, pp. 573-596.

Shnitnikov, A. V. 1969. *Inter-centennial variability of general humidity components* (Vnutrivekovaja izmenchivost komponentov obchej uvlaznennosti) (in Russian). Nauka Publishers, Leningrad.

Sievers, E.W. 2003. *The Post-Soviet Decline of Central Asia: Sustainable Development and Comprehensive Capital*. Taylor & Francis, London, New York, 264 p.

Small, E.E., F. Giorgi, L.C. Sloan, S. Hostetler. 2000. The Effects of Desiccation and Climatic Change on the Hydrology of the Aral Sea. *The Journal of Climate*. Vol. 14, pp. 300-322.

Sorrel, P., S.-M. Popescu, S. Klotz, J-P. Suc, H. Oberhänsli. 2007. Climate variability in the Aral Sea basin (Central Asia) during the late Holocene based on vegetation changes. *Quaternary Research*. Vol. 67, pp. 357–370.

Tarasov, P., J.W. Williams, A. Andreev, T. Nakagawa, E. Bezrukova, U. Herzschuh, Y. Igarashi, S. Müller, K.Werner, Zh. Zheng, 2007. Satellite- and pollen-based quantitative woody cover reconstructions for northern Asia: Verification and application to late-Quaternary pollen data. *Earth and Planetary Science Letters*, Vol. 264 (1–2), pp. 284-298

UNEP. 1992. *World Atlas of Desertification*. Nairobi.

Varuschenko S.I., A.N. Varuschenko and Klige R.K. 1987. *Izmenenija rezima Kaspijskogo morja i besstochnih vodoemov v paleovremeni* (Variations of the Caspian sea regime and of closed lakes in palaeotimes), (in Russian), Nauka, Moscow, 239 pp.

Vinogradov A. V. & Mamedov E. D. 1991. The Aral Sea in the Holocene. In: *Aral Crisis* (Eds B. V. Andrianov, N. F. Glazovskiy, L. M. Levina & L. A. Timoshuk), pp. 66–75. (in Russian).USSR Academy of Sciences, Moscow.

White, R.P. and J. Nackoney. 2003. Drylands, people, and ecosystem goods and services: A Web-Based Geospatial Analysis, World Resources Institute, http://pdf.wri.org/drylands.pdf.

5

Survival Strategies Facing Effects of Climate Change in the North Cameroonian Drylands

Wakponou Anseleme, Monique Mainguet and et Frederic Dumay

Abstract

This paper is the synthesis of researches of quaternary on the Chad lake basin, our field investigations, our previous palaeogeographic research underlined by the sédimentological analyses, ^{14}C *radiocarbon and thermoluninescence dating. The objective is to show that after the effects of geological fluctuation, paleoclimatic oscillations, the Chad lake basin experiences since historical times an increasingly drying up due to the combination of geographic factors such as its close position in the Sahara desert, centre of the climate dynamics and human activities. This situation frequently generates recurrent live conditions to which man has to face.*

Introduction

Cameroonian drylands belong to the Chad basin. This area close to the Sahara desert, centre of climate dynamics has always deeply undergone climatic changes, that are oscillations of the recent Quaternary, historical times fluctuations marked by droughts and the actual drying-up. The field work extends from the upper Benoue river basin (9° N) to the actual south bank of Chad lake (13° N) (figure 1).

The objective of this paper is to show that after the influences of palaeoclimate oscillations, the Chad lake basin since historical times is submitted to an increasingly drying-up condition. This status frequently generates live conditions resulting in an environmental crisis always responsible of social difficulties that the society has to solve for its survival.

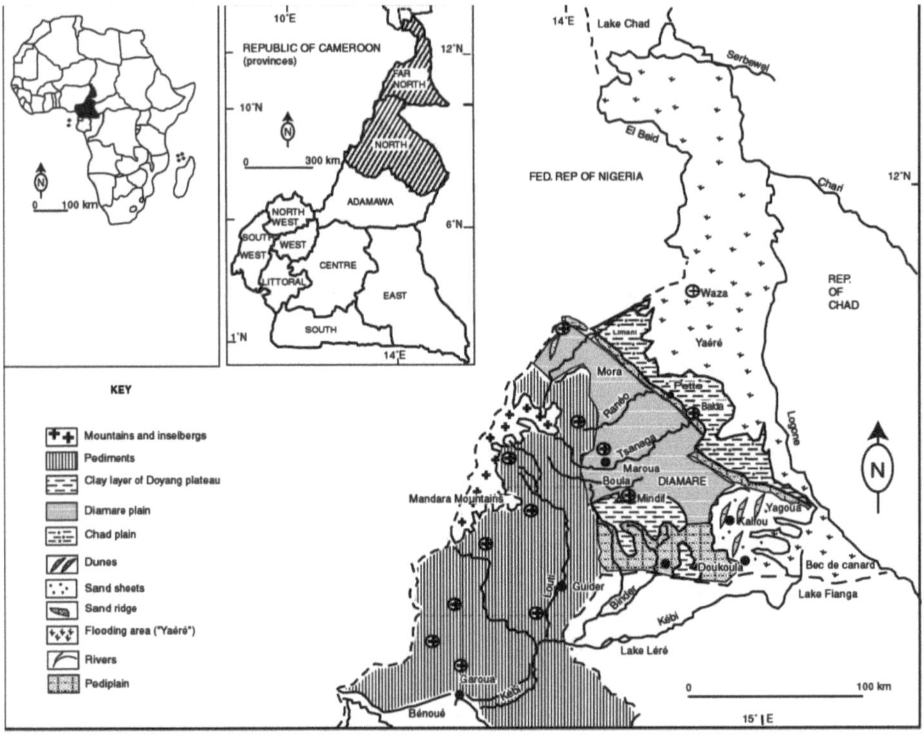

Fig. 1 : Location map of the study area with some geomorphic aspects of North Cameroon

From the Recent Quaternary Paleoclimate Oscillations to Present Day Variations in Chad Lake Basin and Its Southern Limit

The Chad lake basin is the most studied area of Central Africa. Indeed, numerous palaeoforms and palaeoformations that it contains: erg, sand dunes, sandridges, pediments, ferruginous tropical soil patches ... have been mysteries during long times as for their origins and their ages (Wakponou, 2004). Furthermore, fluctuations of the level of the Chad lake resulting from its closed basin and its borders have to be correlated with the different geomorphic activities, fluvial activities generated by the precipitation rates, the vegetative cover which all constitute good indicators of the climate evolution in the area.

We propose on the table 1, a chronology of the quaternary in eight episodes clarifying the climate and the geomorphologic evolution of the Extreme-North of Cameroon. The evolution is nowadays dependent on human actions, responsible of the degradation (desertification) of the environment (Wakponou, 2004).

Historic droughts are well-known. Indeed, reports of the early explorers in Cameroon at the end of the 19th and at the beginning of the 20th Century (Peterman, 1854 ; Barth, 1860 ; Freydenberg, 1907) quoted by Beauvilain (1985 and 1986) and recent investigations mention those droughts in 1830, 1850 – 51, 1893, 1908, 1912 - 1914, 1939 – 45, 1955, 1968, 1972, 1974, 1980, 1983 - 1984 and 1985. The figure 2 gives the rainfall fluctuations in the Cameroonian dry areas responsible of the climate fluctuations.

The Cameroonian sudano-sahelian and sahelian ecosystems as everywhere are prone to water resources shortage. 7 to 8 months of the dry season with severe insolation (2750 to 3000 h per year) and hight temperatures (28° average per year) and intense evaporation (more than 2000 mm per year) are depended. The hydric deficit shown by a weak climatic rate (0,21 in Maroua-Salak : N 10° 18' -E 14° 14') (Suchel, 1987), generates a quasi general drying-up, worsened by hot and dry continental Harmattan winds. These winds (NE - SW), loaded by siliceous desert dusts called dry hazes, are predominant (60 to 70 %) from January to March, in the heart of the dry season. Thermal contrasts are pronounced. While diurnal temperatures of the hottest months (March, April) exceed 35°C and even 41°C towards the Chad lake, months of December and January experience cold nights : 10°C in Maroua-Salak, 10°C in Mokolo (N 10° 45' - E 13° 49') in the Mandara mountains and 9°C in Guider (N 9° 56' - E 13° 57'); this situation is obviously due to powerful polar urges in the heart of winter in high and mean latitudes. The shortness and the precariousness of rainfall, seasonal thermal contrasts are the climatic characteristics of the area.

Table 1 : A proposal of the quaternary paleoenvironmental chronology of the Chad lake basin.

Ages BP (OBDY Lab –No., radiocarbon dating TL Age (Thermoluminescence dating) Reference and source of proxy data	**Paleoclimatic interpretation**	**Location & archaeological stages**		**Vegetation**	**Morphodynamic dynamics processes**
		Chad lakebasin	**Northern Cameroon**		
100 ± 160 BP (OBDY 578), (recent ?) charcoal at 20 cm depth 1.170 ± 150 BP (OBDY 801), charcoal at 65 cm depth 2.780 ± 90 BP (OBDY 577), charcoal at 170 cm depth 3.330 ± 1.000 TL and 3. 550 ± 760 TL, sand on the top of dune 4.260 ± 35 BP (OBDY 869), fossil wood at 5 m depth	Short alternations of humid and dry periods		Actual, recent (Hervieu, 1979)	Expansion of forest	8 : recent fluvial dynamics : acceleration of soil erosion, outcrop of the basement complex; continuation of downslope pedimentation; phyto-instability on the slopes by human activities; filling of river beds; beginning wind erosion and deflation.
6.000 ± 240 TL (Durand *et al.*,1981). 8.000 BP (Pullan, 1965) 10.000 BP (Grove *et al.*, 1964) (aeolian 8.720 ± 420 BP (OBDY 379), nodular limestone	Short episode of tropical humid climate	Nigero-Tchadien (Servant, 1983)	Sub-actual (Hervieu, 1979)	forest	7 : major lacustrine episode : inducing lacustrian bar fine sand, coarser grain) and filling up

Contd.

					of the clay blow-outs; formation of limestones concretions within the clayey soils.
7.000 BP Diatomite layer, aeolian sands (north of Maga Chad lake) (Servant, 1983) ; Lake regression around 6.300 ± 200 BP 14.500 ± 1.690 and 15.360 ± 2.040 TL, Pette, red sands at the base and 3.330 ± 3.350 TL at the top of a dune 11.940 ± 750 TL, red sand 8.980 ± 1.150 TL and 8.240 ± 1.120 TL, red sand on the top of a dune 10.640 ± 3.930 TL and 8.670 ± 2.140 TL red sand on the top of dune	Dry period Sub-humid- sudanian rainfall 800 mm per year Saharo sahelian arid to semi-arid period; rainfall below 400 mm per year	Kanemian (Servant, 1983)	Bossoumian (Hervieu, 1979)	forest regression formation of an open and discontinuous vegetation	6 : aeolian episode : disappearance of the lakes (Servant, 1980); negative sediment budget, appearance of longitudinal sandriges, temporaly drying up of the Chad lake;formation of blow-outs; pedogenesis of the sand of Kalfu sandsea; red palaeosoil formation; deposit of palaeo sandsea in Cameroon extending to the south to Doukoula (N10°08′ E 14°59) towards Kolofata (N11°11E14°02) at the northwest.
34.460 ± 5.690 – 32 950 ± 5.160 BP (TL), palaeosoil at the base of sand dunes at Pétté	Dry period Tropical humid climate .	Upper Ghazalian (Servant, 1983)	Peskeborian (Hervieu,1979)	dense vegetation (forest ?)	5 : loess Episode: Doyang layer (Mainguet and

Contd.

According to Mathieu (1980), Pleistocene Lake episode: Mega Chad lake 400-420 m (Pias, 1967); According to Servant and Servant -Vildary (1978), many lakes within the Chad lake basin					Wakponou, 2002); fluvial episode (Marliac and Gavaud, 1975) ; silt-clay layer ; kaolinite reworked by Aeolian action; High evapotranspiration of the small lakes ; discordant sedimentation of silt and clay on the Continental Terminal (CT).
Middle stone Age Advanced Acheuleen (Marliac and Gavaud, 1975) , Paleolithic stone tool encrusted duricrust 50 000 BP.	Dry/sub-humid with two contrasting seasons : rainfall estimated between 600 and 800 mm	Lower Ghazalien (Servant, 1983)	Douroumian (Hervieu, 1979)	More open discontinuous vegetation cover	4 : Pisolitic duricust formation with a coarsed vacuolar structure
Plio - PleistoceneTertiary, Miocene ?	Semi-arid	Pre-Ghazalian (Servant, 1983)	Pre-Douroumian (Hervieu, 1979)	Sparse Vegetation cover	3 : Pedimentation on inselberg footslopes
Eocene, Oligocene	Long arid period			Sparse Vegetation cover	2 : Continental Terminal deposits
Mesozoic	Tropical humid climate			Dense Vegetation cover	1 : Humid tropical weathering of the Precambrian basement

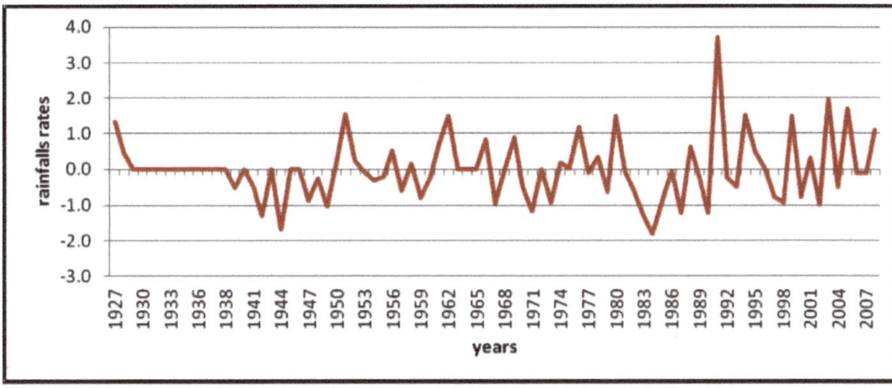

Sources of data : Archives de la SOCOTON

Fig. 2 : Fluctuations of rainfall in Maroua during 82 years, from 1926 to 2008.

Manifestations and Climate Change Perceptions

The sahelian cameroonian zone has 90% rural activities. The empirical perception of the farmer is summarized as " it's no longer raining as before ". The qualitative referents to define empirically climatic changes are therefore: hydric deficits, permanent agricultural calendar disruptions and recurrent grasshoppers invasions. Thus, decrease of rainfall is the main characteristic of the climatic changes.

The sketch of inter-annual isohyets in the Extreme-North of Cameroon is given in figure 3. Annual rainfall means increase from 1000 mm at Guider: N 9° 36′ - E 13° 57′, altitude 340 m in the Sudano-Sahelian area in the south to 400 mm around the Chad lake borders in the sahelian -semi-arid zone in the north. The length of the rainy season decreases in the same way: from 6 months (April – September) to three months (June, July, August). The last two months experience the higher part of rains with above 50 % of the total rainfall. Rainfall increase from the East (Maroua : 800 mm ; N 10° 37′ - E 14° 18′, altitude 380 m) to the West, from the Diamare plains to the Mandara mountains (Mokolo : 1 000 mm ; N 10° 45′- E 13° 48′, altitude 800 m) due to the altitude.

In fact, the rainfall fluctuations results more from the great temporal irregularities and the bad space distribution than from global deficits. This situation leads farmers to uncertainty as for the beginning of the rainy season for seedling or its end that gives way to the beginning of dry season sorghum cultivation ("*muskwari*") and the decline of the agricultural yield.

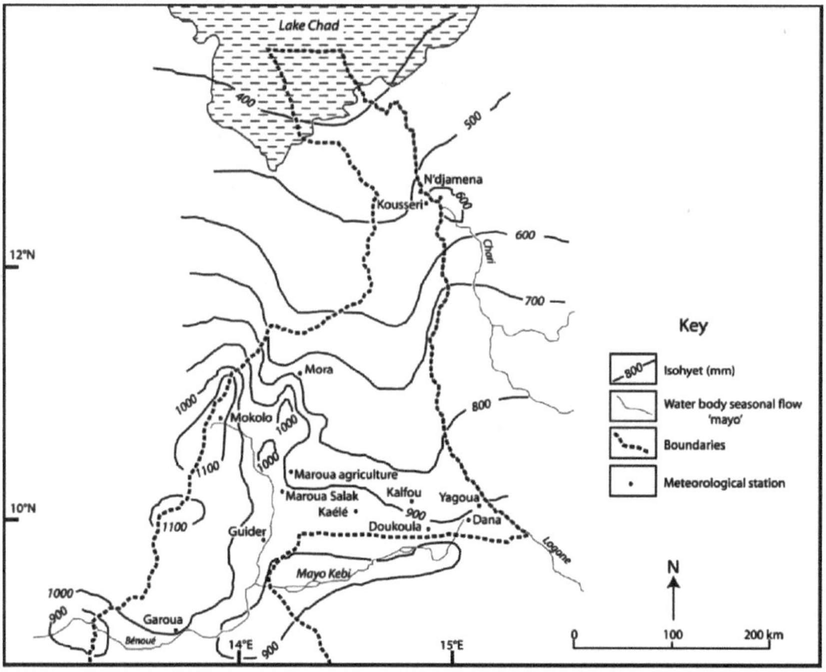

Fig. 3 : Sketch of inter-annual isohyets in the Extreme-North of Cameroon

Drying-up generates four types of droughts:

- Climatic drought: 5 months (November - March) to 8 months (October - May) dry season with ETP > P;
- Hydrologic drought: waterbody flows only during 8 months (April - November) to 4 month (June - October); that is why the rivers are called "*mayo*" (seasonal flow);
- Edaphic dryness: - regosoils, rocky or granular soils which intensify the infiltration and are inappropriate soils to the agriculture because of very weak capacities of water reserves, very deficient physical properties, very high risks of erosion and degradation (" *hardés* ") and hydromorphic vertisoils on tropical black clays in low areas (" *karals* ") which favour the evaporation;
- Agricultural drought: years with deficient rainfall.

The sahelisation could be the consequence of the change of water flow rate: first of all rainfall and therefore, the evolution of surface water and underground water. Indeed, rainfall is characterized by pronounced irregularity. Bare soils and denudation of the soil, accelerate water runoff and increases flow of large river courses as well as it reduces infiltration abilities which could have supplied to the underground water.

The irregularity of rainfall occurs often early into the sudden rain interruptions in the agricultural season (*deficit rains*). The annihilation of the vegetative growth due to the water deficit which is very necessary for agriculture, conveys the agricultural drought. This situation misleads farmers for sowing period, predisposing the populations to famine. On the other hand, environmental drought causes sicknesses such as meningitis, but drought related diseases are those caused by water shortage for human consumption without paying attention to the quality. The populations are thus submitted to hydric endemic sicknesses: onchocercose, amoebiasis, typhoid, bilharziasis, ascariasis, ankylostomiasis, dysentery, cholera, malaria, dracunculose or Guinea worms, etc.

Droughts also create food scarcity which memory transmitted from generation to generation. Food shortage of 1912-1913 was called "*skoum nodoumo*" i.e. ("the great hunger") by Kotoko (fishing population surrounding the Chad lake and Logone river). The hunger of 1921 was named "*ankra ahkouk*", i.e. "to disavow his brother" by the Choa Arabs (stock-breeder population around the Chad lake), (Saïbou, 2001). Human crisis is simultaneous to environmental crisis. Frequent droughts cause migrations towards viable areas. Indeed, frequent farmers - stock-breeders conflicts occur around the Chad lake to control black hydromorphic vertisoils (*karals*), fertile lacustrine deposits and /or residual ponds resulting from the Chad lake narrowing due to the dryness. The abilities of water retention and consequently green grass untill late in the dry season make those lands special areas for dry season pastures but also for the cultivation of dry season sorghum (*muskwari*). The last conflict between Kotoko of Tchede village and Musgum of Sifna village for farmland in the area which lasted from January 4th to January 15th 2007 and made 119 deaths, (News paper « L'œil du Sahel » N° 227 of February 13th, 2007). That's why Homer–Dixon (1999) wrote, « *Scarcities of critical environmental resources – especially of cropland, fresh water, and forests contribute to violence in many parts of the world…* ».

In this dry environment, even if conflicts are ethnic, they deal with ecological and economical context marked by natural resources scarcity. Frequent trans-boundaries tensions between Cameroonians and Nigerians or between Cameroonians and Chadians are also to be linked to this situation.

The main sign of climate drying-up is the desertion of the cotton cultivation in the traditional area by the Cotton Development Company (SODECOTON). Indeed, this company relocated its shelling factories from the Sudano-Sahelian domain to the Sudanian domain southwards.

Simultaneously, cotton farms are shifting in the same direction. That is a great disagreement to the very beginning of cotton basin peasants.

Survival Strategies to Face the Climate Drying

In the Cameroonian Sudano-Sahelian and the Sahelian zones, only the so called large cities have running water (Maroua, Mokolo, Yagoua, Kousséri and recently in 2005, Mora). Peripheral quarters are not connected. Even in these urban centres, 80 to 90 % of the population are farmers or craftmen, others are merchants, that is why just a few minority can subscribe to the National Water Company of Cameroon (*La camerounaise des eaux*). Rich merchants have their own water towers and some families have private wells. The majority of population buys drinking water from retailers (Figure. 4).

Fig. 4 : A young water retailer in Maroua (December 2007)

Facing the environnemental degradation because of the climate drying, people living in this vulnerable area have, since millennia and this before the so called modern hydraulic facilities, build know-how and developed adapted techniques to obtain enough water for the consumption, the household, the agriculture and the breeding.

The access to water for daily consumption : drinking, bathing, dishes, washing..., depends on the topographic location:

- In the Mandara mountains where small dams built downhill rocks cracks "*Thaal* " are dry, when water reserves collected from gutters or under rocky slabs during the rainy season are exhausted, when wells in pockets of regoliths are dry despite the creation of stillwater reaches to feed them, women and adolescents are constrained to daily migrations up to 20 km by foot. It is worth bearing in mind that during these critical moments (full dry season: February - April), the use of water is limited to basic needs such as drinking and cooking.

- In the nearest piedmonts, women and adolescents scrape out, the bottom of small excavations in regolith pockets over days (Figure. 5).
- Downhill in low lands (plains, glacis...), the sedimentary cover becomes thicker and is a good water-bearing layer. People access to this water by wells.

Fig. 5 : A woman is collecting water in a small excavation of regoliths (Dec. 2007)

In valleys, people get to the underflow by funnels, dug in coarse sand of dry river beds. On the other hand, during the first three months of the dry season (October - December), residual pools remain in seasonal flow river beds (*mayo*). Those residual pools are preferential meeting point between men - adults and children (for daily water consumption, dish, washing, swimming, household, fishing in some cases), the cattle and mosquitoes in search of humidity (Figure. 6). In the flood plain of Chad (*yaéré*), populations organise their survival with the water of residual pools, of the Logone river and the water of the irrigation canals in the ricefields.

Fig. 6 : Young women washing up the dishes and clothes in the mayo Danay (Dec 2007)

Aware of their suffering, population joint Common Initiative Group (GIC) or others associations to benefit from assistances in their search of water, notably those of:

- Non Governmental Agencies (*ONG in French*): Cooperation for American Relief Everywhere (CARE); United States Agency for International Development (USAID); German : " Gesellschaft für Technische Zusammenarbeit (GTZ) "; Dutch : Dutch Geotechnic Institute (ING) or Chinese ;
- Projects financed by the United Nations Development Program (UNDP) ; the Food Agricultural Organization (FAO); the Islamic Bank for the Development (BID), and the Cameroonian government: Special Fund for Rural Action (FSAR), the Project of the Development of the Mandara Mountain Region (PDRM), the Integrated Development Mission of Mandara Mountains (MIDIMA), the Waza-Logone project, the Underground Water Program (PES), the Emergency Hydraulic Program for Villages (PUHV); the Community Initiative Project for the Development of Karhay (PCIDK)…. ;
- The social action of the Company of the Development of Cotton (SODECOTON) ;
- Projects held by the World Environment Fund (FEM) such as the Water, Ground, Tree project (ESA);
- The valuable actions of religious congregations, such as the Bethleem Foundation of the Italians Missionaries in Mouda.

All these agencies contribute to build cemented wells and borings for drinkable water, to create stillwater reaches for the ground water and the underflow refilling and to improve cultivation techniques by keeping running water in farms. These techniques depend on the topography.

In the mountains and on inselbergs (isolated mountainous hill) slopes, the techniques of dry stone terraces have been perpetuated from one generation to another by ancient populations (Mandara or Wandala, Mafa, Kapsiki, Matakam, Mofu, Zoulgogue, Ouldeme...). (Figure. 7)

Fig. 7 : Terraces in the Mandara Mountains

In less sloppy sectors runoff erosion is fought by stone ridges perpendicular to the general slope of the field, by the association of plants, strip of grasses like *Brachiaria* and *Pennisetum,* by the association of crops and trees (agroforestery) such as *Azadirachta indica, Acacia senegal, Ziziphus mauritania, Acacia albida* (Figure. 8).

Fig. 8 : Stone ridges associated with strip of grasses.

Long before the discovery of under Vegetation Cover System Cultivation (*Agriculture sous Couvert Végétal: SCV*) and the policy of tree plantation extolled by Non Governmental Organisations and public authorities, farmers have permanently preserved some trees in their farmlands. Indeed, they have secular knowledge on the usefulness of trees of their land. If *Acacia albida* for example is a leguminous and gives an aerial forage during the dry season, others have edible fruits, bring shade or are used in the traditional pharmacopoeia, in the cosmetic, in the constructions, in rope weaving, in the biological insecticide manufacture or are used as firewood. Furthermore, branches or straw cover fight splash erosion and preserves the humidity in the soil at the beginning of the dry season. A cover of 15% of corn remnants reduces the erosion of 75% as compared to a bare ground, (Ketcheson and Stonehouse, 1983).

To restore barren ground known as *"hardé"* by the runoff control, farmers make deep ploughing, to facilitate the penetration of roots and to maximise macro porosity of soils. Hillocks or ground ridges favour the infiltration and the filling of the underground water (Roose, 1994). This technique rehabilitates *karal* soils i.e vertisoils and soils with vertic similarities on tropical black clay in the depressions and the flat sectors. Those black cotton soils are the out of season's sorghum areas, locally called *muskwari.*

The last innovation of famers facing climatic drought is the adoption of new crops with short vegetative cycle, high yield, having a good resistance to phytosanitaries diseases and to attack of insects. We can name improved groundnut and bean (*niebe*) species; their vegetative cycle passes from 120 and 90 days to 75 and 60 days respectively, thanks to research of the Agronomic Research Institute for Development (IRAD).

Herders have built dams before farmers and created artificial ponds to water their cattle up to 3 - 4 months after the end of the rains in September. This practice compensates risks due to natural residual pools in river beds which are the main drinking troughs at the beginning of the transhumance (November - December).

In an ultimate effort to face climate drought, people tries to control the water so as to struggle against food shortage. Therefore, Cameroonian authorities have ordered in 1979 the building of a dam in Maga for fishing and rice cultivation. This dam has changed the lake into a very large swampy land (*yaere*) which offered the dry season pasture for the four country members of the Chad lake Basin Commission (CBLT): Chad, Niger, Nigeria and Cameroon. As Wijikan et al (1984) wrote, concerning Africa as a whole, droughts are appalling in dry areas because agriculture encroached on ecological fragile lands in the past reserved to nomadism.

Facing endemic diseases, public authorities react punctually by strengthening the epidemiological supervision, purifing the areas, awaring populations...and by vaccination.

Conclusion

The Extreme-Nord-Cameroon on the edge of the northern African Sahel, has undergone probably with a light temporal, quaternary paleoclimate oscillations influences known by the Sahara desert. Now a days environmental dynamic is ordered by a general dryness. This situation which can be justified by natural conditions is exacerbated by the increased use of natural resources by man that jeopardizes every day of his existence. As in all semi-arid zones, the question of water resources, its safeguard and its uses is a major challenge. Since millennia, people living in this vulnerable area have built know-how and developed adapted traditional and/or modern strategies to get to the water for consumption, household, agriculture and breeding.

In these areas, inter-seasonal climate variabilities more than changes lead to months of water excess and months of water deficit due to loss. Suggested strategies for by hydraulic facilities are :

- underground dam in river beds to retain the underflow;
- the artificial refill method, notably by driving water excess of the rainy season in the artificial aquifers made by the thickness of hundreds meters of sediment deposits;
- the promptness in the fulfilment of hydraulic projects;
- the cessation of using water as a political blackmail;
- new relationships between populations and their ecosystem by a better way of managing natural resources.

All these solutions would improve life conditions in case of natural excesses.

References

Beauvilain A., 1985. Remarques sur la situation pluviométrique actuelle au Nord - Cameroun. Rev. de Géogr. du Cameroun. Vol. V, N° 1, Yaoundé, pp. 26 - 34.

Beauvilain A., 1986. Les variations du niveau du lac Tchad. Rev. de Géogr. du Cameroun, Vol.VI, N° 2, Yaoundé, pp. 121 - 137.

Durand A. et Mathieu P., 1981. Evolution paléogéographique et paléoclimatique du bassin tchadien au Pléistocène Supérieur. Rev. Géol. Dyn. Géogr. Phys., N° 22, Vol 4, pp. 97 - 109

Grove A. T. and Pullan R. A., 1964. Some aspects of the Pleistocene palaeogeography of the Chad basin. African Ecology and Human Evolution. Ed H. C. Howel, F. Bourlière & Cie, London, pp. 230-245.

Hervieu J., 1969. Le Quaternaire du Nord-Cameroun. Schéma d'évolution géomorphologique et relation avec la pédogénèse. Cah. ORSTOM, Péd.,VIII, N° 3, Paris, 172 p.

Homer - Dixon Th. F., 1999. Environmental, scarcity and violence. Princeton University Press, Princeton and Oxford, New Jersey, 253 p.

Ketcheson J. W. And Stonehouse D. P., 1983. Conservation tillage in Ontario. Journal of Soil and Water Conservation, Vol. 38, pp. 253 - 254.

Mainguet M. et Wakponou A., 2002. Les derniers souffles sahariens à la lisière du Monde sahélo-soudanien. Rapport de Mission de terrain 09 - 30 janv. 2002, Extrême-Nord-Cameroun, 32 p. et annexes.

Marliac A. et Gavaud M., 1975. Premiers éléments d'une séquence paléolithique au Cameroun septentrional. Ass. Sénégal et Quatern. Afr., Bull. liaison, Sénégal, N° 46, déc. 1975, pp. 53-66.

Pias J., 1967. Chronologie du dépôt des sédiments tertiaires et quaternaires dans la cuvette tchadienne (Rép. du Tchad). Note présentée par Camille Arambourg. C. R. Acad. Sci., Sér. D., Paris, pp. 2432 - 2435.

Pullan R. A., 1965. The recent geomorphological evolution of the south central part of the Chad basin. Samaru Res. Bull. N° 50, pp. 115-139.

Roose E. 1994. Introduction à la gestion conservatoire de l'eau, de la biomasse et de la fertilité des sols. Bulletin Pédologique de la FAO, N° 70, 1ère édition, Rome, 420 p.

Saïbou I., 2001. Conflits et problèmes de sécurité aux abords sud du lac Tchad. Dimension historique (XVIème - XXème Siècle. Thèse de Doc. Univ. de Yaoundé I, 381 p.

Servant M., 1983. Séquences continentales et variations climatiques : Evolution du bassin du Tchad au Cénozoïque Supérieur. T. et D. ORSTOM N° 159, Paris, 573 p.

Suchel J.B., 1972. La répartition des pluies et les régimes pluviométriques au Cameroun. Travaux et document de géographie du tropicale n°5 Université Fédérale du Cameroun, CEGET-CNRS, Talence, 287 p.

Wakponou A., 2004. Dynamique géomorphologique des basses terres soudano-sahéliennes dans l'Extrême-Nord-Cameroun. Thèse de Doctorat NR, Laboratoire de Géographie Zonale pour le Développement, Univ. de Reims Champagne-Ardenne, 229 p.

Wijikan A. And Timberlake L., 1984. Natural disasters: Act of god or Acts of man? Earthscan Publication, London.

6

Climate Change Impacts and Small Scale Farmers : Perceptions and Coping Strategies in Southern Zambia

Emmanuel Mutamba and Richard Chintu

Abstract

Increased human activities such as agriculture, mining, and manufacturing have resulted in rapid increase of green house gases (GHGs) in the atmosphere. Atmospheric accumulation of the GHGs results into global warming and climate variability or change. Climate change due to enhanced greenhouse effect arising as a result of human activity is considered a major global environmental threat to mankind today. Perhaps nowhere in the globe are people more vulnerable to climate change than in Sub-Saharan Africa (SSA). Most livelihoods in SSA are climate-sensitive due to over dependency on land and other natural resources and are thus perched on the brink of disaster due to climate change.

Climate change is now undoubtedly affecting millions of people across Africa and it actually threatens to wipe-out the past and current global efforts made to tackle poverty unless urgent and robust decisions and actions are taken. This is also a wake-up call for the world to adopt alternative development models that would ensure massive cuts in GHG emissions to avoid a potentially irreversible change. Scientists have warned that unless the global community, especially the developed countries, take genuine steps now to reduce their emissions; many more millions of people in the developing world will be negatively impacted by climate change.

Zambia like many other third world countries has not been spared from the adverse effects of climate change. Already, uncertainty and variability in the weather pattern and intermittent episodes of severe floods and droughts are posing unprecedented threats to agriculture and food security in parts of Zambia, especially the southern region which even before climate change concerns assumed international debate, had been vulnerable to drought episodes.

Climate change experts have predicted that though developing countries like Zambia have had minimal contributions to the overall GHGs build up, they are likely to suffer the most due to their poverty which make them highly vulnerable to climate related shocks and stresses. Farming communities in SSA now have no option but to adapt to climate change, by identifying and applying strategies and practices that will increase their livelihoods resilience to impacts of climate change.

In southern Zambia, which lies in a low rainfall ecological zone, lack of access to income and credit facilities was cited by some farmers and local NGOs as the main factor inhibiting adaptation. The study also showed that reduced household sizes, farming experience, access to water and alternative energy (alternative to fuel wood and charcoal which contribute to deforestation and green house gases (GHGs)), land tenure rights, off-farm activities, telecommunication (email, radio, mobile phones, TV), reliable transport, and access to health, sustainable agriculture, profitable agri-business and agro-extension services were the main factors that could enhance adaptive capacity among the vulnerable rural communities. Thus, the government should design policies aimed at improving some of these factors where they are poor and introducing them where there are none.

This research reports adaptation behavior of farmers in the face climate variability or change through in-depth interviews, and focus group discussions (FGDs) with small scale farmers in southern Zambia and other agricultural practitioners and experts around the country. In total, over 150 interviews with experts, individual farmers and farmer groups were made in Siavonga, Monze, Chirundu and Lusaka districts. Farmers were carefully selected with the assistance of local government and NGO staff in the respective study areas.

The survey which was conducted between September and October 2009 collected a large range of data. However, this study used principally the section of the survey on perceptions of climate change, adaptations made by farmers, and challenges or barriers to adaptation. Open-ended questions were used to ask farmers and development NGOs whether they had noticed long-term changes in mean temperature, mean rainfall, the incidences of disease out breaks, and vegetation cover over the past 20 years. Questions about adaptation and the constraints to adaptation were also posed. The questionnaires also included questions to inform about the potential impact of climate variability and change on household vulnerability and farm production.

The study adopted a participatory bottom-up approach that sought to investigate actual climate change related adaptations at the farm and community levels, as well as the factors that appear to be driving them. Based on the case studies in Siavonga in Southern Zambia, this paper also attempted to capture and share farmers' awareness and perceptions of climate variability and change, and how they have adjusted their farming practices in response to the perceived changes.

The study concluded that the country's climate-sensitive sectors of agriculture, fisheries, forestry, water, energy and human health have already been adversely affected by climate change and this has had a negative impact on economic, social and environmental aspects of the country. The southern part of Zambia has particularly suffered recurrent droughts. The country has in the past few years also experienced severe flooding in certain parts,(including parts of southern Zambia) thereby causing havoc to local communities. The floods have often resulted in damage of agricultural crops, infrastructure such as roads and bridges and a number of rural people have been displaced resulting in disruption of livelihoods.

Introduction

"Farming has now become completely different and difficult. Sometimes rains start in late November or later instead of mid October as in the past. When it comes, it rains and rains continuously for almost the whole day, which is bad for our crops. This time we had a mix of persistent droughts and floods. The weather patterns have gone completely mad here. When we expect a drought, there is too much rain and floods and vice versa. The rains destroy houses and wash away fields, roads and bridges. It is getting even worse each year!" Says, 50 year old Dickson Siangoma Malundu - The Headman of Malundu Village, Lusitu, Siavonga

Climate change refers to any change in climate over a long period of time, whether due to natural variability or as a result of human activity. It refers to changes in the climate as a whole, not just one single element of the weather.

Climate change is expected to have serious adverse environmental and socio-economic impacts on livelihoods of millions of poor small scale farmers in Sub Saharan Africa, Zambia included. In particular, the poor rural, whose livelihoods largely depend on the natural resources, are likely to be the worst hit by the adverse impacts of climate change. The extent to which climate change impacts on the rural poor is largely dependent on the nature of adaptation strategies used in response to climate change.

Zambia like the rest of SSA currently faces many challenges - poverty, a high disease burden, rapid population growth, food insecurity, floods in some areas and limited access to water in many others. Climate change will either exacerbate or modify the current types of threats or it may introduce new threats altogether which may affect natural and human systems independently or in combination with other determinants to alter productivity, diversity and functioning of ecosystems and livelihoods *(IISD 2003)*. Climate change is likely to drive the majority of the rural Zambian farmers into acute desperation and destitution, as household assets are lost and public infrastructure such as roads and bridges are destroyed and such natural resources as land, water and forests dwindle and government public resources are constantly diverted from development sectors to deal with emergencies. Since the dawn of time, the earth has experienced periods of cooling and warming, with mean temperatures remaining relatively stable. These changes were minimal and stable due to the energy balance between land, sea and the atmosphere. However, such human activities as combustion of fossil fuels, industrialisation, urbanisation and deforestation have led to significant build up of GHGs in the atmosphere. One of the most important negative consequences of atmospheric GHG accumulations is that most of the heat that would otherwise escape from the earth is bounced back and retained, causing a steady rise in the average global temperature with time. In fact according to IPCC, eleven of the last twelve years (1995–2006) are among warmest years since records began in 1850. The warming trend over the previous century was reported as 0.6°C in the IPCC's Third Assessment Report (TAR) published in 2001; it is now 0.74°C. A constant global warming trend may cause sea levels to rise and submerge surrounding low lying areas thereby displacing populations and disrupting natural ecosystems, let alone human development and livelihoods.

The research on which this presentation is based was conducted in Siavonga, Monze, Chirundu and Kafue using a "bottom-up" approach, which sought to gain insights on climate change from the farmers based on household and community surveys. Farm-level data were collected from 135 households in Monze and Siavonga between September and October 2009. The study examined how farmer perceptions compare with scientific observations on climate. The study also examined the strategies being used by small scale farmers to cope with the effects of climate change.

Concepts of Climate Change Adaptation, Vulnerability and Mitigation Explained

Climate change has moved beyond being an environmental challenge to one that threatens poverty reduction and development initiatives around the world. The global scientific community has confirmed that the world is now locked into a pattern of change and the opportunity for preventing *any further* warming (IPCC 2007). Thus, societies must respond by both minimizing further warming *(by reducing the concentration of greenhouse gases in the atmosphere)* and finding ways to adapt to such impacts of warming as shifting precipitation regimes, more frequent and extreme weather episodes, and sea-level rise.

Adaptation

Adaptation is a process of adjustment to or coping with new or modified circumstances in order to prevent further damage and/or to recover from damage. In the context of climate change, adaptation is understood as the actions people take in response to, or in anticipation of changing climate conditions in order to reduce adverse impacts or take advantage of any opportunities that may arise (Tompkins and Adger 2003). The need for, type and scale of adaptation depends on the kind of change taking place, as well as the vulnerability of people and natural systems to this change.

Vulnerability

In terms of climate change, vulnerability refers to the susceptibility of people and the natural systems to the harmful consequences of climate extremes and variability depending on their adaptive capacity and the sensitivity.

Mitigation

Mitigation refers to efforts to alleviate or reduce the negative impacts of an effect. In the context of climate change this may entail efforts to stabilize atmospheric GHG concentrations by cutting emissions at their sources using strategies that effectively reverse deforestation and forest degradation and to adopt environment friendly and sustainable development pathways as opposed to the wide-spread conventional fossil based development. However, if current global policies and strategies to mitigate climate change and related steps towards sustainable development remain unchanged, global GHG emissions will continue to grow over the next few decades.

Reducing peoples' vulnerability to climate change is closely linked to the poverty reduction agenda, since poverty is both a condition and determinant of vulnerability. Many of the world's poor are already vulnerable to climate risks due to factors such as settlement on marginal lands.

High dependence on climate-sensitive livelihoods, and limited access to or unavailability of resources to respond to shocks and stresses (ADB, 2003). If people do not have the resources to deal with today's stresses, then they are unlikely to be able to deal with the additional stresses associated with climate change, a condition known as the *'adaptation deficit'* (Burton 2004). Adaptation to climate change must start with reducing this deficit. To this end, many strategies currently used for development and poverty reduction have an important role to play.

Our research in Southern province in Zambia revealed that many parts of the province are already experiencing one form of climate change related phenomenon or the other. Farmers mentioned such trends as changes in rainfall amount and distribution, leading to an increased frequency and intensity of such severe climatic events as floods and droughts including out breaks of diseases and pests. It is evident from the testimonies of older citizens of southern Zambia that the province which is one of the driest parts of Zambia will be even getting warmer and drier with time.

Findings and Conclusion

Agriculture and food security situation amidst climate change

Sub-Saharan Africa is currently known to be the most food-insecure region in the world. Climate change is likely to aggravate the situation if robust measures are not put in place. Some areas may become inhabitable for certain types of organisms because of being too hot, too wet, too dry or intensely infested by devastating pests or/and diseases.

Already some rivers have dried up in southern Zambia and this alone has had a very devastating effect on livestock production in the province which was once teeming with various types of livestock especially cattle. Currently the livestock populations have dwindled and most of them look strangely emaciated because of dehydration, starvation due to shortage of natural fodder, heat stress and a whole range of pests and diseases. Equally, the livestock owners appear to be in varying states of worry and panic over an impending loss of their most valuable asset, cattle locally known as *"Ng'ombe"*. Cattle occupy a very high position as a household asset in the Tonga culture to which most indigenous people of southern Zambia belong. Cattle is a traditional symbol of wealthy and prestige which is also used as a proper and respectable payment for dowry by men who are marrying. Cattle have also long been widely used as draught power for ploughing crop fields in the province. The dwindled, diseased and weakened cattle population will have a telling effect on food security in the province and beyond as less and less land will be cropped. This negative effect on livestock, humans and food security clearly illustrates that climate change is real and its effects would be more disastrous countrywide if effective adaptation strategies are not devised and widely disseminated and adopted among the vulnerable communities.

Some implication of climate variability on people's livelihoods in Southern Province

Considering current indications and literature on climate change, southern province which is the driest part of Zambia with rainfall ranging from 400 to 800 mm per annum, one can wisely predict that the area is likely to become drier with time. The climate crisis has forced some farmers and agricultural experts to turn and look up to irrigation and drought resistant crops as a form of adaptation to the escalating water stressing conditions in drier areas. However, there is a potential danger that producing food using irrigation and drought resistant crops would also become increasingly unattainable with the progressive drying up of natural open water bodies and the inadequate recharge of the fast depleting ground water due to insufficient rainfall, increased run-off and over-extraction.

Until the recent past, fishing also used to be an important source of livelihood for the people of southern province because of the major rivers from the north of Zambia such as the Zambezi and the Kafue rivers (and its flats) that traverse the area. However, fishing as an economic activity in the area is facing threats from climate change as some of the once water-swollen tributaries of the two major rivers are now barely flowing or have dried up completely. In addition there is a strong concern that the Kafue flats may be shrinking in terms of their total area and water content as more and more of the wetland area is being reclaimed for agricultural production. Apart from maintaining a high aquatic biodiversity as a habitat, such wetlands as the Kafue flats offer an important ecological service in terms of water balance, storage and supply by acting as natural sponges or sinks for trapping and storing water in the wet season and gradually releasing it into perennial rivers and streams during the dry season. Although naturally the Kafue flats are seasonally recharged with water by rainfall, in the face of climate change; drought, over-extraction of water and unsustainable land practices may lead to water depletion and thus reduced biodiversity through loss of the aquatic habitat.

Irrigation farming and its implications on the local livelihoods

In southern Zambia, as elsewhere in semi arid parts of SSA, water is quickly becoming the limiting resource that affects many agricultural decisions in the face of climate change. As water becomes scarce and begins to lower or disappear, most small-holder farmers begin to farm closer to water sources, often planting directly in the slightly drier or wetter parts of river or dam beds. Some farmers locate their gardens too close to such water sources as rivers and lakes in order to lessen labour for irrigation. This in itself is a potential source of environmental disaster as far as floods, erosion and river siltation are concerned. A number of farmers actually complained of their entire fields being washed away or submerged by water-swollen rivers and lakes in the wet season. This is because they establish their fields on the river and lake banks in the dry season when the water levels are low. However, at the peak of the rainy season the water levels rise and cover the cultivated areas on the otherwise relatively higher and drier banks. Most households depend on streams as a source of water for domestic use. Given this scenario, it is observe that increased use of livestock manure, fertilizers and other agro-chemicals in the fields that are close to natural water systems, may pose a danger of nitrate and pesticide pollution of such fresh water systems. This situation may be harmful to the various forms of life and the environment in general. Additionally, the erosion and flooding causes household farm area to shrink while rivers become wider and shallower as soil deposition onto the river-bed escalates. The combination of soil, climatic (high rain-fall) and topographic (steep gradients) conditions and inadequate soil conservation practices render some of the cultivated lands in the study area very susceptible to degradation by erosion. There is a yearning need to disseminate sustainable farming practices which emphasize soil and water conservation so that agricultural intensification around water bodies and wetlands does not compromise environmental quality, life, safety and sustainability in the face of climate change.

The commonest irrigation technique used in wetland gardens by small scale farmers in southern Zambia is manual using buckets or cans and streams or dug wells as sources of water. This trend was consistent in the entire study area, except in some sites were the Diocese of Monze and Harvest Help Zambia have introduced the use of treadle and motorised pumps. Mechanised irrigation poses some big challenges in that over extraction in the upper parts of the river might cause water shortages for farmers whose fields are located along the lower parts of the river. This might be a source of conflict in the future as water shortages become acute.

"In Lusitu area in Siavonga, acute shortages of irrigation water have become common in recent years as the dry season progresses" Say Fr. Spencer Nyendwa of the Diocese of Monze

Although most parts of the study areas receive abundant water as a result of frequent flush floods from a network of rivers, streams and lakes in the rainy season, the availability of water in space and time does not always coincide with uses and needs. There is no simple technology available to the majority of the small scale farmers in the area to regulate, store and transfer the abundances of water supply during the rainy and flood episodes in the wet season to meet agricultural and domestic demands during droughts and in the dry season. If the excess flood water in the rainy season could be prevented from running-off and completely draining away into streams, availability of irrigation water during droughts and the dry season would be improved.

In some areas of Siavonga and Chirundu districts utilization of wetlands for dry season gardening, and water supply for livestock and domestic purposes was observed. However, this intensification must seriously take the issues of degradation and sustainability into account. The current trend of population increase and the progressive drying of the area due to climate change, might result in intense competition for the use of wetlands by different stakeholders and then issues regarding property rights, water rights and land tenure structure on wetlands would become critical. It is important therefore to have adequate information on the dynamics of these issues to enable local communities and policy makers make informed decisions concerning the sustainable future use of the fragile ecosystem in the face of climate change. Therefore, there is a need to characterise access to wetlands and related production systems and to develop recommendation domains and policies regarding their future use for small-scale irrigated gardens and livestock in view of the climate change.

Building farmers' resilience against climate change through agricultural sustainability

Climate change is now manifesting itself with great intensity, and the current global financial crisis, has just accentuated the vulnerability of farmers, and others whose livelihoods are closely linked to the climate and natural resources. Initiatives are urgently needed to help small scale farmers and other vulnerable groups to protect and promote agricultural production. Simple, inexpensive actions could be taken, such as setting-up effective meteorological and early warning systems, improving agricultural extension services so as to increase crop yields, and establishing local networks of information exchange between communities.

The International Center for Living Aquatic Resources Management (ICLARM) and the World Fish Centre has shown that integrated agriculture-aquaculture (IAA) farms can produce several folds the income generated prior to introduction of IAA. By retaining water on the land, ponds have enabled farms to continue food production and balance their losses in seasonal upland agriculture which is highly susceptible to climate change and variability. IAA farms have also been known to produce a diverse range of food, more consistently throughout the year and more food in drought years than non-integrated farms. By controlling water flow across the landscape and increasing plant cover *(especially trees, shrubs and grasses),* IAA also reduces soil erosion and increases soil nutrient and organic matter content. The types of intervention used in IAA *(e.g, terracing, stone lining, contour ploughing, storm drains, inter-cropping, hedgerows, boundary planting, pond construction, improved fallows)* have been tested in many places and found to be extremely effective in improving agricultural sustainability. In such a system, sustainable agricultural practices such as agroforestry and IAA are used in a complementary manner. The potential of agroforestry in increasing the productivity of IAA needs to be widely tested and promoted in the study areas to enhance food security and income generation for the small scale, resource-poor rural households to improve their resilience and adaptability to climate change.

Fish farming can enhance local people's resilience and adaptability to climate change (Photo : R. Chintu)

Common climate change effects, stakeholder interventions and farmers coping strategies in Southern Province

In the recent past extreme weather-related events such as flooding have frequently destroyed some infrastructure used to store or transport agro-produce from production areas to markets in Southern Province and elsewhere. For instance, in the wet season of 2007, terrible flush floods hit Lusitu, Lusangazi, Lufua and Musaya areas sweeping away some crop fields, residences, household property roads and the Lusitu Bridge which links Siavonga district to other urban towns. This unprecedented amount of destruction acted as a huge disincentive for farmers who could potentially produce more food.

Development organisations working in southern have been quick to realised that dealing with the impacts of climate change calls for measures that will effectively minimise losses or take full advantage of the available adaptation opportunities. It is clear that government, non-governmental organisations (NGOs), local communities and all the other stakeholders should play a role in tackling this colossal global crisis. The public must be made to understand the challenges of climate change and to adjust their practices and livelihoods, for example, testing and adopting water harvesting and conservation techniques, suitable crops and animal varieties that are fast maturing, drought and disease tolerant. Increased uncertainty in climatic conditions has caused agricultural experts to advise farmers to diversify their crop and livestock production systems in order to avoid total losses when climatic extremes occur.

Despite the negative effects of climate change the lake shore in Siavonga district has potential to be the bread basket of Southern province because the abundant water, the higher crop growth rates and the rich alluvial soils make the area suitable and attractive for agriculture. The Director of Harvest Help Zambia (HHZ), one of the NGOs working in this area, advised that government and NGOs must collaborate to harness the huge agricultural potential in the area. *"We are looking up to government and other potential stakeholders to assist us translate our ideas into reality"* he added.

The program of development : NGOs in the area are mostly based on agriculture and are targeted at the small scale farm households that are food insecure, with poor agricultural practices and those whose settlements and agriculture are seasonally disrupted by recurrent floods and droughts.

For instance the main *"food for assets"* program by HHZ carries a host of subprograms such as the health (latrine, and refuse disposal pits), education, tree planting and land restoration (promotion of stone lines and storm drains for soil and water conservation). The *"Food for assets"* program is a donation by the World Food Program (WFP) and the Government of Zambia (GRZ). Food is given to vulnerable households to enable them feed and better cope with climate change by producing their own food for the future when supplied with simple irrigation equipment and initial seeds. The farmers in the project areas mostly grow irrigated maize and assorted vegetables for home consumption and income generation whilst taking advantage of agricultural extension and farming inputs from NGOs and government. In addition to the fertilizer obtained on loan, most farmers use cow dung to further improve vegetable production.

"Recently my institution introduced a credit facility to assist in building the resilience of the most vulnerable farm families to the stresses and shocks of climate change in Lusitu and other areas; so far the facility has successfully assisted 10 women-headed households to set up income generating activities (IGAs) to cushion the impacts of drought, floods and crop failure", said Fr. Spencer Nyendwa of Monze Diocese, Southern province.

He added that microfinance may play a role in filling the market gaps left open by traditional banks and some state-run development programmes, which have been unwilling or unable to effectively provide financing for the poor who do not have collateral. Thus, government and other development players must seriously consider the role of micro-financing in increasing resilience of small holder farmers' to shocks and stresses of climate change.

About 6,000 resource poor households eke out their meagre livelihoods in this area which is threatening to turn into a real desert. In the near future, the impacts of climate change are also likely to have a more devastating effect in the highly degraded area of Lusitu than elsewhere in Zambia. In the recent past occasional heavy downpours of the *"climate change related"* erratic rainfall has already been causing heavy soil erosion from extensively deforested land. For instance, in 2007 widespread massive destruction of such infrastructure as buildings, roads and bridges was partly attributed to the drastically degraded soil structure which impedes rain water infiltration causing heavy run-off and floods.

In 1958 during the construction of the Kariba Dam, the indigenous residents of Chief Chipepo and Simaamba areas were forced to relocate from Gwembe Valley to Lusitu in Chief Sikongo area which was very sparsely populated at that time. These people being pastoralists brought along them their scores of livestock. They also started clearing forests to give way to crop production. Both livestock and crop production led to rapid land degradation with time as human population grew, thus drastically reducing the carrying capacity of the area.

Typically degraded areas in Lusitu; *(Photo : R. Chintu)*

This suffices to indicate that poor resettlement policies which do not take into account the effect of the changing climate and the carrying capacity of the land may breed disaster. Ironically, the power grid which transports power from Kariba to Lusaka traverses the present day home of the "displaced" Gwembe people; "Lusitu", but there has not been any attempt by government to provide power for consumption by the local people.

Development organisations working in Lusitu in Siavonga district identify it as being one of the most typically degraded lands in southern Zambia. Development staff from the NGOs admitted that currently the situation has been made worse by the challenges of climate change coupled with inadequate public extension staff to collaborate with them. Mr. Kasenzi, Director of HHZ intimated that the tree planting efforts in Lusitu were seriously frustrated by lack of water to raise seedlings in the dry season to transplant in the rainy season.

Development organisations in the south are trying to find ways of reforestation to reverse the progressive and widespread land degradation in Lusitu. The NGO development facilitators and the local farmers have been experimenting on vegetative tree propagation to circumvent the water problems which poses a challenge to raise tree seedlings from seed. The study revealed a number of indigenous plants including the *Mukwa* tree species that are able to grow vegetatively from stems.

A vegetative established live fence (shorter stakes) along side a dead fence of logs infested with termite (L) and (R) A feeble attempt of fencing a garden against wild life and livestock; *(Photo : R. Chintu)*

Dry season wetland gardens booming with assorted vegetables in Siamatik (L) and (R) Cow dung made ready for application *(Photo : R. Chintu)*

The off-season maize and vegetable production is important to Lusitu farmers as a means of supplementing rain-fed agricultural production which is also increasingly becoming vulnerable to the vagaries of climate change. This form of agriculture which is quickly taking root in Siavonga is important not only for its income, but also as a safety net against the often challenging climate-susceptible rain-fed agriculture. Mr. Kabudula's testimony *"In 1994, I first started with manual irrigation using simple hand buckets. In 1996, I managed to buy a treadle pump from the income I realized from selling vegetables. A few years later, I bought a motorized pump which made irrigation much easier. The market is good most of the time except in the rainy season when there are too many fresh vegetables in the urban market and thus I resort to selling this produce locally*".

Watering gardens with buckets is labour intensive for farmers (L) and (R) A woman farmer and some traders negotiating for a fair price in a tomato garden in Siamatika *(Photo : R. Chintu)*

Rain water harvesting (ground water recharging) techniques

One area where HHZ has scored success is in ground water recharging techniques in the semi arid areas of Siavonga. The technology involves digging of large troughs measuring tens of meters wide and at least two meters deep to trap rain water which might otherwise have been lost as run-off to stream or lower lying areas. Once the rainy water collects in the troughs, it slowly sips down into the subsoil thereby recharge the ground water levels. Whereas two years prior to introduction of this technique, farmers could dig wells deeper than 20 metres without finding any water, now the wells yield water even at depths as shallow as three meters.

A water-well located in a rain-water harvesting trough offers good access to ground water *(Photo : R. Chintu)*

Weirs and dams are part of the water harvesting and conservation techniques introduced in the communities *(Photo : R. Chintu)*

Small scale farmers' perception of usefulness of forests and trees in the face of climate change

In view of climate change and the effects of deforestation, tree planting and management has never been as important and urgent challenge as it appears today in the minds of the farmers in Southern Province. The farmers in the study area readily acknowledged forests and trees as part of the most important but quickly disappearing natural resources. They noted that people everywhere in rural Zambia depend on trees for an enormous range of vital products and services for local communities. Trees provide materials to build homes and furniture, bridges, poles, works of art, musical instruments, toys, and hundreds of other everyday household articles. Trees also bear fruits and provide a variety of useful medicines, pest poisons, dyes, and oils. Many families in the study area confirmed deriving a major part of their cash income from the sale of medicines and various other natural forest related products such as honey, mushrooms and fruits. Forests are also a principal source of energy in rural Zambia through the use of fuel wood. Farmers are also aware that even trees that do not supply food themselves may support their resilience to climate change by contributing to soil fertility and conservation, reducing incidences of droughts and floods, providing livestock fodder, or by providing shelter or shade to people, animals and crops and creating favorable micro-climates in the general ecosystem.

They also know that trees in the forests and in the other land-use systems regulate stream-flows and precipitation, thus preventing devastating floods, soil erosion and providing habitats for enormously diverse populations of plants and animals, both micro and macro, whose importance is just beginning to be understood. Communities also appreciate forests as being vital to society for recreation and leisure, cultural and religious pursuits. However, the continued over-grazing and conversion of Miombo woodlands into agricultural production and increased demographic pressure among other factors have greatly contributed to widespread deforestation in the study area. It is also acknowledged that indigenous plant species are slow growing and take decades to regenerate into sizable trees once they are cut down. Annually, distances from which, women and children *(whose task is to fetch fuel wood)* collect fuel wood are increasing because of the fast receding forests due to rampant deforestation. Thus, to the farmers of Southern Province, the dwindled forest areas combined with climate change only symbolize the dawn of danger, hopelessness and misery!

Indigenous knowledge and climate variability

Almost everywhere in southern Zambia, farmers confirmed that they have the ability to predict the weather, through, the careful observation of nature, without the use of a weather station. In several villages, farmers are able to identify changes in the behaviour of plants, insects or animals *(changes in colour, shape, bearing, period of maturation, migration, reproduction, nesting places, and so on)* that tell them whether the rainy season will be early, or short, or whether a drought will be eminent. Systems could be set up through which individual farmers, or communities could share this information. They could then prepare by selecting short-cycle varieties, for example, or planting crop in wetlands if a drought is expected.

The Way Forward

The poor farming communities in parts of southern province, assisted by government, NGOs, donors and other stakeholders are painstakingly trying to better adapt to climate change. There is, however, need for the key development actors to partner and work closely together in devising initiatives that would help to consolidate existing knowledge and generate new information on climate change to help affected communities to make informed decisions on what could be possibly done beyond individual or household actions. Equally important is the need for development partners and the other stakeholders to deliberately and proactively mainstream climate change in development plans to try and avoid emergency situations or the desperation and ineffective responses when the emergencies inevitably occur.

In the face of climate change the challenge facing development partners is to respond appropriately to the development challenges of

Southern province; notably poverty, land degradation, low agricultural productivity and income. Since these problems are interlinked, they require integrated and robust solutions which address the concerns of vulnerable communities simultaneously. Greater investment, by funding partners and government in risk reduction and building community resilience to climate change is urgently needed. The Zambian government, donors and other stakeholders need to explore the possibility of targeted insurance products that respond to the peculiar nature of risks that small scale farmers face. Programmes such as flood mitigation, water harvesting and rural infrastructure development require more serious policy considerations, both for their risk reduction and productivity enhancing benefits. Microfinance and other financial services to the poor and vulnerable communities so that they can engage in productive activities to help them build assets, stabilize consumption and protect themselves against risks should be seriously considered as a strategy to build the resilience of communities to climate change. In order to address the more complex structural problems of poverty and vulnerability in the face of climate change, it is important to consider microfinance alongside a range of other interventions such as sustainable land use, health, education, market and clean water access and renewable energy technologies coordinated with national growth strategies that promote a combination of short-term development gains and longer-term view to sustainable socio-economic transformation.

References

ADB, 2003. African Development Report.

Burton, Ian. 2004. Climate Change and the Adaptation Deficit. Occasional Paper 1, Adaptation and Impacts Research Group (AIRG), Meteorological Service of Canada, Environment Canada, November 2004.

IISD, 2003. Livelihoods and Climate Change. Combining disaster risk reduction, natural resource management and climate change adaptation in a new approach to the reduction of vulnerability and poverty : A Conceptual Framework Paper. Prepared by the Task Force on Climate Change, Vulnerable Communities and Adaptation.

IPCC.2007. Climate Change 2007 : Synthesis Report.

Tompkins, Emma L. and Adger, W. Neil. 2003. Building resilience to climate change through adaptive management of natural resources. Tyndall Centre for Climate Change Research, Working Paper 27.

7

Desertification Status in the Democratic Republic of Congo

Ernest Manganda Iloweka

We let you know that the type of people in a given region is closely linked to the kind of resources found in the concerned region; and which are necessary for human beings to survive.

In fact, more than 100 countries possess arid lands, where more than 40% of the global population lives. The soils of these regions present too much fragility and they are too degraded that they can't allow people to survive.

Let's emphasize that in Africa, two third of the total area are arid zones. Thus, for Africans, the type of their struggle for adaptation to the context of their land, illustrates their ingenious capacity to model their hostile surroundings and to leave marks of their cultures all over prehistory and history.

Social organizations in those arid areas are closely linked to modes of exploitation and valuation of resources (water and land). However, human productions in those areas are doubly menaced; on one side by the action of humans who exploit resources for their immediate needs, and on the other side, by climate adverse effects during long periods of drought. It means that climate change and violent meteorological phenomenon worsen circumstances. That leads to environmental degradation and biodiversity reduction; alarming deterioration of vital conditions; and intensification of poverty, both in rural and urban areas; particularly among women and young. Consequently, there are non durable (viable) agricultural practices, overexploitation, overgrazing, excessive and illegal forestry exploitation; destruction of forests and bushes by non-controlled fires, pillages and conflicts with weapons; reduction of vegetation, due to the scarcity of rain; of water in general and menaces of heaps of sands.

We, unwillingly, notice intensification of poverty, decline of standard of living; violence, infectious sicknesses; sexually transmissible, STD/HIV. Women are the most menaced, as they are invisible managers of the environment.

Factors of degradation in arid areas can be classified in two aspects; namely natural factors and anthropic activities.

Anthropic Activities

- Industrial activities create migration of people urban areas.
- Excess of use of biodiversity resources makes up the most important menace on the ecosystem of arid zones. In these regions, sedentary life is discouraged by the scarcity of rain and vegetation.
- The process of sedentarization started since several decades, increasing difficulties of agriculture and colored by very rapid pauperization of rural zones.
- We notice migration of herds of animals even if they are not accompanied with the shepherd. This has a negative impact on the vegetation and raise conflicts, on one side, between wandering and sedentary peoples of arid zones on the matter of land property; on the other side, between animal raisers and farmers because of animal wanderings.

Natural Factors

Aridity is encouraged by mechanisms at the global scale. Aridity is either reinforced, or attenuated by the following local factors.

- The level of evaporation, the permeability of soils (sandy soil, rocky soil or clayey soil). The relief also plays a role, with the fact that encourages accumulation, drain or infiltration of water from the rain. It's important to emphasize that the United Nations Convention to Combat Desertification constitute the frame of references for all strategies of sustainable management in arid areas. This convention is a juridical commitment of great range; taking into account the peoples' concerns; notably those related to their daily life, with consequences of degradation of lands.

To guarantee the strategies of the struggle to combat land degradation, the convention has recommended the composition of programs to be implemented on different scales, on the basis of long term strategies and priorities.

The protection of ecosystems and the rational management of natural resources goes, since today, through the amelioration of the poor populations standard of living. That must be done at the level where population remain constrained to a non-sustainable exploitation of natural resources. Consequently, a new approach in the environment management necessitates concentration on poor population. The activities to be undertaken must focus on poverty alleviation and the vulnerability of local communities. Such a new orientation of activities on the environment allows the transfer of benefits from services which are linked to the exploitation of ecosystems like water, soil and all other natural resources. They need to improve their standard of living; particularly for the most vulnerable categories, namely women and children. That will allow them to gather, reunite necessary conditions for the durability and the diversity of these resources.

Economic Power of Great Arid Zones

1. Arid zones possess non-pollutant and inexhaustible sources of energy such as:
   a. Solar energy: It can respond to the need of energy in arid areas, instead of woods; at a very low cost or free of charge.
   b. Wind: In arid zones, winds are frequent and constitute an important source of energy; able to produce electricity and allow machines like mills to work.
   c. Biogases: It can be obtained in all areas where agriculture and ranching are practiced together.
   d. Landforms: arid areas offer stones on a hard soil, convenient for comfortable constructions and beautiful houses.
   e. Water: In arid zones, water is more necessary than anywhere else over the world. Water in rivers and lakes evaporates quickly. To overcome this barrier, extraction of water from the underground is made and canalized at the places it must be used, in the oasis.
   f. Petroleum: In Africa, most of the arid areas possess petroleum under the ground; making an important source of energy, necessary for the economic development of these areas.

g. Biodiversity : It constitute an identity and wealth for arid zones. Several species of savage animals are integrated into cultural life of populations. For example, by furnishing necessary elements for decoration. The biodiversity is an important source of plants giving various medicines as well as more other wealth; transmitted from one generation to another.

Constraints

a. Important climate changes are causing great hydrological disequilibrium, with very negative impact on agricultural productions.

b. Human activity : Desertification is caused by the interaction between physical, biological, political, social, cultural and economic factors released by man. For example: Fires in the bush and forests degrade the environment.

c. The lack of investments hinder local population to realize their programs. There are many barriers to cross to convince public, national and international investors. For example: the lack of security related to land ownership of producers, the weakness of the civil society organization, the feebleness of the state regulations and the high fluctuation in purchasing.

This initiative can be considered as a general frame for programs and activities to undertake in southern countries, regional and international; in collaboration with northern financial institutions, national and international. This supposes collaboration between national and international financial organizations all over the world; for the implementation of sustainable development programs; in partnership with concerned parties, such as institutions of information and civil society organizations; at national and international scales.

Conclusion

The world is in constant mutation. The natural surrounding is evaluated, under the influence of climate changes; the environment degradation; the appearance of new diseases and ravaging beings.

The economic context is under demographic pressure, appearance or disappearance of markets, due to high concurrence. That implies permanent investigations and researches. Particularly in developing countries, where it can make profit of limited financial and human resources.

Researches need to orientated towards an economic and independent agriculture calling biological techniques and showing favor to local resources. It shows particular interest to agricultural products under climate change conditions and good techniques of productions on good soils.

To really serve agriculture, researches must take experiences from generations of famers. In a pragmatic way, following are the suggestions :

- Support the elaboration and the use of national policies and programs in what concerns agricultural techniques and those related to crops.
- Mobilize and request the intervention of every member of the community to realize collective actions concerning the struggle against desertification, preferably in the frame of PAN.
- Organization of capacity building of rural communities on how to use energy from the Sun, the wind and biogas, etc.
- Organization of collective actions to create projects (for example a project of reforestation) with families, farmers, animal raisers and political responsible.
- To inform pupils how to combat desertification in schools.
- To vary animal and agricultural products for better use of earth resources and avoid overproduction of one product.
- Bring people to discover the importance of water in arid zones.
- Drip irrigation of cultivated areas.
- Organization of mechanisms for follow up of actions by PAN concerning the struggle against desertification.
- Invite the international community to support efforts in arid zones to apply strategies of integration of arid areas in the general and global policy of the environmental and biodiversity protection;
- Supply technical assistance to reinforce human and institutional capacities, along with teams and equipments of follow up and alert in case of catastrophes due to meteorological and climatic degradations.
- Promote a type of trade allowing people in arid areas to survive without excessive exploitation of their lands.

On the basis of the above analysis, donors and funding agencies need to work in close collaboration and coordinate efforts of actions promoting sustainable development of arid zones, taking into account, researches and experiences carried out in these areas.

Concerned governments and populations must be aware of this situation, mobilize energies to coordinate efforts.

In this case, there should be a focus on the "Development in African countries".

8

Desertification and Adaptation of Producers' Endogenous Agricultural Practices in Burkina Faso

Poda J.N., Belem M and Dibloni O

Abstract

Burkina Faso is a landlocked country in West Africa with a dry tropical climate characterised by a short rainy season and a long dry season. The annual rainfall varied between 300mm in the northern part of the country to 1200mm in the meridian part. Since many years, the degradation of climatic conditions is observed. Therefore, the annual rainfalls of the 1971-1981 period were inferior to those of the precedent periods, and also the deterioration of the climatic conditions led the limits of some ishoteques "isohyète" to move southward, more than 50km farther. Furthermore, the decennial rainfalls since 1920 were subjected to a latitudinal displacement to the south. For example the ishoteque "isohyète" 500mm which were located at the North of the 15°N outside Burkina Faso during the years 1950-1960 moved gradually year by year and reached the 14°N during the years 1970-1980 inside Burkina Faso.

The dryness and degradation of environment are related and could become irreversible with effects on the survival and welfare of populations. It results in:

- *The temporal and spatial irregularity of rains resulting in rainfalls inferior to 20mm (the ones useful for rainfall crops), and in the increase of week rains,*

- *The increase in erosion factors (strong winds, inundations, increase of temperature),*
- *The expansion of vectorial-transmitted tropical diseases (malaria for example) and recurrent food shortages.*

Thus, the image prevailing is often the accelerated precariousness of vegetal and animal production conditions, the high incidence of poverty resulting in a vulnerability of rural populations. Therefore, due to the gradual increase in the hostility of environment, populations have developed adaptations as the sole alternatives to survive (adoption of adaptive endogenous techniques of soil and water conservation instead of migration).

Upon this central question of adaptation to the process of desertification, the answers of populations are derived from their daily experiences since several years. These experiences enriched by lessons learned on recurrent droughts are less documented at scientific level. They are related to:

- *Water conservation and its optimisation for plant ("Zaï" or "demi-lunes" techniques, dry season crops...),*
- *Struggle against erosion and protection of soils (straw mulching, protecting valves with stone, vegetal or both and allow soil fertility restoration...)*
- *Adaptation of cultural practices (agroforestry, top slopes weeding, utilisation of organic manure...),*
- *Adoption of agricultural innovations (certified seeds along with the short cycle ones, crops diversification ...).*

Poverty which is one factor of environment degradation will amplify with desertification. Environment changes will affect more and more the food production systems, resulting in the increase of malnutrition and famine. Therefore, to reach of the first Millennium Development Goal which aims to reduce 50% of population proportion (with a daily income less than 1 USD) requires documentation on convenient endogenous practices to face desertification, and researches to improve traditional systems with the support of communities.

Introduction

Burkina Faso, as an illustration of tropical Africa, appears at the crossroads of climate contrasts. It is a landlocked country with no direct access to the sea. The climate is a tropical dry one with two distinct seasons: a rainy season and a dry one. The annual rainfall varies between 350mm in the far north and 1,200mm in the southern part. The average temperatures vary between 30-33°C maximum generally in March, April and May and 13-14°C in December and January.

As far as rainfall is concerned, isohyet fallback phenomenon is already noticed and it appears as isohyets withdrawal of more than 200 km towards the south. However, the frequency of exceptional events such as floods and droughts might grow in number. So the city of Ouagadougou has noted on September 1st, 2009 an exceptional rain of 263 to 300mm in some areas with 10% of the population made homeless; such heavy rains date back to more than a century.

So in Burkina Faso, like elsewhere in Sub-Saharan Africa environmental degradation related to climate is not and will not be a new fact; it is getting worse with human activities, which leads to lower productivity and will worsen with climate changes. In this context, the biosphere reserve of the Hippopotamus Pond in Burkina Faso would in the framework of Séville (UNESCO, 1996) strategy implementation, reconcile biodiversity and biological resources preservation with their sustainable use.

Populations, illiterate in majority, need real life patterns including simultaneously sustainable preservation and development in the climate change context. These patterns can only be applied if they are taken from the populations' ingenuity and if they take into account all the social, spiritual, economic needs of the society and use a good scientific basis. The adaptation consists in adjusting inside a human system, in response to a climate stimulus and its current or expected effects including climate variability and extremes; so, it is a process taking root in socialisation, social and political learning and expresses itself through systems and decisions to face climate stresses.

The Sustainable Management of Marginal Drylands (SUMAMAD) project in this phase II gives to stakeholders in Burkina, the opportunity to assess and impulse traditional strategies for water and soils preservation and for production.

The documentation of some endogenous knowledge and technologies of the populations that we would like to validate in scientific area, in the framework of five-year implementation of SUMAMAD project, appears as a negotiable alternative, a response to climate uncertainties and demographic pressure.

The presentation shows the impacts of climate changes and endogenous adaptations.

Methodology of Research

The methodology of research in SUMAMAD Project implementation in Burkina Faso is an eco-systemic approach using qualitative and quantitative approaches. They are based on the whole process of research on main principles namely participation, gender and interdisciplinary aspect.

For this presentation, data were collected by means of a certain number of tools such as (i) discussions and surveys, meetings, workshops in and outside the Hippopotamus Pond biosphere reserve. Documentation and information about the RBMH site have been presented below.

Results and comments

The scope of climate changes

The group of intergovernmental experts, in their last report on climate evolution (GIEC, 2007) asserted that in the course of 21st century, global warming in Africa will be above the global level. The average increase in temperature over the period 1980/99 and 2080/99 might reach 3 to 4° C all over the African continent i.e.1.5 more than the global level. This increase will be less significant in the coastal and equatorial areas (+3°C) and more significant in the Western part of the Sahara (+4°C) where Burkina Faso is located. The climate all over West Africa including Burkina Faso is governed by West Africa Monsoon (WAM) whose wanderings condition life for 300 million people living in a 7.8 million square kilometre area.

Faced with such a situation, the population on the field are on the cutting edge of climate changes impacts and their adaptations have been survival alternatives as time goes by.

Variability and climate change indicators in Burkina Faso

Since the early 1970, drought through its spatial spread, its length and seriousness (up to 50% of rainfall deficit over the 1970-1990 period) has been incomparable since climate series exist. The indicators are violent storms, strong irregularity of rains, disturbance in the length of the various seasons in the year, time-lag between sowing periods, progressive disappearance of biodiversity, lower crop yields, modification of fodder system, mainly the disappearance of the most grazed plant species, noticeable changes in landscape physiognomy and wildlife disappearance. At the social level, the population have experienced poverty in households and more malnutrition.

Demographic evolution and climate change

In Burkina Faso, the general population as well as the urban one has nearly doubled in every 20 years. To that considerable demographic growths are associated more exploitation of the nature and more urbanisation (Déjoux 1988, FAO 2003).

Among the solutions adopted by the populations, now extremely vulnerable, there are migrations whether local, regional, international, rural-rural or rural-urban. The seasonal but also final migratory influx concerns all age groups, men as well as women. In such a context, there are more risks in the vicinity of the Hippopotamus pond biosphere reserve which, for decades, has been a settlement area for the rural migrations from the northern and central regions of the country.

Climate change and crop yield

Rainfall determines sowing date, plants growth and yield. As regards rainfall in Burkina Faso, Albergel (1987) has identified three types:

- Increasing daily precipitations amounting to less than 20 mm;
- Daily precipitations amounting between 20 mm and 40 mm and constant in time;
- Daily precipitations amounting over 40 mm with increasing number.

The disadvantages that might result from Carbonnel (1983) and Albergel (1987) analysis could be (i) variability of season start and end for rain-fed crops (ii) degradation of farming lands, lower soil fertility.

Populations have developed water management techniques for plots and trees, crop diversification, etc. So, zaî, half-moons, stone-belts makes farms better for crop production. To this can be added out season and irrigated crops.

Climate change and pastoral system

Climate changes and their consequence pertaining to the bad spatio-temporal rainfall distribution have significant impacts on pastoral system. As regards the pastoral aspect, data show overgrazing from 20 to 50%, the reason for sensible reduction of ground cover productivity, resulting in unproductive glacis.

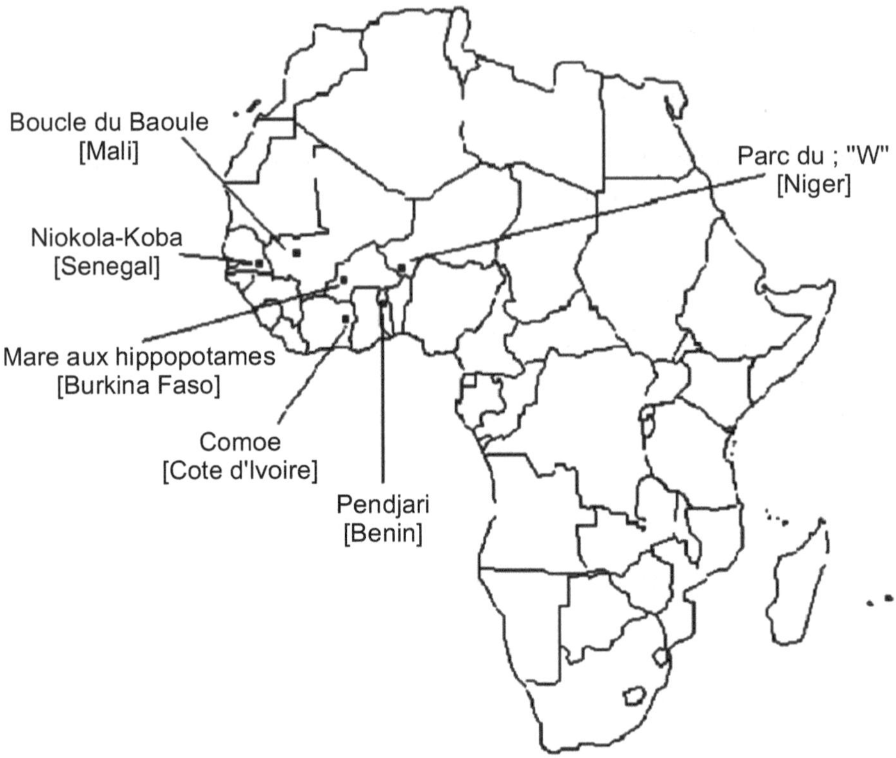

Biosphere Reserves in dryland Africa

The considerable temperature variations impact on fodder and deteriorate to its poor quality thus contributing to a progressive increase in lignin content and decreasing appetence for it. Another phenomenon is the proliferation of non-grazed species like *Sida cordifolia* commonly known as Green Aids, *Cassis,* and so on which, because of their adaptation capacity to new ecological conditions and non-appetence by animal during their vegetative phase, are more and more invading grazing land to the detriment of the other appreciated species under disproportionate exploitation because of animal pressures leading to significant qualitative and quantitative imbalance of fodder. The consequences on livestock are feed deficit, their lower productivity, fecundity and their vulnerability to diseases. Populations have developed strategies like regional and international transhumance, fodder plants producing, fodder cutting and conservation. Destocking and fattening should be encouraged.

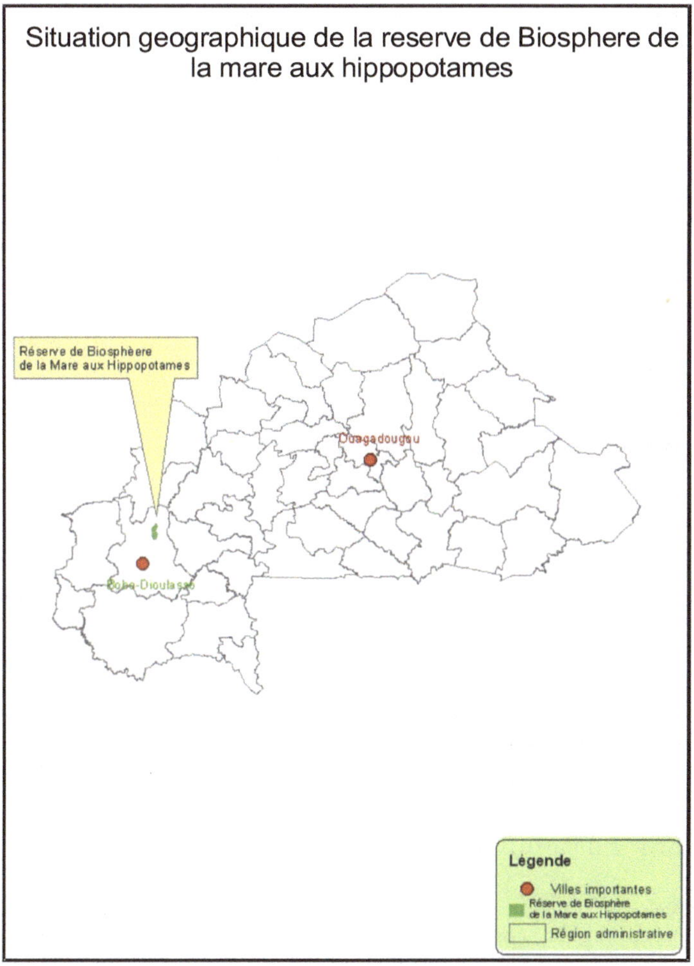

Climate change, food security and health

The lower plant and livestock production compromises people access to safe, nutritive and sufficient quality food to satisfy their basic food needs. So, many households, self sufficient in the past, are now getting into conjunctural or temporal food insecurity.

Serious consequences are foreseeable: more food and nutrition vulnerability with more growth troubles hindering harmonious development for children, more cardio-vascular and respiratory diseases.

As regards malaria related to rains and bacterial meningitis related to harmattan, their spatial distribution is related to the latitude and rainfall. A longer season might impact on these two diseases.

Conclusion

Burkina Faso like all Sahelian countries is faced with drought and natural resources degradation. This situation leads to migration from degraded zones in the north to better zones in the south. It is in this last region particularly favourable to agriculture that Burkina Faso is testing the concept of biosphere reserve within the reserve forest of hippopotamus pond which was made biosphere reserve in 1987.

The sustainable management of resources seems to be now the mobilising theme to understand the range of climate adaptations at stake in arid countries. From now on, at local level as well as in all arid countries, climate change challenges can be addressed not as fatalism but in the angle of sustainable development.

SUMAMAD team and MAB committee of Burkina Faso are then called upon to respond appropriately from real life and local populations' endogenous practices with the view to taking appropriate and concrete actions to control the adverse effects of climate changes in Burkina Faso.

These adaptation responses that will result in profit will help a balanced and sustainable management of ecosystems and biological diversity for future generations. The shared experiences among countries with arid climate in the framework of SUMAMAD could help build a set of approaches on natural resources rational use and exploitation as a capital for local populations and for mankind survival

References

Albergel. J. 1987.- Génèse et prédétermination des crues au Burkina Faso: du m^2 au km^2 Etude des paramètres hydrologiques et de leur Evolution. Ed.ORSTOM. Collection Etudes et thèses, Paris 1988 : 336pp

Balland P. 1992. - Sécheresse, vies animales et végétales dans les cours d'eau, les barrages et les estuaires. *Sécheresse*, 3 (1) : 45-56.

Bremond P. 1990. - Application des techniques électrophoréitiques à deux aspects de la biologie des populations de schistosomes africains : caractérisation des parasites et de leurs hôtes intermédiaires. Détection des schistosomes hybrides. In Act. conf. internat. OCCGE. schisto. *Niamey 30 janvier-2 février* 1990 : 182-189.

Capblancq J. et Toureng J. N. 1978. - Hydrochimie de la rivière lot - *Ann. Limnol.*, 14 (1-2) 1978 : 25-37

Carbonnel J.P. 1983. - Evolution climatique récente en Haute Volta I.Un paramètre pluviométrique peu utilisé: la pluie journalière moyenne annuele. *Not. Doc. voltaïques*, 14 (3-4) Juillet : 1- 20.

GIEC. 2001. Incidence de l'évolution du climat dans les régions : Evaluation de la vulnérabilité Afrique rapport du Groupe d'Experts Intergouvernemental sur l'Evolution du Climat Variabilité climatique en Afrique de l'ouest (Niger – Progrès N° 007 janvier 2008)

Githeko A. K. ; Lindsay S.W.; Confalomen U.E.; Patz J.A. 2001 - Changement climatique et maladies à transmission vectorielle: une analyse régionale ; *Bulletin OMS N° 4* ; 2001 ; 62 - 70

Guinko S. 1984. - Végétation de la Haute Volta - Thèse de Doctorat es Sciences Naturelles U.E.R. Aménagement et Ressources Naturelles Université de Bordeaux III, 2 vol. 394p.

Harouna MOUNKAILA Changement climatique, démographies et migration in Niger - Progrès N° 007 janvier 2008, P 34 - 3.

Hubert N'DJAFA OUAGA Perception paysannes de la variabilité et des changements climatiques et les pratiques locales d'adaptation. in Niger - Progrès N° 007 janvier 2008) P 41 - 42

Illya MIKO (IRAN) Changement climatique et pastoralisme au sahel in Niger - Progrès N° 007 janvier 2008, P 44 - 46

Jean Bernard DUCHEMIN, Ramatoulaye LAZOUMAR et Isabelle JEANNE Impact du changement climatique sur la santé in Niger - Progrès N° 007 janvier 2008) P 28 - 30

Levêque C. 1975. - Mollusque des herbiers à *Ceratophyllum* du lac Tchad: Biomasses et variations saisonnières de la densité. Cah. ORSTOM, *Ser. hydrobiol.*, vol.IX (1) : 25-31

Moumouni ASSANE DAGNA et Boureïma ALPHA GADO Changement climatique et sécurité alimentaire in Niger - Progrès N° 007 janvier 2008) P 31 - 33

O.M.S. 1993. - Lutte contre la schistosomiase. Deuxième rapport du comité O.M.S. - d'experts - *Sér. rap. tech.* N?830 : 100pp

Paperna I. 1972. - Habitat selection and population chandes of buninid snails in the Volta lake Ghana. Bull. *IFAN*, XXXIV, A (4) : 23-57.

Poda J. N., Sawadogo L. 1994 - Hôtes intermédiaires et prévalence de la bilharziose au Burkina Faso. *Sci. Tech*, Vol. XX (2) : 54-67.

Poda J. N., Sellin B., Sawadogo L., Sanogo S. 1994 - Distribution spatiale des mollusques hôtes intermédiaires potentiels des schistosomes et de leurs biotopes au Burkina Faso. *OCCGE INFO.* N°101 : 12-19

Poda J. N., Sellin B., Sawadogo L. 1994 - Dynamique des populations de *Bulinus senegalensis* Müller, 1781, dans une mare temporaire de la zone nord-soudanienne du Burkina Faso. Rev. Elev. Méd. *Vét. Pays Tropicaux* 47 (4) : 375-378

Poda J. N., Sawadogo L., Sellin B., Sanogo S. 1995. - Evolution des hôtes intermédiaires de schistosomes dans les périmètres irrigués du Burkina Faso : cas de *Bulinus truncatus rohlfsi* Clessin, 1886, (A paraître dans la rév. Agro. Afric, Abidjan - Côte d'Ivoire)

Poda J. N. 1996. - Typologie des biotopes des hôtes intermédiaires potentiels des schistosomes au Burkina Faso. *Annales de l'université du Bénin* (Lomé, Togo) *Série Sciences*, Tome XII, 1996, pp. 71-84.

Poda J. N. 1995. - Structure des tailles et accroissement de deux gastoropodes pulmonés: *Bulinus truncatus rohlfsi* Clessin, 1886, et *Bulinus senegalensis* Müller, 1781, dans deux sites de la zone nord-soudanienne du Burkina Faso. (En cours dans les Ann. de l'Université de Ouagadougou).

Sur le plan plus global, le rapport du le Groupe d'experts Intergouvernemental sur l'Evolution du Climat (GIEC) de 2001 Incidence de l'évolution du climat dans les régions : Evaluation de la vulnérabilité Afrique.

Tounao KIRI Changement climatique et gestion intégrée des ressources en eaux in Niger - Progrès N° 007 janvier 2008, P 25 - 26

9

Status of Wetland Degradation in Banni and Its Impact on Maldhari Community in Arid Kachchh, Gujarat

Jagruti Shah

Abstract

The arid Kachchh district, located at the western most part of the Gujarat State in India, is known for its large extent of area under wetland habitat, which is represented by numerous seasonal water bodies and coastal habitats. Vast stretch of grasslands also available in the northern border of the district, the Banni (23^0 19′ to 23^0 52′ N; 68^0 56′ to 70^0 32′ E), which encompasses an area of 2,617 km^2. There are many depressions in Banni, locally known as Dhandh (shallow lakes). During monsoon, rainwater gets collected in these depressions and form shallow lakes. About 34 smaller and bigger size Dhandhs are present in the Banni. Banni communities are highly dependent on the Dhandhs for various purposes. Due to the erratic rainfall in arid Kachchh, Maldharis is highly unstable for livestock in search of fodder and water during drought years. During the normal rainfall years, livestock from neighboring talukas of Kachhh and even from other districts depends on this grassland and Dhandhs. During last decade, the Banni grasslands have seen serious environmental changes due to the invasion of an exotic plant Prosopis juliflora and increasing the salinity are the major issues of Dhandh degradation. Thus, the Banni is gradually losing its capacity to sustain human and livestock population.

Main objectives of the present paper is to examine the local perceptions of wetland and grassland change, socio-economic status, and factors that cause degradation of the wetland and vegetation resources in the Banni grassland.

Introduction

In the context of arid and semi-arid environments, wetlands are transitional areas that are permanently, seasonally or occasionally waterlogged with fresh or saline water, including both natural and man-made areas that support characteristic fauna and flora (Mermet, 1986; Dugan, 1990). These environments support distinctive vegetation and provide habitat for many wetland dependant animal species. Wetlands in arid regions characteristically have very high rates of evaporation. The duration of inundation is therefore highly variable depending on inflows and seasonal evaporation rates (Goodrick, 1984). Arid and semi-arid regions receive very low and erratic rainfall and experience frequent drought. Livestock farming is the predominant source of their economy. During drought period, most of the wetlands act as the natural fodder bank. These wetlands are saline or non saline and are very dynamic systems subject to natural (seasonal) as well as man induced changes which produced alterations in wetland ecosystems. Many arid wetlands are very shallow water bodies but have great importance for biodiversity. This situation is true in Kachchh especially in Banni.

The salinity of arid wetlands varies from fresh to saline. The salinity of many playas and salt lakes usually increases as the wetland dries. Inland saline wetlands around the world. i. e., closed lakes, dry lakes, salt pans, inland sabkhas and playa-lakes, have been studied using the classical approaches of sedimentology, limnology, or geochemistry (Castaneda and Herrero, 2005). Detailed description of these environments requires specific diagnostic tools and indicators. The Kachchh District probably supports one of the largest networks of wetlands in the Gujarat State. Arid Kachchh district covers maximum area under wetlands. Based on the distribution, the wetlands of Kachchh can be divided into three major types; the central inland wetlands (man-made), coastal wetlands in the southern and western region and saline wetlands (Ranns and Banni) in the northern limit. Major parts of the Great Rann of Kachchh, Little Rann of Kachchh and Banni grasslands, for the substantial part of the year support shallow wetlands of larger dimensions. Wetlands in this arid region are critically important for sustainable management of natural resources.

Banni Wetlands

Banni (23^0 19′ to 23^0 52′ N; 68^0 56′ to 70^0 32′ E) located in the northern border of the Kachchh district, which encompasses an area of (2,617 km^2) (GUIDE, 2009) . Banni is administratively protected as grazing land, soils are poorly drained and saline, and thus are generally beyond the margin of profitable crop production. Whereas the introduction of cooperative dairy development is responsible for major improvements in milk production and collection, infrastructure and services in the 46 Banni villages still remain poor (Bharara, 1993).

Flooding during the monsoon isolates the Banni from mainland of Gujarat. In Banni, which is predominantly a flat area, have many depressions, where rainwater gets collected during monsoon.These shallow water lakes are locally known as *'Dhand'*, *'Jhil'* or *'Talav'* range from <2 ha to > 3500 ha in size. There are four types of wetlands in Banni, *Kar* (smallest wetland), *Chhachh* (bigger wetland then Kar), *Thathh* (bigger wetland then chhach) and *Chhari* (biggest wetland). All these wetlands are supporting large numbers of waterfowls. After monsoon only 4-5 month water is drinkable after that more salty and not useful for livestock also except Chhari dhandh.

Material and Methods

The approaches for data collection are as follows:

a) Secondary information

b) Geographic Information System (GIS) and remote sensing (RS) techniques (Garg, Singh and Murthy, 1998) and

c) Field based methods for socio economic study

Details of Satellite Data

Indian remote sensing satellite IRS – 1C and 1D data is used in the present study which has spatial resolution of 23.5 m. The interpretation has been done at 1:50,000 Scale.

Data Used and Interpretation

In the present study, post monsoon data was used. The study area is falling under 2 scenes of cloud free satellite data of October 1997 and October 2007 Path (89 & 90) and Row (55).The data was georeferenced with Survey of India (SoI) topographical maps of 1964 to 1986 on 1: 50,000 scale and prepare a False Color Composites (FCCs) map of Banni. A preliminary interpretation was undertaken through visual interpretation of different colors tone in the imagery for marking the wetlands, which was represented by a light to dark blue to black with smooth texture and irregular shape were mapped. These preliminary interpreted maps were refined through ground truthing. ERDAS IMAGINE (9.3) and Arc GIS (9.3) image processing software were used for these purposes.

Following methods were used to assess the pressure on wetlands/ dhandhs by the local communities :-

1. Group and team dynamics (discussion and meetings)
2. Interviewing and dialogue (pre-designed questionnaire method)
3. Visualization and diagramming (Participatory Natural Resource Appraisal (PANR) (Chambers, 1981; Chambers, 1994; Collinson, 1981; Mukherjee, 1997)

In the questionnaire method, information from the respondents pertaining to wetlands were gathered on major items, like general information and dependency. Secondary data on rainfall pattern and livestock population were collected from various agencies.

Results and Discussion

The wetlands with a spread of 10 ha and above were digitized from satellite data of 1997. There are 19 wetlands/dhandhs covering 22,584 ha (GUIDE, 2008). (Map 1). As per example Chhari-dhandh, Servo-dhandh, Vakrio-thathh, Khirjog-dhandh, Abdha-jheel, and Luna-dhandh etc. good waterbodies were present in Banni.

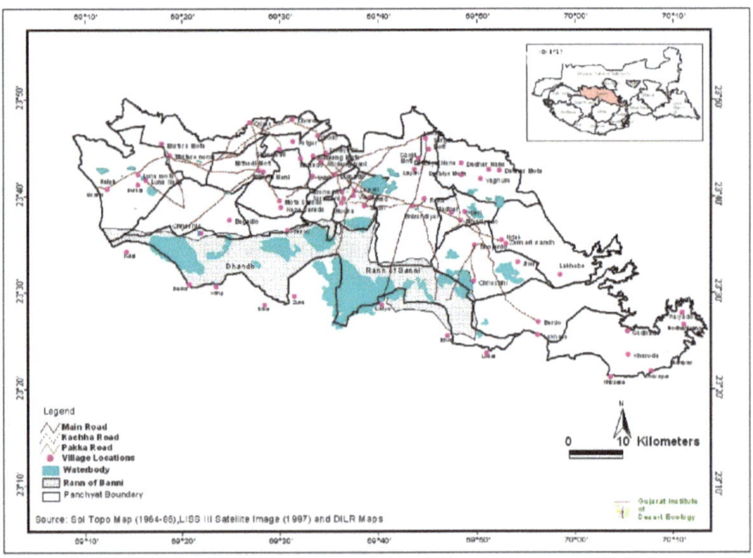

Map 1 : Wetland Distribution in Banni Grassland

Issues in Banni Related to Wetland

Banni is an unique and dynamic ecosystem where grassland coexists with wetland ecosystem. It is known for its large extent of area under wetland habitat, which is represented by numerous seasonal water bodies and saline habitats. They supports a rich and unique assemblage of fauna and flora. People's perceptions show that now the palatable grasses and native woody tree species have become scattered and water pools dried out. Some issues have been identified for wetland degradation (Figure 1).

Grazing | *Microcystis* Algal Growth | *Prosopis* encrochment

Fishing | Tree Cutting | Livestock Use

Fig. 1 : Factors causing Wetland Degradation in Banni

Overgrazing

Arid and semi-arid regions receive very low and erratic rainfall and experience frequent drought. Livestock farming is the predominant source of their economy. During drought period, most of the wetlands act as the natural fodder bank. This situation is true in Kachchh and especially in Banni. The degradation of Banni grassland is through overgrazing by local and migratory livestock.

Sedimentation

Rapid increase in livestock population and over-grazing in and around the wetlands is expected to enhance soil erosion and thereby, creates sedimentation problem in the wetlands.

Irregular Rainfall

The Kachchh region and Banni being a part of arid to semiarid climatic domain. It is characterized by irregular, short but high intensity rainfall pattern. Rainfall pattern at Kachchh region is as follows (Figure 2).

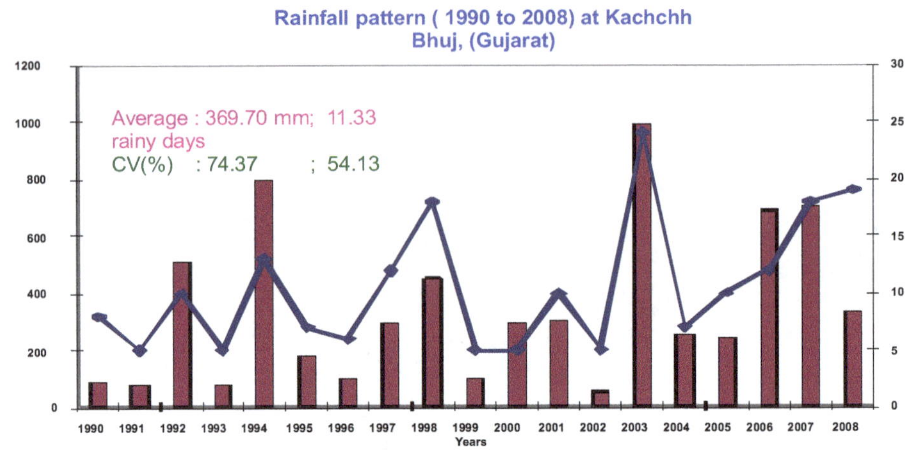

Fig. 2 : Rainfall pattern of Kachchh region

Changes in Species Composition

Local people of Kachchh and Banni have observed a change in the surrounding habitat from grassland to woodland (mainly dominated by *Prosopis juliflora,* locally called Gando Bavar) during their lifetime.

Current Scenario

Shah, (2009) reported that remote sensing data of 2007 showed 29 wetlands covering an area of 21,600 ha whereas data of 2009 showed 34 small and large size wetlands with the water spread over 12,027 ha (Shah, Patel and Vijay kumar, 2011). This result shows that in the decade (1997 - 2007) and during a two year period (2007-2009), wetland number has increased but the water spread area has decreased due to natural or manmade activity. Chhari dhandh is the biggest dhandh present in Banni and it covers water spread area of nearly 3695 ha. In normal rainfall years, livestock from neighboring talukas of Kachchh and even from other districts depends on this Banni grassland and dhandhs.

The Banni supports a large number of Muslim pastoral community, collectively known as "Maldhari". The Maldhari herd is composed of camel, sheep, goats, cattle and buffaloes. They produce ghee (purified butter), wool and handicrafts. In Banni, cattle are owned by Maldharis and now they keep more buffaloes than cattle. Information was gathered mainly on jat (Maldhari) communities from Banni.

During discussion with local inhabitants it emerged that the Banni communities are heavily dependent on the natural grassland and wetlands for various purposes. Earlier, most of the ponds had dense vegetation cover including palatable grass species and fodder trees. Now the grasses and woody tree species have become scattered and water pools also dried out.

Availability of grazing lands around the villages is not sufficient for livestock round the year. Due to the erratic rainfall in Banni and Kachchh, Maldharis remain in search of fresh fodder and water during drought years. This results in heavy pressure on the natural resources like grasslands and wetlands.

This change in vegetation have effects on the livestock population: cattle have declined, since these animals thrived on the Banni grass; on the other hand, the number of buffaloes has increased. Overall, steady increase in the livestock population has been recorded from Banni.

Table 1 shows that during the last 25 years, livestock (Cattle and Buffalo) population has increased. Census data shows that in Banni, livestock population (57,073) is relatively high as compared to the human population (16,783) based on 2006 data.

Table 1 : A Comparative Panchayat wise data of Livestock population in Banni

Banni Panchayat	Cattle and Buffalo		Goat and Sheep	
	1982	2007	1982	2007
Bardo	2238	3475	247	10
Bhirandiyara	5523	4436	2548	1788
Bhitara	2211	4626	0	1025
Bhojardo	2096	2803	139	1576
Chachla	1270	2370	0	0
Dadher Moti	1113	2763	784	588
Dadher Nani	997	0	1133	0
Dhordo	0	0	0	0
Dumado	0	0	0	0
Dethiya	897	0	482	0
Gorewali	2748	4335	6116	4881
Hodka	3286	4508	2403	1660
Luna	2939	1881	789	1713
Mishariyado	1589	5397	1127	845
Mithdi	2976	259	27	799
Raiyado	788	773	6	109
Shervo	574	1006	0	3435
Udhmo	352	0	716	0

Problems in the Banni Dhandhs

Since the 1950s, the Banni has experienced major changes that have largely degraded the quantity, composition and productive potential of its plant population, having serious repercussions on livestock production. While significant proportions of the Banni were reported that most native trees, and associated native shrubs, forbs and grasses have disappeared. The compound influence of climate, grazing, and the introduction of non-native plants have brought irreversible changes to its plant composition.

Loss of plant cover, deteriorated soil structure, and the construction of check dams in Kachchh, used to control flooding and provide drinking water, have curtailed the natural movement of detritus and silt that periodically brought fertility to the Banni (Saxena, 1993).

Today, the grazing land and wetland vegetation has been overgrazed by livestock, and that is a major problem being faced by these Dhandhs. Banni at one time was considered the largest grassland of its kind in the Asia, but it has fallen upon sad times in the last decades. The degradation of Banni area has been linked to increasing salinity, illiteracy of its inhabitants, growing human and livestock population and a large scale growth of *Prosopis juliflora. P.juliflora*, an evergreen plant is one of the most invasive species of arid and semi arid regions across the world and also found in Banni grassland. Due to the existing dry conditions, there has been a spread of *Prosopis juliflora* in the Banni grassland, which is adversely affecting the health of the wetland habitats as well as other terrestrial vegetation species and also domestic animals. One of the major threats that is cutting of woody vegetation is also reported from wetlands. In Chhari and Luna dhandh like wetlands had recorded freshwater fishing by a commercial or individual fisherman. The Banni dhandhs face natural threats like erosion and sedimentation, droughts and scanty rainfall.

Due to this misunderstanding, unawareness and overexploitation of its natural wealth, during the last few decades tropical wetlands have been destroyed and degraded.

Problem Faced by Maldharis

Banni is classified under forestland but is currently in the administrative control of Revenue Department. The local villagers have customary settlement rights without ownership rights. Today, livestock production on the Banni is principally maintained by free grazing on depleted sources of grasses, sedges, and forbs, on the leaves of *Suaeda fruticosa* (a native salt-tolerant tree species), and on the pods and seeds of *Prosopis juliflora* which pastoralists often collect and boil before feeding. But some of the Maldharis are not used to prepare boiled food for cattle, and then cattle spread the *Prosopis* by their dung.

During and post-monsoon seasons, livestock are grazed within the Banni, moving between man-made wells and tanks in which freshwater has been trapped during the rains. During the dry season, maldharies are forced to move cattle outside Kachchh District.

The pastoralists are a good source of information about animal feeds which are largely unknown to the farming community. They have identified several trees, bushes, grasses and weeds which improve milk yields, fat percentages in the milk and also reproduction (Rangenekar, 1992a; Rangenekar, 1992b). Pastoralists are knowledgeable about livestock, their behavior, feeding habits and production characteristics (Rangenekar and Rangenekar, 1992; Rangenekar, 1993). But now, there is a need to collect the traditional knowledge from experienced maldharies and circulate this knowledge to other peoples and help to stop more degradation of the natural resources. Some major problems are as follows.

- Migration in search of fodder
- In drought years, selling livestock as well as animal products like wool,
- Buffalo seems to be the most common livestock purchased or sold by Maldharis,
- In summer less water availability for livestock whereas livestock need more water during summer compared to winter season,
- No proper facility for storage of milk.

The study reveals their perceptions of the environmental changes, and their priorities for conservation of the natural vegetation in the region. Involvement of communities will help to formulate focused development plans that have more likelihood of success (Joshi, 2009).

Conclusion

This research work is crucial for the study of the evolution of these wetland habitats in a decade. Natural habitat has been lost through gradual regeneration of invasive species like *Prosopis juliflora*. This has resulted in significant loss of wetland area, degradation of remaining natural resources and a consequent decrease in the native land use type and species composition.

Future Steps to Overcome Wetland Conservation

In this context, while high resolution satellite imageries are very useful data sources for delineating wetlands, the ecological surveys of wetlands, which constitute sampling of different bio-physical attributes. These are critical for their management planning. The information gathered from this study is expected to be useful for the management and development of wetlands, which provide benefits to both the wildlife and mankind living in and around.

There is a need to study the wetlands of Banni and come up with a proper strategy for stopping future degradation.

- Development of grassland for fodder security
- Generating employment for the villagers
- Alternate livelihoods like value added products from *Prosopis juliflora*
- Detailed socio economic study needed for level of usage of dhandhs and its economic influence
- Detailed geo hydrology study is essential for the conservation of the wetlands

Inland saline wetlands pose unique opportunities for research, education, aesthetic appreciation and experience managing the interface between nature and man. Currently, steps have to be taken to preserve the eastern saline wetlands like Banni Dhandhs.

Acknowledgements

The author is thankful to Dr. V. Vijaykumar, Deputy Director, GUIDE, Bhuj for providing necessary facility. The author also acknowledge to Dr. S. Somasundaram for useful suggestions to prepare this chapter.

References

Bharara LP. 1993. Socio-economic aspects of Banni pastoralists and factors effecting Changes in their perceptions toward the future. In: Cincotta, R.P. and Pangare G. (eds) *Pastoralism and Pastoral Migration in Gujarat.* Institute of Rural Management Anand, India.

Castañeda C and Herrero J. 2005. The water regime of the Monegros playa-lakes established from ground and satellite data. J. Hydrol 310, 95-110. 285

Chambers R. 1981. Rapid Rural Appraisal, Rationale and Repertoire. IDS discussion paper No. 155. Institute of Developmental Studies, Sussex.

Chambers R. 1994. Participatory Rural Appraisal (PRA): Analysis of Experience. World Development. 22 (9):1253-1268.

Collinson MP. 1981. The exploratory Survey Content, Methods and Detailed Guidelines for discussion with farmers in CIMMYT Eastern Africa Economic Programme. Farming System Newsletter.

Dugan PJ. *Wetland conservation: A review of current issues and required action.* IUCN, Montreux. (ed) 1990.

Garg JK; Singh TS and Murthy TVR. 1998. Wetlands of India, National-Wide Wetland Mapping Project-Project report. Space Application Centre, Ahmedabad. Pp. 239.

Goodrick GN. 1984. *Wetlands of North-Western New South Wales. Occasional Paper No. 6.* NSW National Parks and Wildlife Service, Sydney.

GUIDE. 2009. Integrated grassland Development in BANNI Kachchh District, Gujarat. Progress Report. Gujarat Institute of Desert Ecology - GUIDE. Pp.35.

GUIDE. 2008. Study of Wetland Habitats in Kachchh District and Suggesting Stakeholder Driven Management Strategies. Final Report. Gujarat Institute of Desert Ecology - GUIDE. Pp. 278.

Joshi PN. 2009. Local perceptions of grassland change and priorities for conservation of natural vegetation in Banni, Gujarat. News from Dry net, No.4.

Mermet L. 1986. Terres et eaux: Approaches techniques pour conserver et mettre en valeur les zones humides. Centre d.Etudes des Systemes et des Technologies Avancees (CESTA), Paris.

Mukherjee N. Participatory Appraisal of Natural Resources. Concept Publishing Co. Delhi. 1997.

Rangenekar S. 1992a. Women in livestock production in rural India. *Proceedings of 6th AAAP Animal Science Congress,* Bangkok. 23-28 November 1992.

Rangenekar S. 1992b. Studies of indigenous knowledge, work sharing and perceptions of livestock production by women from pastoralist families. In: Cincotta, R. and Pangare, G. *Pastoralism and Pastoral Migration in Gujarat. Institute of Rural Management,* Anand, India.

Rangenekar SD and Rangenekar DV. 1992. Involvement of women and children in goat keeping in some villages of Gujarat and Rajasthan. *Proceedings of Vth International Conference on Goats,* New Delhi. 2-8 March, 1992.

Rangnekar DV. 1993. Traditional livestock production systems among pastoralists: Their perceptions of the production system and attitude to change. In: Cincotta, R.P. and Pangare G. (eds) *Pastoralism and Pastoral Migration in Gujarat.* Institute of Rural Management Anand, India.

Saxena SK. 1993. Present status of the Banni Grassland. In: Cincotta, R.P. and Pangare G. (eds) *Pastoralism and Pastoral Migration in Gujarat.* Institute of Rural Management Anand, India.

Shah JP. 2009. Study of selected wetlands of Kachchh with special emphasis on landscape ecology and socio economic status using GIS and remote sensing. Unpublished Thesis at Hemchadracharya North Gujarat University, Patan.

Shah JP; Patel P and VVijaykumar. 2011. Landuse Landcover Classification of Banni Grassland *Proceeding of Symposium on Banni Grassland* 4th – 5th March 2011 at Gujarat Institute of Desert Ecology, Bhuj-Kachchh.

10

Geo-Environmental Hazards and the Future Planning of the Coastal Area between the Two Nile Branches Rosetta and Damietta for Sustainable Development, North of Nile Delta, Egypt

Mohamed Sabry Mohamed El-Bady

Abstract

The balance between resources and population constitute the solid base for drawing the future map of sustainable development in the study area in all developmental sectors such as agriculture, industry, tourism, and commercial sectors. Geoenvironmental hazards such as erosion, sea level changes, sand dune encroachment, siltation of the navigation channel of the Damietta harbor, salinization of the soils, the negative effects of the protective structures and bad management of the coastal areas must be studied in details before making the future planning for any area. The main geomorphologic units in this area are: beaches, sand dunes, wetlands, urban areas, reclamation lands and lagoons.

The rate of erosion from 1984 to 2000 in the Rosetta beach is -117.6 m/yr, while it is -17.6 m/yr in the Damietta beach. On the other hand, the rate of erosion in the period from 1984 to 2005 is -104.6 m/yr and -11.4 m/yr in Rosetta and Damietta beaches, respectively. Since, Nile Delta is a typical wave dominated delta (Coleman et al., 1981) and Tide is a semi-diurnal microtidal regime with a maximum tidal range of 50 cm (UNESCO/UNDP, 1978), the erosion and accretion are very dangerous hazard in the area.

Due to the reclamation of vast areas of the coastal dunes, the lagoons and wetlands, the new reclaimed lands are added to the agricultural deltaic land. Thus, the coastal sand dunes, lagoons and wetlands are decreased. The main urban centers have expanded, and new urban centers (Damietta harbor, the New Damietta city and Mansoura city in near future) have been constructed at the expense of vast areas of lagoons, coastal dunes, and backshore flats. The siltation of the navigation canal of the Damietta harbor, salting of the soil (new and old) and sand dune encroachment are very important hazards in the study area. All the previous geoenvironmental hazards and others threaten the sustainable development in the study area.

Introduction

Egypt is the second most populous country in Africa. Its population is over than 70 million and is growing at 1.66% per year (World Bank, 2003). Now, the population is about 80 million. The projected population by the year 2025 is from 85 - 95 million (UN, 1979, UNEP, 1988). In the coastal area, living in less than 3 m, it could be over 12 million (about 7 million by 2020, east of Rosetta Nile, (ARICON, 1989). It is located between 22° to 32° North and 24° to 37° East. The coastline of Egypt extends for more than 3500 km along the Mediterranean sea and Red sea.

The study area (Fig. 1) has about 155 km length and about 10 km wide, the coastal area of study including El-Burullus lagoon and many cities such as Rosetta, El-Burg, Baltim, Gamasa, New Damietta city and Ras El-Bar.

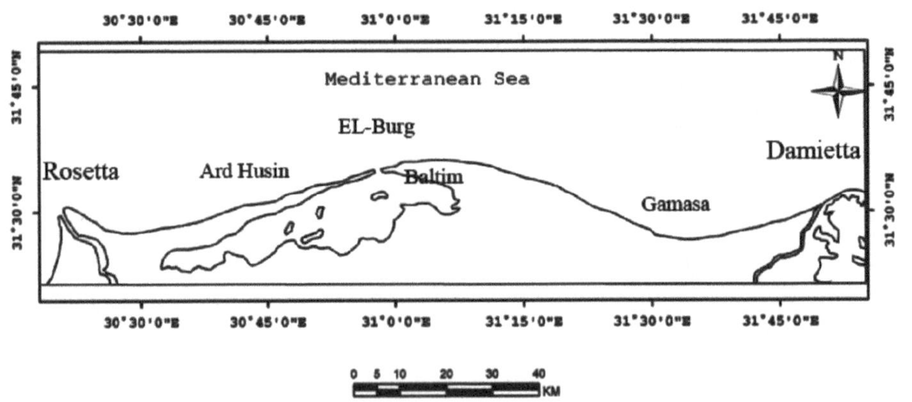

Fig. 1 : Location map of the study area

The Nile delta has received considerable attention by scholars because of its important role in daily life and development aspects of Egypt. The beaches of the Nile Delta are backed by coastal flats followed by coastal dunes and three lagoons (from east to west, Manzala, Burullus, and Idku). Assessment presumed that by the year 2100 the coastal areas with elevation below 1 m above sea level will be submerged (Stanley, 1988)

The field of environmental studies is usually concerned with the impacts of both human and natural phenomena that affect the earth's environment as climatic and geologic hazards. Intense storms and winds, oceanographic factors such as waves, ice and sea level changes, together with extreme precipitation are called climatic hazards. On the other hand, the geologic hazards include a lot of phenomena such as earthquakes, volcanoes, erosion, accretion, land instability and sand dune encroachment.

The Nile Delta coast, which constitutes about 300 km, hosts a number of highly populated cities such as Alexandria, Port Said, Rosetta, and Damietta. The Nile Delta with its coastal front on the Mediterranean is considered vulnerable to the impacts of climate change. In addition to the expected rise in sea-level the shoreline erosion, stresses on fisheries and salt water intrusion in groundwater create major challenges. These factors also produce stressful effects on water and agricultural resources, tourism and human settlements. The northern Egyptian lakes, which constitute about 25% of the total Mediterranean wet lands and produce about 60% of the fish products, are also highly vulnerable to the impacts of climatic changes. From the geographical point of view, the major delta cities such as Rosetta, Baltim, Gamasa, Damietta and Port Said lies below 2 m elevation (comprising a zone of 30-55 km wide). This zone is very important due to its high level of agricultural and fish production.

Materials and Methods

Three satellite images 1984, 2000 and 2005 (Fig. 2, 3, 4) are used for detection the geomorphologic units with time. The satellite image showing the change of each geomorphologic unit in years 1984, 2000 and 2005 also show the shoreline positions in these years. Some representative samples have been collected from beaches, dunes and soils according to the field observations. Sedimentological and chemicals analyses are used to study the collected samples.

Geoenvironmental Hazards

Following are the main environmental hazards that are facing south of the Mediterranean Sea including the study area.

1. The coastal erosion and sea level changes.
2. Sand dune encroachment.
3. Salting of soils.
4. Drying of many parts in Burullus lagoon for cultivation and/or farm fishing.
5. Siltation of Damietta Harbor navigation channel.
6. Irregular distribution of the populations on the coastal plain.
7. Pollution.
8. The negative effects of the coastal protection projects.
9. Management of the coastal areas.

Erosion and Sea Level Changes

Coastal erosion

Due to the increasing erosion, many of the world's low-lying coastal regions will be exposed to potential inundation. Deltas that are deteriorating because of sediment starvation, subsidence and other stresses are particularly susceptible to accelerated inundation, shoreline recession, wetland deterioration, and interior land loss (Biljsma *et al.*, 1996; Day *et al.*, 1993). These encouraged many author such as Nicholls *et al.* (1999) to predict that the sea level rise will cut over 22% of the coastal wetlands of the world by 2080. Such losses would reinforce other adverse trends of wetland loss resulting primarily from direct human actions as estimated by Detr (1999) to be about 40% of 1990 values by the 2080s. Stabilization scenarios developed by Detr (1999) show a large reduction in wetlands to be 6-7%, compared with unmitigated emissions (13%). Before the construction the Aswan High Dam the growth of the Delta was 12 feet/yr., this is a small rate if compared with other rivers (Mississippi, 250 feet/yr; Volga, 1000 feet/yr). It is due to the fact that the river Nile flows in the arid zone for almost 2500 km before reaching the Mediterranean Sea. About 40 years ago and up to date, the Nile Delta beach has been suffering extreme erosion.

Historically, the study of the published maps reveals that the Nile Delta coastline was steadily advancing since early 18th century (UNDP/

UNESCO, 1978). The average advance along the Rosetta and Damietta promontories was 30 m/yr and 10m/yr respectively (Sestini, 1992) with total progradation of some 3 - 4km between 1800 and 1900 (Lotfy and Frihy, 1993). However, the last century has been marked by intensive erosion with dramatic changes in the delta shorelines. Such changes are generally attributed to the construction of dams and barrages along the main Nile course, prominent among them being the Aswan High Dam which had closed the Nile course in 1964 (Stanley and Wingerath, 1996). Erosion may also have been caused in part by climatic factors and the reduction in the Nile floods (Sestini, 1989). Sestini (1992) suggested an overall reduction of 25% in the Nile discharge by 1910. This is in fact due to the decrease in monsoonal rainfall over East Africa (Rossignol - Strick, 1983). From 1910 to 1965 the Rosetta promontory receded by 2.5 km with the destruction of the first lighthouse, while the Damietta promontory receded by about 2 km (Sestini, 1992). From 1970 to 1990, the rate of shoreline retreat had increased along the Rosetta promontory, being 106 m/yr, while being 10.4 m/yr at Damietta promontory (Frihy and Komar, 1993). The human impact on the coastal areas and the construction of several structures along the coastline may also play a considerable role on the changes in the oceanographic parameters and hence accelerates erosion with abnormal patterns (El-Asmar 1994, 1995 and 2000). Such coastal erosion and shoreline retreat may represent another potential environmental hazard that may threaten and impede the developmental activities in the study area between Damietta and Rosetta.

The rate of erosion attracted the attention and many of authors such as Fanos (1995) which mentioned that the rate of erosion ranges from -18 to -230 m/yr in the period from 1900 to 1989 at Rosetta beach.

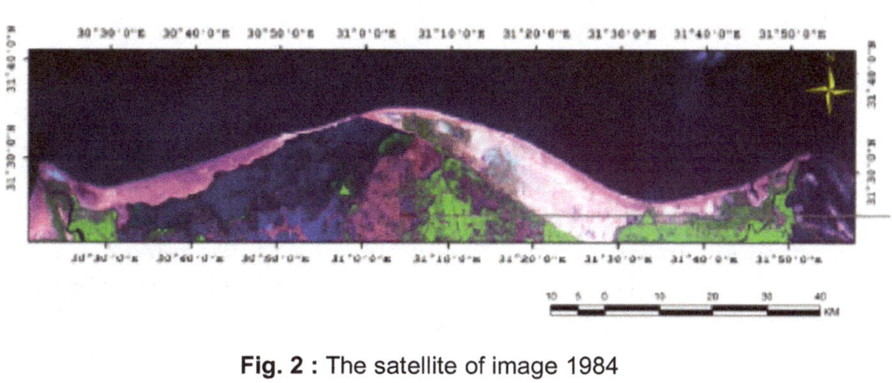

Fig. 2 : The satellite of image 1984

Fig. 3 : The satellite of image 2000

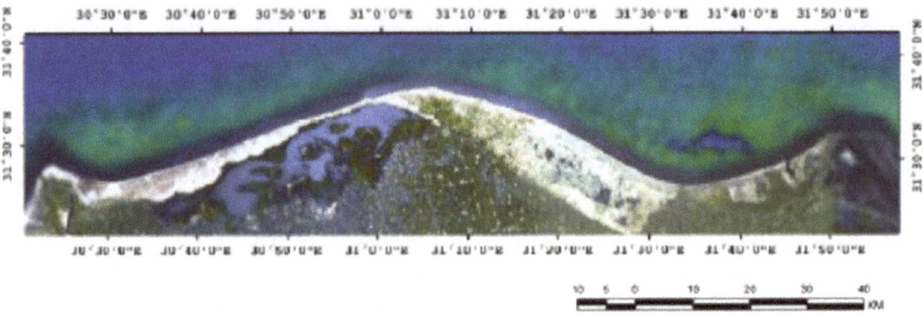

Fig. 4 : The satellite of image 2005

Ahmad (2002) detected the rate of erosion in the period from 1984 to 2000 as - 118.6m/yr. The rate of erosion in the period from 1909 to 1989 of the west and east Burullus have been estimated by El-Fishawi (1994) as -1.3 and -3.8 m/yr, respectively. On the other hand, White and EL-Asmar (1999) recorded the rate of erosion along the Nile Delta beaches in the period from 1984 to 1991 as -113.8m/yr, -15 m/yr, -5.1m/yr, -5.0 m/yr, and 2.1m/yr for Rosetta, Damietta, Baltim, west and east of Burullus, respectively.

Present study is concerned with the measurement of the rate of erosion in two period at the two promontories Rosetta and Damietta. The first period from 1984 to 2000 and the other from 1984 to 2005 is based on satellite images and field observations. The rate of erosion from 1984 to 2000 in the Rosetta beach is -117.6 m/yr, while it is -17.6 m/yr in the Damietta beach. On the other hand, the rate of erosion in the period from 1984 to 2005 is - 104.6 m/yr and -11.4 m/yr in Rosetta and Damietta beaches, respectively. This means that, the rate of erosion decreases due to the continuous construction of protection structures. Utilizing the same satellite images and geomorphologic maps, the relatively high rate of erosion along the beach of the study area (especially Rosetta and Damietta promontories) can be observed. (Fig.

5). The values of rate of erosion in Rosetta and Damietta promontories compared with other previous authors e.g. Frihy and Komar (1993) and White and El-Asmar, (1999) are given by table (1).

Table. 1 : Comparison in the rates of erosion estimated for Rosetta and Damietta Promontories

Authors	Studied Period	Erosion in Rosetta (rate by m/yr)	Erosion in Damietta (rate by m/yr)
Frihy & Khafagy, (1991)	1909 to 1983	53 - 58 m/yr.	18 - 28 m/yr.
Frihy & Komar, (1993)	1971 to 1990	106.00 m/yr.	10.4 m/yr.
White & El-Asmar, (1999)	1984 to 1990/1991	113.8 m/yr.	15.00 m/yr.
Ahmed, (2002)	1984 to 2000	118.6 m/yr	----
Present study	1984 to 2000	117.6 m/yr.	17.6 m/yr.
Present study	1984 to 2005	104.6 m/yr.	11.4 m/yr.

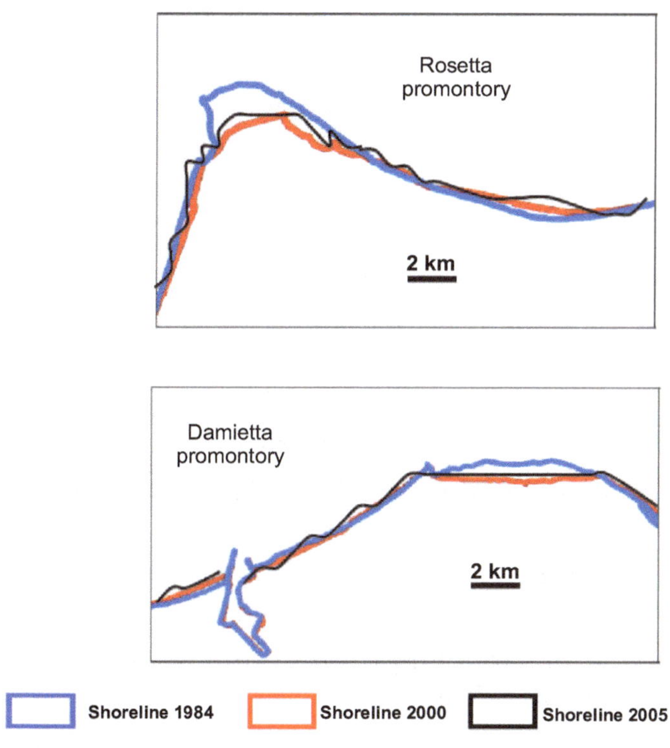

Fig. 5 : The shoreline changes detected in years 1984, 2000 and 2005 of Rosetta and Damietta promontories

The erosion and accretion of the beaches in the study area between Rosetta and Damietta have been discussed. The accretion and erosion phenomenas are local and affected by the continuous protective structure projects in or near the beaches of the study area. The main sites of erosion (Fig. 6) are recorded in Rosetta beach, east Burg El-Burullus Village in the beach front to El-Bannaen and El-Marazka villages, east the Baltim Resort, and in-between the detached break water in Ras El-Bar and Baltim, and to the east of Damietta Harbor (Plates 1,2). On the other hand, the accretion observed in east Rosetta, the front of the detached break water in Baltim and Ras El-Bar, Gamasa, New Damietta, west of the Damietta Harbor and some sites in Burg El-Burullus Village.

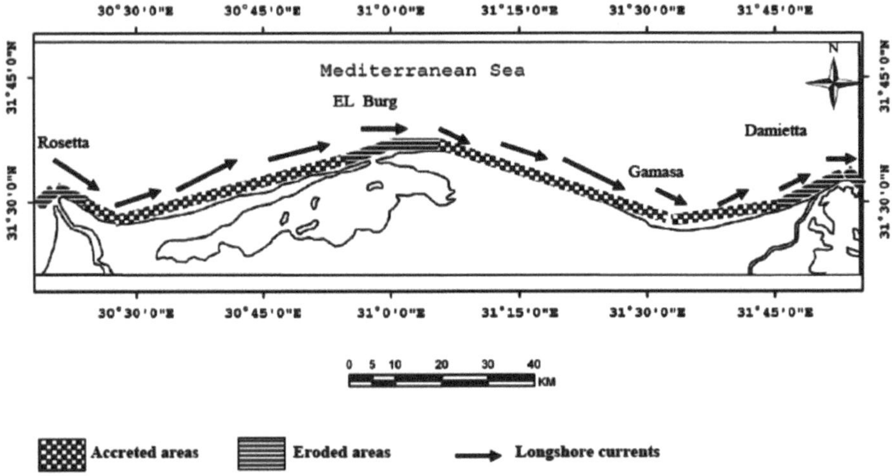

Fig. 6 : The areas of accretion and erosion in the study area between Rosetta and Damietta branches (Modified after Frihy, 1994).

Following are main reasons of erosion that threaten the coastal plain in the study area,

- Dams along the river Nile such as Aswan Dam, High Dam and different barrages decrease the arrival of sediment to the sea.
- Reduction in shoreward moving sand from the shelf because it has been exhausted, due to steepness.
- Increased loss of sand shoreward from beach by increasing wind drifting.
- Reduction in sand supply from dunes because of inland migration or stabilization.
- Reduction in sand supplies from dunes due to the removal of sand dunes by human.
- Sea level rise.

From the geomorphological and sedimentological studies, different conservation structures are proposed besides the present protective structures in the Nile Delta to overcome the erosion problems (Fig. 7). According to the locations of the erosion and accretion (Fig. 6) in the Nile Delta, the new proposed structures are as follows:-

- In Rosetta, two jetties on both sides of the promontory must be erected to protect the Rosetta outlet as in Damietta, the western jetty must be longer than the eastern because the current comes from west to east. This proposed conservation structure is very important because the present protection structures do not finish the erosion in the Rosetta promontory.
- In Burg El-Burullus, the western jetty of Burullus inlet must be longer, and the Basaltic sea wall must be extended to the east to reach the Baltim resort to overcome the erosion between El-Burg and Baltim. In Baltim, another number of detached break water must be built eastward to reach to Kitchener drain.
- In Damietta Harbor and Ras El-Bar beach, a jetty must be constructed to the west of the western jetty of harbor. It must be longer than the western and the detached break water must be increased in quantity to reach to the eastern jetty of Damietta Harbor (Fig. 7).

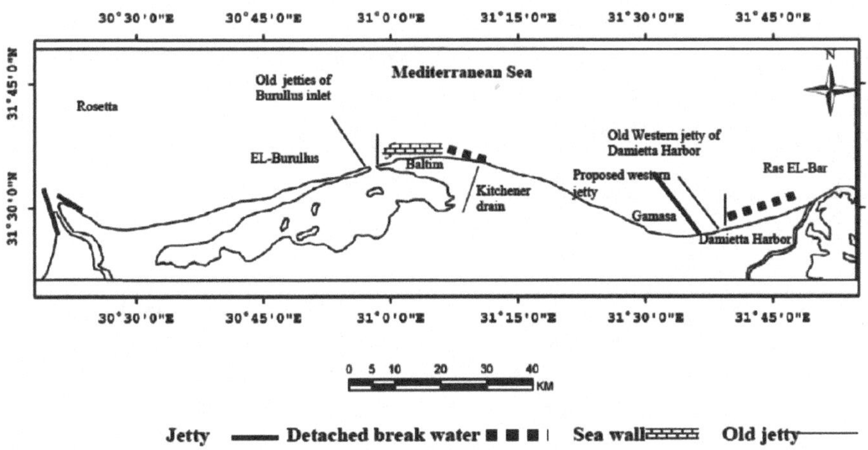

Fig. 7 : The type of proposed conservation structures in the study area between Rosetta and Damietta

Major reasons of erosion are : (i) The decrease of the sediments discharged by the Nile to the coast due to construction of many barrages and dams (Table. 2) along the Nile course especially the Aswan High Dam in 1964. The Nile discharged about 85- 90 million tons of sediments into the sea in the late summer of 1964 (Sestini, 1992), (ii) Eastward littoral drift, where it drives about 1,000,000 m^3 of beach and near shore sands to and beyond north Sinai, (iii) The offshore wind, where it continuously remove a large amounts of sand beach (Sestini, 1992), (iv) Construction of a numbers of protective structures along the Nile Delta coast, where they may cause accretion in areas and erosion in the other along the coast.

Table. 2 : Chronology of Major Works on the Nile course (Shaplan, 1997)

Year	Work (project)
1843 - 61	Construction of Delta barrages
1890	Repair and renovation of Delta Barrages by the British
1902	First Aswan Dam
1902	Assuit Dam
1903	Zifta Barrage
1909	Isna Barrage
1912	Aswan Dam raised
1925	Sennar Dam on the Blue Nile
1930	Hammadi Barrage
1933	Aswan Dam raised again
1937	Jebel Auliya Dam (on White Nile, south of Khartoum)
1951	Edfina Barrage
1954	Owen Falls Dam completed (at outflow from Lake Victoria)
1958	Sennar Dam raised
late 1950's	Hydro-power station installed at Aswan Dam
1960-1970	Construction of Aswan High Dam

Sea level changes

The major lowering of sea level occurred at the end of Pliestocene from about 35000 to 18000 years B.P. (Coutellier and Stanley, 1987) and continued in lowering between 20000 to 15000 years B.P. (Frihy and Stanley, 1987). The sea level has been risen from 14000 to 7000 years B.P. (Nageb, 1992), then lowered from 3000 to 2500 years B.P. (Morner, 1971). The sea level rise or the submergence of delta due to the increase of the sediment load was studied by El-Askary and Frihy, (1984) and Coutellier and Stanley (1987). For the last 100 years, the rise in sea level was 1.0 to 1.4 mm/yr. (Butzer and Hansen, 1968).

According to Milliman (1992) the changes of sea level over geologic time scales occur in two distinct ways: tectonic changes related to the rate of sea floor spreading and climatic changes related to the glacial events. In addition, warmer water expands more than colder water for a given change in temperature. The geographical distribution of sea level change results from the geographical variation of thermal expansion, changes in salinity, winds, and ocean circulation. Sea level also changes when the mass of water in the ocean increases or decreases. This occurs when oceanic water is exchanged with the water stored on land. The major land store is the water frozen in glaciers or ice sheets. An increase in global temperature may cause a rise in sea level due to melting of glaciers, ice caps and ice sheets and due to the thermal expansion of water (Milliman, 1992).

Plate (1)

Plate (1) : The erosion in the study area and protection structures

Photo (1): The erosion and sea wall in Rosetta

Photo (2): the erosion after the Basaltic sea wall in El-Burg

Photo (3) Erosion in Baltim beach before the detached break water

Photo (4) Tombolo formation in Baltim resort

Photo (5): The erosion of Baltim shore in 1996

Photo (6) The same shore changed into accretion after construction of detached break water in 2004.

Photo (7): The erosion of Baltim shore in 1996

Phot (8) The same shore changed into accretion after construction of detached break water in 2004.

Plate (2)

Plate 2 : The erosion in the study area and protection structures

Photo (1) The erosion in Baltim between detached break water

Photo (2) The erosion east of Baltim resort

Photo (3) Sea wall was constructed by rip-rap of basaltic rocks and the jetty along the western side in Ras El-Bar

Photo (4) The concrete sea wall in Ras El-Bar

Photo (5) The eastern jetty in Ras El-Bar

Photo (6) The Eastern sea wall and eastern jetty in Ras EL-Bar

Photo (7) The detached break water in Ras El-Bar

Photo (8) The third old groin (damaged) in Ras El-Bar

Photo (9) The erosion in between the detached break water in Ras El-Bar

Photo (10) The Family beach in Ras El-Bar with shells and shell fragments (eroded)

Nakicenovic et al (2000) predicted a rise in mean global sea level of 12 cm to 88 cm from 1990 to 2100, while IPCC (1996) expected a rise of 18 cm by 2030 and 50 cm by 2100 in the Mediterranean region.

Based on tide gauge data, the rate of global mean sea level rise during the 20th century is in the range 1.0 to 2.0 mm/yr, with the main value of 1.5 mm/yr (IPCC 2001). Grenier (1990) predicted the submergence of three major deltas around the world including the Nile Delta. Map (Fig. 8) shows the possible submerged coastal areas along the Nile Delta due to the rise in sea level by 1 and 3 m respectively. Khafagy *et al.* (1992) estimated that for a 1-m sea-level rise, about 2,000 km^2 of land in coastal areas of the lower Nile delta may be lost to inundation. Outside the delta, erosion is expected to be quite limited.

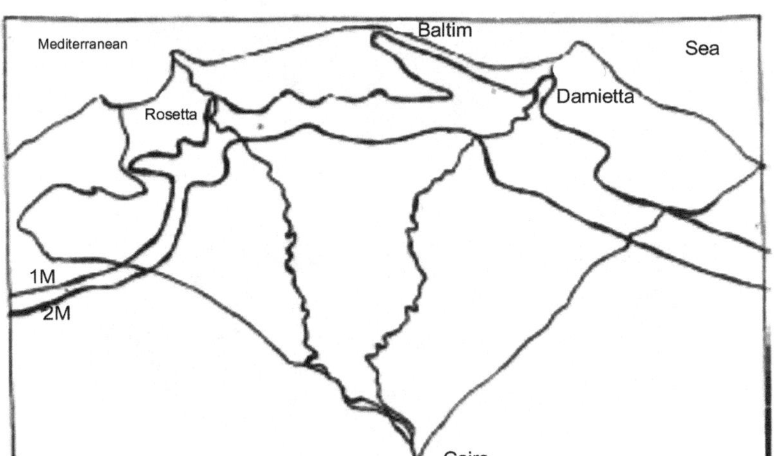

Fig. 8 : The possible submerge of the coastal zone along the Nile Delta due to the sea level rise (after Grenier, 1990)

BBC News (1997) shows the amount of land that would be lost in the Nile Delta if the world's oceans rose by 1.5 meters (Fig. 9).

Fig. 9 : The Effect of the rise in sea level on the Nile Delta (BBC News, 1997)

Figure (10) shows the land of northern Nile Delta that would be lost if the sea rose by 0.5 m and 1.0 m, where Most of a 50 km wide land strip along the coast is less than 2 m above sea-level and is protected from flooding by a 1 to 10 km wide coastal sand belt only, shaped by discharge of the Rosetta and Damietta branches of the Nile.

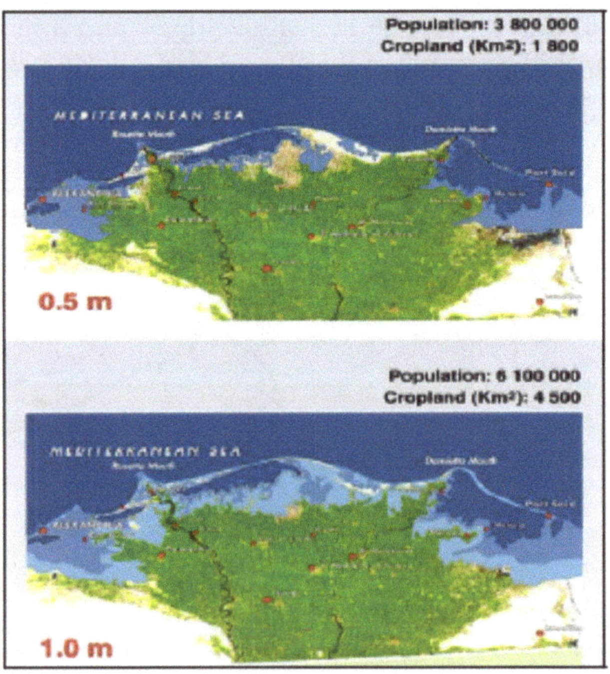

Fig. 10 : The effect of sea level rise (by 0.5 and 1.0 m) on the Nile Delta (UNEP/GRID, 2002)

Generally the sea level now will cause many change in the coastal area e.g:-

When the Mediterranean sea level in the study area rises or the coast of the Nile Delta (study area) subsides according to many authors such as (IPCC, 2001c ; UNEP/ICSU/WMO, 1985 ; Khafagy *et al.*, 1992 ; Broadus *et al.*, 1986; Milliman *et al.*, 1989; Sestini, 1989; Ante, 1990; El-Raey, 1990; El-Sayed, 1991; Milliman, 1992; Stanley and Warne, 1993 and others) new wetlands will occur and accordingly the greenhouse gases emissions will increase , especially CH_4 and CO_2.

When the sea level rises, the agricultural lands will be converted into wetlands and the loss of agriculture lands will lead to a degeneration or interruption of social and economic services. The loss of agricultural soils will decreased the GHG emissions.

Rising of sea level (Figs. 9, 10 and 11) would destroy the weak parts of the sand belt, which is essential for the protection of lagoons and the low-lying reclaimed lands. Most of sand dune belts adjacent to the shoreline are as in Burg El-Burullus, Baltim and El-Nakhel areas (Fig 12).

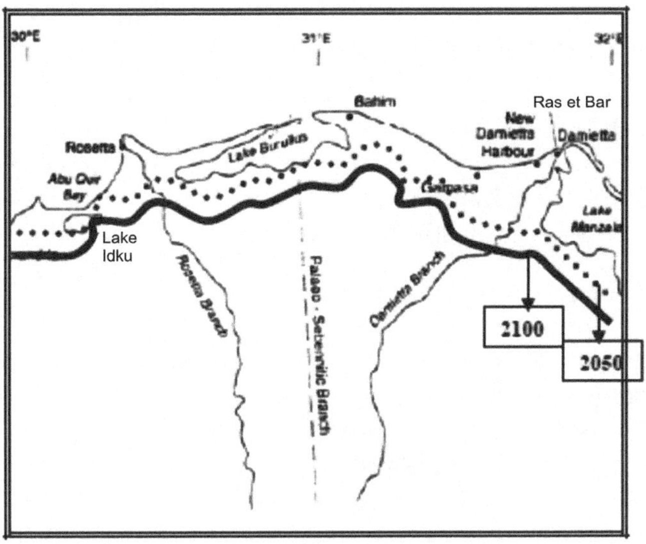

Fig. 11 : The possible submergence areas of the coastal zone along the Nile Delta due to the sea level rise in two years 2050 and 2100 by using the rates of Milliman, 1992 and IPCC, 2001.

Fig. 12 : The Effect of sea level rise (by 0.5 and 1.0 m) on the Nile Delta (UNEP/GRID, 2002)

The erosion of Nile Delta beaches is discussed by many authors such as (Frihy et al, 1998; Khafagy et al, 1992 ; El-Fishawi, 1994; Frihy and Khafagy, 1991; Frihy and Komar, 1993; Ahmad,1991, 2002 ; White and El-Asmar,1999 and others). The erosion in the study area beaches may be due to the continuous development activities in the coastal area (local and global areas) that leading to some climatic changes or/and tectonic changes where the Nile delta subside by -4.8 mm/yr (Milliman, 1992).

When the sea level rise by 1.5 mm/yr (IPCC,2001) and the Nile Delta subside by 4.8 mm/yr (Milliman, 1992), the sea level rise by about 20 cm in 2025, about 25 cm in 2050, about 52 cm in 2075, and about 67 cm in 2100, where many coastal areas well disappear (Fig. 11).

Sand Dune Encroachment

As mentioned before, the sand dunes in the study area are classified into two categories, the old fixed dunes and recent mobile dunes. The main directions of wind blow are N, W, and NW directions, cause the movement of mobile dunes (Barchan and longitudinal). Sand movement may impede the urban, resorts (Plate 3, Photos. 1, 2, 3, 4,and 5), reclamation and agricultural activities, and threatening the international coastal highway (Plate 3, photos. 6 and 7) by sand accumulation that decrease the economic activities of the present and proposed projects. The sand dune encroachment has increased with time especially in urban areas (Plate 3, photos. 4 and 5). The palm trees in the study area suffer from destruction by sand dunes (Plate 3, photo. 8 and 9).

Several methods have been used for dune fixation among which are the tar spraying (Plate 4, Photo 1), wind fences (Plate 4, Photo. 2), and stones (Plate 4, Photo 3). Wind fences are not costly and nonpolluting if compared with stonewall and tar spraying, which prevent the land utilization in agricultural purposes.

Different methods have been used for measuring the sand dunes movement in the study area such as Howard *et al.* (1978), Fryberger *et al.* (1984) and Livingstone (1989) methods. The dune movement at El-Burg Village reached up to 3 m/yr, while no movement was detected at Baltim lighthouse. El-Asmar (2000) recorded that the movement is related to the wind direction and calculated the movement ranges from 3-8 m/$y_{r.}$ at Gamasa. Now, most Gamasa dunes are removed due to development activities such as reclamations, industrial and tourism activities.

Salting of Soils

There are three types of soils namely saline soils, saline-sodic soils, and normal soil in the study area. Saline soils contain large quantities of neutral salts which interfere with plant growth. The samples 37, 41, 53, 50, 29 and 19 (Fig. 13) belong to these type of soil, where SAR < 13 and electrical conductivity is high, greater than 4 dS m^{-1} (4 mmho/cm). The saline soils are recorded mainly south El Brullus Lagoon due to the bad drainage, shallow subsoil water and the nature of soils that cut mostly from the lagoon and surrounding sabkhas.

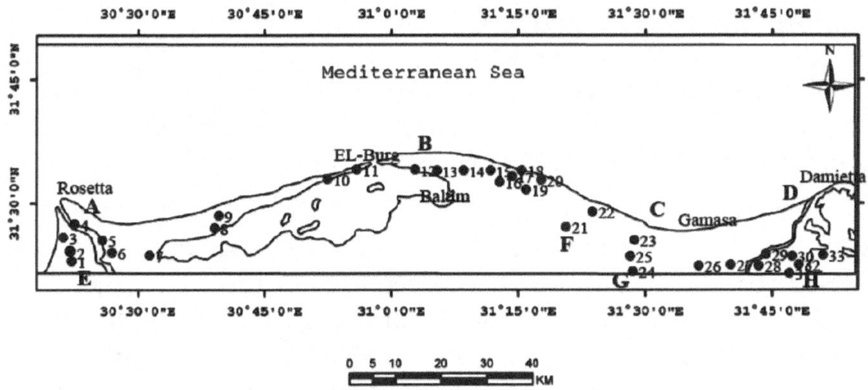

Fig. 13 : The location of the soil profiles in the study area.

Saline-sodic soils contain appreciable quantities of both neutral salts and sodium (SAR > 13 and > EC 4) as in samples 42, 24, 51, 48 and 18 (Fig.13) belong to this type that were recorded near the Nile branch and near the lagoon.

Normal soil, in this type the SAR is less than 13 and EC also less than 4 mmhos-cm as in all other samples. The main sources of salinization are the rise of saline water table, bad drainage, salt water intrusion, the finer texture of the low lying areas as well as the use of brackish water in irrigation.

The artificial fish breeding ponds as well as pars of salt production (Plate 4, Photos 4 and 5) represent one of the environmental hazards for soil salinization. They may cause rise and contamination of ground water by salt water.

Thereby, the governorate must dug a good network of drainage to decrease the water table, decrease the salinity of irrigation, addition of agricultural gypsum to decrease the salinity of soils and planting salt-tolerant crops.

Drying Many Parts of El Burullus Lagoon

With the end of 20th century, the Egyptian Governorate planned different agricultural projects to cut lands of El-Burullus Lagoon especially in the western and southern parts. It dig canals for irrigation and drain and for transportation of excess water to the lagoon (Plate 4, Photo 6). This encouraged a lot of people for cutting parts of the lagoon for fishing farms. The dried land was characterized by fine sediments, saline soil, low water table, and relatively low topography. Thereby, these parts may be covered by sea water when the sea level rises in near future.

Siltation of Damietta Harbour Navigation Canal

With the erection of Damietta Harbour (10 Km west of the Damietta Branch) two jetties have been constructed in 1982 to protect the access channel and the Harbour entrance against littoral drift.

Their length was increased in 1985 to become 1300 m for the western jetty and 600m for the eastern jetty. The accretion occurred west of the western jetty and the erosion occurred east of the eastern jetty. The navigation channel suffered from siltation processes. Due to the poor selection of Damietta Harbour site and the location of the jetties, a lot of sea sediments are accumulated in the navigation canal. These sediments cause a lot of problems for ship navigation. Thereby, the authority of the harbour are paying more than 25 million Egyptian pounds for dredging the sediments.

For combating the siltation in the navigation channel, there are two proposal. The first proposal, construction of another jetty (Fig. 7) with about 2.5 km length to the west of the western jetty of the Damietta harbour to partially prevent the transportation of sediments coming from the western eroded areas. The second proposal, the old jetties of harbour must be lengthened. To cover the coast of dredging, the dredged sand can be used as a source of heavy economic minerals, builds, and nourishment the eroded beaches.

Defective Distribution of Population on the Coastal Plain

There is no plan for distribution of populations along the coastal plain, where, they are concentrated mainly in the western parts around Rosetta, in central parts around Baltim and El-Burullus, and in the eastern parts around Ras El-Bar. There are some patches of fishing and/ or farmer villages with bad distribution. The planner must be redistributing the peoples in all the area especially after the construction of international road.

Pollution

The pollution in the coastal area is divided into the air pollution, marine pollution, and land pollution. The air pollution has increased after the construction the international road and the increasing of the industrial activities and individual factories. The marine pollution occurred near the Damietta Harbour and in the outlet of the drains. The land pollution resulted from the increasing human activities and from the human and animal waste.

The Negative Effects on the Coastal Protection Projects

There are many protection projects that have been constructed from the beginning of the 1990 till now in Ras El-Bar, Baltim, El-Burg, and Rosetta. The detached break waters in Ras El-Bar cause beach accretion in the front of the detached break water and also cause the erosion to the west (from familie's beach to the eastern jetty of Harbour). In Baltim , the detached break waters cause the accretion in front of the detached break water and cause erosion in between the detached break water forming isolated islands (Plate 4, Photo 7). The detached break water and other protection structures caused accretion at one place and erosion at another. All the protection projects transport the problem from one place to another and do not solve the problem totally. From the results of sedimentological and mineralogical analyses and from the field observations during four years of the studied area, some of protection structures are proposed to solve problems caused by the old protection structures.

Mangement of the Coastal Areas

The coastal area is very important due to its high mobility and consensual change with any change in climate and human activities. The coastal area between Rosetta and Damietta has many different geomorphologic units affected by each other.

The recent development activities as agricultural and urbanization are concerned mainly with sand dunes. These sand dunes especially near the shoreline as in El-Nakhel area is the first defense for the beach, where it is the main source of sand to the beach and it has heavy minerals that increase the stability of the shore. On the other hand the sand dunes are the storage of high economic rare heavy minerals and using this dune in agriculture will decrease their economic potentialities. Thereby, before using these sands for any purpose, the economic minerals must be extracted.

Now, El-Nakkel dunes have been removed (Plate 4, photo. 8) and will be used in many other purposes. It is removed also in Baltim (Plate 4, Photo 9) to level the lands for building. Before any conservation of the beach, the planner must take in the consideration of the previous studies and the side effects of any project on the surrounding ecosystem.

Plate (3)

Plate 3 : Sand dunes encroachment

Photo (1) : Sand movements threaten Baltim resort from the sand dunes

Photo (2): Sand movement threaten the Baltim resort from the sea

Photo (3): Sand dunes threaten the builds in Baltim.

Photo (4): Sand dune around the build in El-Shehabia in 1996.

Photo (5): Sand dunes around the same build inEl-Shehabia in 2005.

Photo (6): Sand dune movements on the international road in El-Nakhel area.

Photo (7): Sand dune movement on international road in El-Burg.

Photo (8): Sand dune movement effect on the palms in El-Nakhel area.

Photo. (9): Sand dunes around the palm trees in El-Hammad

Plate (4)

Plate 4 : Salinization, drying of lagoon, negative effects of protection structure, and poor management of coastal area

Photo (1) : Stabilization of sand dunes by spraying

Photo. (2) : Stabilization of sand dunes by stone in El-Burg photo. (3) : Wind fence and reclamation for cultivation

Photo (4) The fish breeding artificial ponds adjacent to international road

Photo (5) The salt production adjacent to the international road

Photo (6) Reclamation activities adjacent to the Burullus lagoon by drying

Photo (7): Isolated detached break water as island in Baltim.

Photo (8) Remove the sand dunes in El-Hammad area

Photo (9): The plain area prepared to building in the Baltim inlet resort after remove the dunes.

The Future Planning for the Study Area

Objectives of the future planning

1. Creating a new attractive areas for human habitation.
2. Redistribution of the population.
3. Increasing the industrial areas.
4. Redistribution of the old industrial activities on the study area.
5. Increasing reclamation and cultivated lands.
6. Increasing the fish farms.
7. Increasing the number of tourism villages and resorts.
8. Increasing the job opportunities.
9. Determination of all natural and human hazards in the study areas and measures to combat.

Steps for the future planning

To make the future planning, we must follow the following steps:

1. Data collection and analysis.
2. Identification of problems and potentialities.
3. Available alternatives.
4. Implementation.
5. Better understanding for the future changes in the area.

How to achieves the complete future planning ?

A complete future planning can be achieved by detecting the suitable planning of all sectors in different fields especially agriculture, industry and tourism.

The agriculture sector

The agriculture sector is the main sector required to build the complete future planning. The poorest can lay claim to only 9.8% of national income, while the richest generates nearly 40%. Consequently, almost one-quarter of the population lives in poverty (World Bank 2002). About three-fifth of the country's agriculture production is in the low-lying delta (Strzepek, *et al.*, 1995). Agriculture contributes 17% of the country's Gross Domestic Product (GDP) and is the largest source of employment, constituting 30% of the labour force (Agrawala, *et al.*, 2004). Egypt is fairly unique in the distribution of its population, land-use and agriculture, and economic activity which makes it extremely vulnerable to any potential impacts on its water resources and coastal zone.

It is known that the sea level rising in the Nile Delta can be attributed to the coastal subduction, melting of ice of both north and south poles, and the construction of Aswan high dam. Thus the low agricultural lands of Delta will be lost due to the coastal inundation. Sestini (1992) showed the land reclaimed from 1960 to 1985 and after 1985 (Fig. 14). This map is based on Beaumont *et al.* (1988) and on the satellite image of 1989, where the Ministry of Irrigation suggested that the 2.8 to 3 million feddans (1 Feddan = 0.42 ha) could be reclaimed in 2000.

For the study area, the problem of salinity is the main issue thus the salinity will treated by washing or by cultivation of a new vegetation loving the salinity. There are two types of lands in the study area, one for reclamation and the other for improving the old agriculture. The proposed areas which must be reclaimed to increase the agriculture lands located in-between Gamasa and Baltim, and the area south the El Burullus Lagoon. They occupy about 280 km^2 (Fig. 15). They are characterized by finer sediments, less saline soils, relatively deep water table and the nearest source of irrigation.

The future planning of the agriculture includes the following:

1. Providing the maximum productive work opportunities
2. Cultivation of new types of vegetation that not cause problems
3. The maximization of land productive value
4. Solution of the problems of soils

There are many patches in the study area which can be used for the fishing farms (Fig.15) beside the coastal lagoon (El-Burullus Lagoon) due to their low topography, heavy saline soil, shallow water table and the presence of source of water recharge. The proposed patches that must be added to the fish ponds occupy about 40 km^2.

Industrial sector

Industry is expected to expand in the next decades by 2.8% to 7.7% annually, according to different economic and political scenarios as well as population and internal market growth (UNEP, 1988) which can lead to increase in job opportunities.

The study area between Rosetta and Damietta with Damietta Harbour as an outlet for export, considered as an ideal field for establishing an integrated production system combining both agriculture and industrial processes. There are two new industrial areas as in Gamasa and New Damietta City. The industrial activities must be tied with natural resource in the study area, thus such types of plants must go with needs of industrial development. Also, the study area is rich in black sands with heavy economic minerals which can be considered as a base for industrial links with a new integrated industry. All proposed industrial projects and activities must be near the sea or the river. The proposed industrial sites occupy about 36 km^2 (Fig.15).

The black sands are the main sources of heavy minerals such as monazite, ilmenite, rutile, and garnet which are used in many industrial activities. They occurred in the eroded beaches (Fig. 6) near to the promontories of Rosetta and Damietta and in local eroded areas in-between as well as in coastal sand dunes near El-Burg village. As mentioned before, the navigation channel of Damietta harbour suffers from siltation problems, where millions of cubic meters silted in the channel each year. To improve the industry sector, two projects have been proposed.

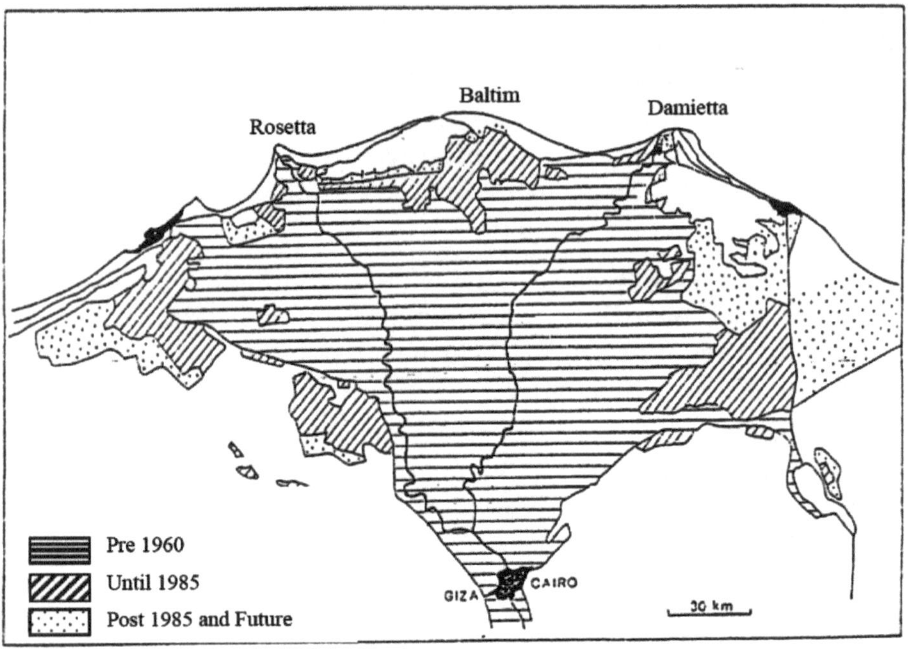

Fig. 14 : Recent and likely future land reclamation projects in the northern Nile Delta (after Sestini, 1992)

The first project is proposed to reuse the dredging sands for the extraction of heavy economic minerals. The proposed team of this project must include members from the Coastal Research Institute, Nuclear Materials Authority, and Department of Geology in Mansoura University. The project can be completed by the following initial steps:

1. Evaluation of the proven reserves of the dredged sands of the navigation channel.
2. Evaluation of the quantity accurately.
3. Determination of all species of economic minerals.
4. The remained light minerals can used for the nourishment of the eastern eroded beaches of the navigation channel (eastern jetty).

The second project is proposed to reuse the sands in the accreted area especially in Gamasa and west of the western jetty of the Damietta Harbour mining in many industrial purpose such as concrete, filling, and beach nourishment. For these uses of sands, mining must make use of sea sand.

There are two ways for mining the accreted sand. The first is to choose the right location for skimmed or excavated. Before starting the excavation the side effects that may occur must be studied, such as change of the sedimentary conditions after the excavation, excavation may cause problems or solve problems. Thus, the location on beach that is subject to accretion and erosion must be studied to conserve the balance on the shoreline and the bottom sediments and morphology must be studied in detail.

Tourism sector

The beach of the study area is extended about 155 km from Rosetta to Damietta. It is characterized by approximately good climate all over the year, thereby it is more attractive for people. It is known that the tourism brings about 7 billion dollar each year to the national income, thus the tourism sector must be developed. The proposed area occupy about 185 km^2 (Fig. 15). The proposed area for tourism in the study area are :

a- Rosetta and Burg El-Burullus sector

It extends from Rosetta tip to 10 km west the inlet of El-Burullus Lagoon (Fig.14). It is characterized by a wide strand plain that increases in wideness from El-Burg to Rosetta. The beach can be used as resort in summer only because of the inundation in winter.

b- Baltim and west of Gamasa sector

The second area is extends from east Baltim to west of Gamasa (Fig. 15). It extends from east of Kitchener Drain to the west of Gamasa Resort. It is characterized by a wide strand plain, flat topography, and few number of small longitudinal and Barchan dunes.

c- Gamasa and new Damietta city sector

It is extends from Gamasa to New Damietta City. It is characterized by moderately wide beach. This area must be used for tourism purpose. It is near to Gamasa City and New Damietta city and near to the different cultivations (Fig.15).

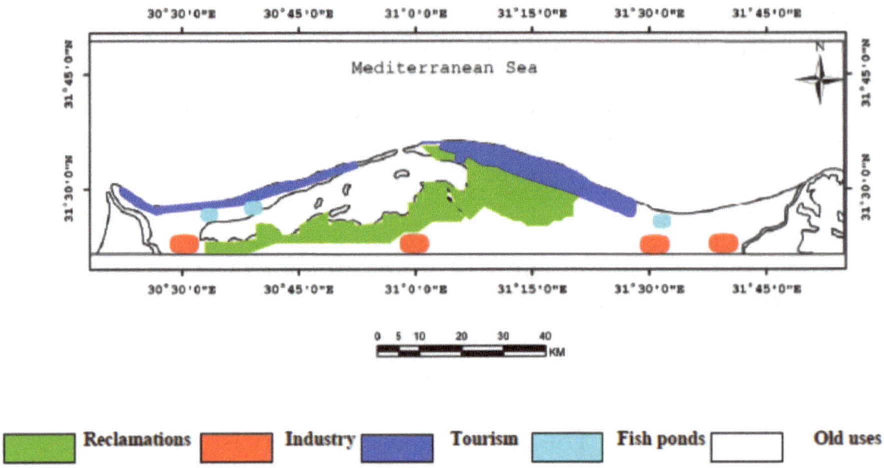

Fig. 15 : The proposed activities in the future, in the study area

Geoenvironmental hazards and its mitigation in future planning

The study area (as any coastal area in the world) is suffering from many problems as mentioned in the geoenvironmental hazards section. This area is considered as attractive for many investments to achieve the sustainable development.

For achieving the right sustainable development in the study area in 21st Century, the coastal area must be subdivided into two subareas from north to south as in the figures (9, 10 and 11). This figure shows the subareas which are subjected to inundation. Thus, the sustainable development is subdivided into two types, the temporary and permanent. The temporary development must be performed in the inundated area (Figs.9, 10 and 11) and the permanent development must be performed to the south of the inundated area.

The sustainable development such as building, transportation, industry, tourism, and agriculture is discussed as follows:

The building

The building, considered the main structure required to perform any development sectors such as industry, tourism, transportation, and agriculture sectors. Where the building change the land utility and human activities. When the number of buildings increase, the population increases, and accordingly the hazards increase. Thereby, planner should consider the impact of the building processes in the study area.

The impacts of building are different according the type of building and their location. If it is residential high rise, then their side effect will be very low, while if it is commercial then the environmental hazard will increase. On the other hand, the industrial construction or equipment generally have dangerous impact on the surroundings ecosystem. Thereby, it must be applicable with environment.

Transportation

The main impact on transportation activity is the increasing GHG especially CO_2. In 1995, the transportation contributed about 22% of global energy related to CO_2 emission, the emission from transport is growing at a rapid rate of about 2.5% annually (IPCC, 2001c).

Many roadlines have been constructed in the study area such as the coastal international road and many other internal link roads between the cities. To mitigate the impacts of transportation, new passenger cars utilizing high grade fuels and new light truck must be used. Navigation in the Nile branches and main canals must be encouraged for low cost transportation. The navigation canal constructed between Damietta harbour and the Nile is very important to mitigate the negative impacts of the other types of transportation.

Industry

It plays an important role in soil, air, and water pollution. With the high expansion of industries, the pollution increases. Thereby, the planner must consider the side effects of the industries on the surroundings ecosystem.

To mitigate the impacts of the industry, the improved technologies must be used in the building or factories or the industrial areas. The environmental impact of every project must be studied before initiation.

Agriculture

Removal of coastal sand dunes for cultivation can accelerate the erosion in the northern Nile Delta. Burning of crop residue may also cause pollution.

To mitigate the impacts of the agriculture in the study area, appropriate crop management practices must be made and natural manures used in cultivations. The coastal sand dunes must be saved on the shoreline from the sea attack.

Tourism

The poor distribution of tourism areas along the coast causes heavy pollution, which leave behind a lot of wastes.To mitigate the impacts of tourism, a redistribution of the constructions and peoples on the coastal area between Rosetta and Damietta can be achieved through the building of a number of villages and resorts. There are a natural hazards in the resort areas such as sand dune movement and coastal erosion that must be mitigated.

The study of all the above previous sectors is required for future planning. It will lead to creation of a new attractive habitation areas, redistribution of the population, increase the industrial areas, redistribution of the old industrial activities, increase the reclamation of land for cultivation, increase the number of fishing farms, increase the number of tourism villages and resorts and increase the job opportunities.

Conclusions

1. The erosion is dangerous in two sites at Rosetta and Damietta promontories.
2. The resorts suffer from the erosion and accretion especially at Baltim, Gamasa and Ras El-Bar.
3. The sea level rise by year with the sinking of the land in north of the study area.
4. Most of the soil north of the study area suffer from salting.
5. Sand dune encroachment affect on the resort especially in Baltim.
6. Sand dune movement on the international roads and on the buildings and ultimately destroy the agriculture.
7. The sand dunes in the area must by stabilized by any way.
8. Many of protective structures must be built.
9. The site of activities must be redistributed.
10. The climatic changes must be taken into consideration.
11. The geomorphologic change detection must be studied with time.
12. The reclamation of land must be increased in the suitable areas.

References

Agrawala, S., Moehner, A. EL-Raey, M., Conway, D., Aalst, M. V.,Hagenstad, M and Smith, J. 2004. Development and climate change in Egypt: focus on coastal resources and the Nile, environment directorate environment policy committee, working party on global and structural policies, working party on development co-operation and environment, Head of Publications Service, OECD, 2 rue André Pascal, 75775 Paris, Cedex 16, France. COM/ENV/EPOC/DCD/DAC(2004)1/FINAL

Ahmad, M. H. 1991. Temporal Shoreline and Bottom Changes of the Inner Continental Shelf off the Nile Delta in the present Century, Egypt. M.Sc. Thesis, Alexandria Univ., 218p.

Ahmad, M. H. 2002. Multi-Temporal Conflict of the Nile Delta Costal Changes, Egypt.Ntional Authority for Remote Sensing and Space Science (NARSS). Littoral 2002, The Changing Cost, EuroCoast/EUCC, Porto-Portugal, ISBN 972-8558-09-0.

Ante, B. 1990. Impact of climate on the socioeconomic structure and activities in the Mediterranean region. In: Changing climate and the coast, Vol.2. U.S. Environmental Protection Agency, Washington, DC, USA, pp. 127-138.

ARICON, 1989. Structural plan of the comprehensive development for north delta area and the northern international road. Cairo Arab Intern. Consultants, Report for Ministry of Development. New Communities Housing and Utilities, 3 vols.(Arabic, Engi. Sum.).

BBC Arabic News 1997. The Effect of a Rise in Sea Level on the Nile Delta, The United Kingdom. http://news.bbc.co.uk/hi/english/special_report/1997

Beaumont, P., Blake, G.H., and Wagstaff, J.M. 1988. The Middle East, a geographical study. Chapter 19, "Egypt: Population grows and Agriculture Development" David Fulton Publishers, London, 511-530.

Bijlsma, L., Watson, R.T., Zinyowera, M., Moss, R., and Dokken, D.J. 1996. Coastal zone and Small Islands. In Climae Change 1995. Impacts, Adaptations and Mitigation of Climaate Change: Scientific technical aanalysis, Contribution to Working Group II to the Second Assessment Report, Cambridge University Press, Cambridge. PP. 289-324.

Broadus, J., Milliman, S., Edwards, D., Aubrex, D.G. and Bable, F. 1986. Rising sea level and damming of rivers: possible effects in Egypt and Bangladesh. In: effect of changes in stratospheric Ozone and global climate, Vol. 4, U.S. Environmental Protection Agency and UNEP, Washington, DC, USA, pp. 165-189.

Butzer, K. W. and Hansen, C.L. 1968. Desert and River in Nubia: Geomorphology and prehistoric environments at the Aswan reservoir. Uni. Of Wisconsin press, Madson, Wisc., 526pp.

Coutellier, V. and Stanley, D.J. 1987. Late Quaternary stratigraphy and Paleogeography of the Eastern Nile Delta, Egypt. Marine Geo. No. 77. pp. 257-275.

Day, J.W., Conner, W.H., Costanza, R., Kemp, G.P. and Meldelssohn, I.A. 1993. Impacts of sea level rise on coastal systems with special emphasis on the Mississippi River deltaic plain. In: Climate and se level change: observations, projections and implications. Cambridge University press, Cambridge, UK, pp 276-296.

DETR, 1999. Climate change and its impacts: Stablisation of CO2 in the atmosphere. U.K. department of the environment, transport and the regions, the met office, Bracknell, UK, 28 pp.

El-Askary, M. A. and Frihy, O.E. 1984. Environmental interpretation of sand grain surface textures in the Rosetta and the Damietta promontories along the Nile Delta Coast. Egypt. N. jahrb. Palaontol. Monatsh., 12: pp. 709 - 716.

El-Asmar, H. M. 1994. Severe coastal Damage along Ras El Bar shoreline, Nortn of Nile Delta: An effect of the construction of the detached breakwater system. *Journ. Geol.* 38(2), 793-812.

El-Asmar, H. M. 1995. Impact of protection structures on physical and sedimentary parameters along the Dammietta coastal Area, Nile Delta, Egypt. *Sedimant. of Egypt* 3: 111-124.

El-Asmar, H. A. 2000. Geoenvironmental studies on the coastal area between Gamasa and Baltim, North of the Nile Delta. *Z. Geomorph.* N .F,44 (1): 59-73.

EL-Fishawi, N. 1994. Relative changes in sea level from tide gauge records at Burullus, central part of the Nile Delta coast. INUA MBSS Newsletter. 16: 53-61.

EL-Raey, M. 1990. Response to the impacts of greenhouse. Induced sea level rise on the northern coastal regions of Egypt. In: Changing climate and the coast, Vol.2. U.S. Environmental Protection Agency, Washington, DC, USA, pp. 225-238

EL-Sayed, M.K. 1991. Implications of relative sea level rise on Alexandria. In: Proceedings of first international meeting, December, 1989, Venice, Italy, pp. 183-189.

Fanos, A.M. 1995. The impact of human activities on the erosion aand accretion of the Nile Delta coast. *Journal of Coastal Research* 11, 821-833.

Frihy, O.E. and Stanley, D.J. 1987. Quartz grain surface textures and depositional interpretations, Nile Delta region, Egypt. Mar. Geol. 77: pp. 247-255.

Frihy, O.E. and Khafagy,A.A. 1991. Climte and human induced changes in relation to shoreline migration trends at the Nile Delta promontories, Egypt. *Catena* 18, 197-211.

Frihy, O. and Komar, P.D. 1993. Long- term shoreline changes and the concentration of heavy minerals in beach sands of the Nile Delta, *Egypt. Mar. Geol.* 115: 253-261.

Frihy, O.E., Dewidar, K.M, and EL-Banna, M.M. 1998. Natural and human impacts on the northeast Nile Delta coast of Egypt. *Journal of Coastal Research,* Vol. 14, No. 3, p. 1109-1118

Fryberger, S.G., El-Sari, A. M., Clisham, T. J., Rizvi, S. A. R. and Al-Hinai, K. G. 1984. Wind sedimentation in the Jafurah sand sea, Saudia Arabia. *Sedimentol.* 31: 413-431.

Grenier, J. C. 1990. The merging of the Nile Delta. V. 136. heavy minerals in beach sands of the Nile Delta, *Egypt. Mar. Geol.* 115: 253-261

Howard, A.D., Morton, J.B., Gad ELHak, M. and Price, D.B. 1978. Sand transport model of barchan dune equilibrium. *Sedimentology*, 25, 307-338.

http://www.worldbank.org/data/dataquery.html. Accessed January 2, 2003

IPCC 1996. Climatic Change 1995-Impacts, Adaptation and Mitigation of Climate Change: Scientific technical analysis. Contribution to Working Group II to the Second Assessment Report, Cambridge University press, Cambridge.

IPCC 2001. Climate Change 2001: Synthesis Report. A Contribution of Working Groups I, II, and III to the Third Assessment Report of the International Panel on Climate Change, Watson, R.T. and Core Writing Team(eds.). Cambridge University, United Kingdom, and New York, NY,USA.

Khafagy, A.A., Hulsbergen, C.H. and Baarse, G. 1992. Assessment of the vulnerability of Egypt to sea level rise. In: Global climate change and rising challenge of the sea. Proceedings of the IPCC workshop, March 1992, Margarita Island, Venezuela. Administration, Silver Spring, MD, USA.

Livingstone, I. 1989. Monitoring surface changes on a Namib Linear dunes. Earth Surface Process Landforms 14:317-332.

Lofty, M.F. and Frihy, O.E. 1993. Sediment balance in the nearshore zone of the Nile Delta cost, Egypt. *Journal of Coastal Research* 9, 654-662.

Milliman, J. D. 1992. Sea level response to climate change and tectonics in the Mediterranean sea pp: 45-57, Modified in 2000

Milliman, J.D., Broadus, J.M., and Gable, F. 1989. Environmental and economic implications of rising sea level and subsiding Delta: The Nile and Bengal examples. *Ambio,* 18.340-345.

Morner, N.A. 1971. Eustatic change during the last 20000 years and a method of separating the isostatic and eustatic factors in an uplifted area. *Paleogeogr., Paleoclimatol., Paleoecol.,* 9: 153-181.

Nageb, M. 1992. Landscape evolution and shallow subsurface depositional sequences of Manzala Lake area, Egypt. Ph.D. Thesis, Fac. Mans. Uni.

Nakicenovic, N., Alcamo, J., Davis, G., de Varies, B., Fenhann, J., Gaffin, S., Gregory, K.,Grubler, A., Jung, T.Y., Kram, T. 2000. Special Report on emissions Scenarios. Intergovernmental Panel Peninsula region.

Nicholls, R. J., Hoozemans, F.M.J. and Marchand, M. 1999. Increasing flood risk and wetland losses due to global sea level rise: regional and global analyses. Global Environmental Change 9, S69-S87.

Rossigonal-Strick, M. 1983. African monsoons, an immediate climatic response to orbital isolation. *Nature* 3030:46-49.

Sestini, G. 1989. Nile Delta : a review of depositional environments and geological history: In: M.K. G. Whatalay and K. T. Pickering (eds) Deltas: Sites and traps for fossi fuels. Special Publication, 41: 99-127. Blackwell Sci. Pupl.,Geol. Soc., London.

Sestini, G. 1992. Implication of climatic changes for the Nile Delta. In: L.J. Jefftic, J.D. Milliman and G. Sestini (eds) Climatic changes changes and mediterranean. 555-601 Edward Arnold, New York.

Shaplan, G. 1997. "Rivers of Discord", St. Martin's Press, New York, 1997

Stanley, D.J and Warne, A.G. 1993. Nile Delta, recent geological evolution and human impact. *Science* 260, 628-634.

Stanley, D.J. and Wingerth, J. 1996. Nile sediments dispersalaltered by Aswan High Dam, the kaolinite trace. *Mar. Geol.* 133: 1-9.

Strzepek, K.M., Onyegi, S.C. Saleh, M. and Yates, D. 1995. An assessment of integrated climate change impacts on Egypt. In As Climate Change: International Impacts and Implications. K.M Strzepek and J. B. Smith (eds.) Cambridge University Press, New York.

UN, 1979. Population distribution and urbanization in selected countries in the Middle East. In: Studies on selected development problems in various countries of the Middle East, New York, 59-78.

UNDP / UNESCO 1978. Arab Republic of Egypt: coastal protection studies. UNDP/ EGY/73?063. Final Report, FNR/SC/OSP/78/230.

UNEP, 1988. The Blue Plan, Futures of the Mediterranean Basin. Executive summary and suggestions for action. Sophia Antipolis, France.

UNEP/GRID, 2002. Potential impact of sea level rise: Nile Delta, Remote Sensing Center, Cairo. Web site: www.UNEP.org.

UNEP/ICSU/WMO, 1985. International assessment of the role of CO_2 and other greenhouse gases in climate variations and associated impacts. Conference Statement, UNEP, Geneva, Switzerland.

White, K. and El-Asmar, H. M. 1999. Monitoring changing position of coastlines using Thematic Mapper imagery, an example from the Nile Delta. Geomorph., 29, 93-105.

World Bank 2002. World Bank Development Indicators. The World Bank, Washington, DC.

World Bank 2003. World Development Indicators: Data and Statistics.

11

Coastal Sand Dune Encroachment as a Main Reason for Desertification in the Coastal Area Between Gamasa and Burg El-Burullus North of the Nile Delta, Egypt

Mohamed Sabry Mohamed, EL-Bady

Abstract

The area of study extends about 65 km from Gamasa City to Burg El-Burullus. The coastal sand dunes are very near to the shoreline in Baltim and El-Burg and far from the shoreline in Gamasa. It is narrow with relatively steep slope northward near Burg El-Burullus and Baltim, while it increases in width eastward to become wide with gentle slope near Gamasa. The coastal sand dunes are very mobile and move toward the agricultural lands, roads (international coastal road), resorts, and other developmental activities. The sand dunes in the study area are classified into two categories, the old fixed dunes and recent mobile dunes. The main directions of wind blow are N, W, and NW directions. It causes the movement of mobile dunes (Barchan and longitudinal). Several methods have been used for dune fixation like the tar spraying, wind fences and stones. Wind fences are not costly and nonpolluting if compared with stonewall and tar spraying, which prevent the land utilization for agricultural purposes.

Introduction

Today over 60% of the world's population lives in a coastal strip or zone (60 Km width) and this number is expected to increase rapidly to 75% by 2025 (UNESCO, 1998). The coastal zone is the interface where land meets the sea and encompasses shoreline environments as well as coastal water (Bartlett, 1994), while Carter (1988) said that it is the space in which terrestrial environments influence marine environments and vice versa of variable width and may also change in time. On the other hand, the coast is the narrow band of the earth which may be covered with water or not or may be covered with a part only. The coastal zone consists of many different ecosystems, like shores, deltas, wetlands, dunes, reefs, lagoons and estuaries. According to Richard and Davis (1996) there are many sources for coastal materials as :

- Sediments eroded from headlands by contemporary waves Its attacks are only important for soft or unconsolidated rocks.
- Sediments may have been brought to shore from offshore zone due to the sea level rise.
- Large rivers deposits high volumes of alluvial sediments in the coastal zone.
- Beach sediments cycling through longshore drift.

Sand dunes may occur where there is a large supply of sand, a wind to move it, and a place in which they can accumulate. Thus, the coastline is an ideal location for these criteria to be met. At places where the saltation process is the means of transport by wind, sand accumulates as dunes in the coastal areas that are not affected by sea water. The size, width, and geometry of sand dunes are much controlled by the geometry of shorelines, sand supply, wind speed, topography and presence or absence of vegetation cover in the coastal area. When the coastal area is characterized by fine texture, the sands are easily to moved by the wind and dune growth is rapid, while the dune growth in the coarse texture beaches takes longer time.

Also sand dune is formed when there are sufficient reductions of wind energy along the direction of sand drift in an active extensive system (Fryberger and Ahlbrandt, 1979). When there is any obstacle such as a rock outcrop or a stand of vegetation, or there is any small depression or gentle dip in an otherwise flat surface, it can fill with sand due to lower wind velocity within the depression (Cooke and Warren, 1973)

Sand dunes are very sensitive to environmental and climatic changes. It is an indicator to paleoenvironments (Livingstone and Thomas, 1993). El-Asmar, (2000) suggested that there is a relation between threshold velocity and mean diameter, where he mentioned that the lowest calculated wind velocity were detected at Gamasa dunes and the highest values were detected at Baltim dunes. He also suggested the wind velocity at both Gamasa and Baltim in the range of 4 to 16 KT (205.6 to 822.2 cm/sec) indicating that these wind velocities are roughly able to transport equal amount of sands with grain diameter in the range of 177 μm to 250 μm to height between 5 to 10 m. The movement of El-Burg dunes is at the rate of about 3 m/yr and the average movement of Gamasa-Baltim dunes is about 8 m/yr.

Fig.1 : The satellite image of 2001 showing the location map

Types of Coastal Sand Dunes

There are three types of coastal dunes, only two main types are predominated longitudinal (linear) dunes and barchan dunes. The third type is the old fixed sand dune, where it is covered partially by recent sand dunes (longitudinal and barchan) or removed by human activities. The majority of coastal sand dunes along the Nile Delta coast occur in the area between Gamasa and Burg El-Burullus that extends about 65 km. The prevailing wind in this area comes from WNW, NW and N and the strong wind occur during the winter (from December to March) especially during December and January with a minor trend from E and NE (El-Fishawi and El-Askary, 1981). The annual frequency shows that wind comes from W to N sector with predominantly NW direction, while the strong storms attack the coast from W to NE (El-Asmar,1994; Tetra Tech, 1984). Sesteni, (1992) said that the months of winter are windiest with dominant wind from NW and WNW, less frequently from N and NE over the eastern Mediterranean coast and in summer. As a result average wind speeds during summer are greater in July. The dominant winds are from WNW and W over central and eastern Mediterranean coast of the Nile Delta.

A) Longitudinal dunes

Before 2002, the longitudinal dunes were recorded mainly west of Gamasa town and near El-Shehabia Villages (Plate 1, Photo. 1, 2 and Figs. 2, 3) where the strand plain is wider from Gamasa to near of Kitchener. They are elongate in shape, more or less straight with continuous serratic crest without breaks with two steep side 1 - 1.5 km long and 10 - 20 m height. The area between Gamasa and Kitchener are characterized by the following features :

1. A relatively wide coastal strand plain.
2. South boundary of vegetation can promote sand accumulation by wind deacceleration.
3. Adequate sand supply due to the wide strand plain and the prevailing wind from WNW and NW. Thus the longitudinal dunes are formed in this area parallel to the wind regime of WNW and NW directions and appear oblique to the shoreline. From 2002 till now, longitudinal dunes near Gamasa are disappeared due to human activities such as reclamation, building, filling and fixation, hence the stabilized longitudinal dunes are still present here. Since 2005, most of the longitudinal sand dunes are of recent formation (small dune) as patches to the west of Gamasa Town and south of the shoreline near the international road (Plate 1, photo. 3 and Fig. 4).

B) Barchan dunes

Barchan dunes are located mainly in area between Burg El-Burullus inlet in the west and Kitchener drain in the east (Plate 1, Photos 4 and 5). They occur in two patterns, isolated and complex. The complex form is recorded at 33 km WSW of Gamasa while the isolated barchan occurs near Gamasa town and in the area between Baltim and Burg El-Burullus. They are located near and parallel to the coast. The barchan dunes have axes nearly normal to the wind regime of WNW and NW directions with 30m height and 50 -300m length. The area between Burg El-Burullus and Kitchener drain is characterized by :

1. A very narrow strand plain between Kitchener and Burg El-Burullus.
2. Relative deficiency of sand supply as the strand plain is narrow and the prevailing wind is from WNW and NW.

From 2002 till now, the small barchan dunes (Plate 1, photo. 6) occur after 2 km from Gamasa and adjacent to the shoreline about 200m far from the sea. These dunes are also recently formed. Most of the barchan dunes in the area from El-Shehabia village to Burg El-Burullus village are recorded in complex form and have longitudinal pattern, parallel to the shoreline (Plate 1, photo. 7 and Fig. 5,6). After 2002, some of Barchan dunes and west Kitchener drain especially in El-Nakhel area and in El-Hammad village are removed and transported by governorate of Kafr El-Shekh for reclamation and cultivation purposes.

C) Old stabilized dunes

The old vegetated dunes have occurred since the last six years (Plate 1, photos 8 and 9). They have either disappeared due to reclamation activities or covered by recent sand dunes. They were described as the lower old dunes by El-Fishawy and El-Askary (1981). They are characterized by low elevation that ranges from 4 to 7m a.s.l, and conspicuous cross stratification and dense vegetation cover. They are also characterized by their relatively higher content of fine grain constituents (silt and clay). These dunes have been dated using Optical Simulating Luminescence (OSL), and have ages of 2.6 ± 0.6 Ka and 1.9 ± 0.4 Ka (El-Asmar, 2000). They possibly represent the first generation dune developed from wind-blown beach sand, which reworked from the old promontory of the Sebennitic Branch.

Sand Dune Encroachment

As discussed earlier, the sand dunes in the study area are classified into two categories, the old stablized dunes and recent mobile dunes. The main directions of the wind blow are N, W, and NW directions, and it cause movement of mobile dunes (Barchan and longitudinal). Sand movement may impede the urban, resorts (Plate 2, for Plate 2; See plate 3 of Article # 10 by same author Photos 1, 2, 3, 4,and 5), reclamation and agricultural activities, and threatening the international coastal highway (Plate 2, photos 6 and 7) by sand accumulation. It decreases the economic activities of the present and proposed projects. The sand dune encroachment increases with time especially in urban areas (Plate 2, photos 4 and 5). The palm trees in the study area suffer from destruction by sand dunes (Plate 2, photo 8 and 9).

Several methods have been used for dune fixation among which are the tar spraying (Plate 3, For plate 3; See Plate 4 of Article # 10 by same author Photo 1), wind fences (Plate 3, Photo. 2), and stones (Plate 3, Photo 3). Wind fences are not costly and non-polluting if compared with stonewall and tar spraying, which prevent the land utilization in agricultural purposes.

Plate (1)

Plate 1 : Coastal sand dunes

Photo. (1): The longitudinal dunes formed west of Gamasa before 2002 (After El-Bady,2001)

Photo. (2): The longitudinal dunes formed in El-Shehabia before 2002 (After El-Bady,2001)

Photo.(3) Small longitudinal dune recently formed after 2002 near Gamasa

Photo. (4) Barchan dune in El-Burg

Photo. (5) Barchan dune between Kitchener and Baltim before 2002

Photo.(6) Small barchan dune recently formed after 2002 near Gamasa

Photo. (7) Chain of complex longitudinal dunes

Photo. (8) Old stabilized dune near Gamasa (after El-Bady, 2001)

Photo. (9): Old stabilized vegetated dune covered by recent dune sand

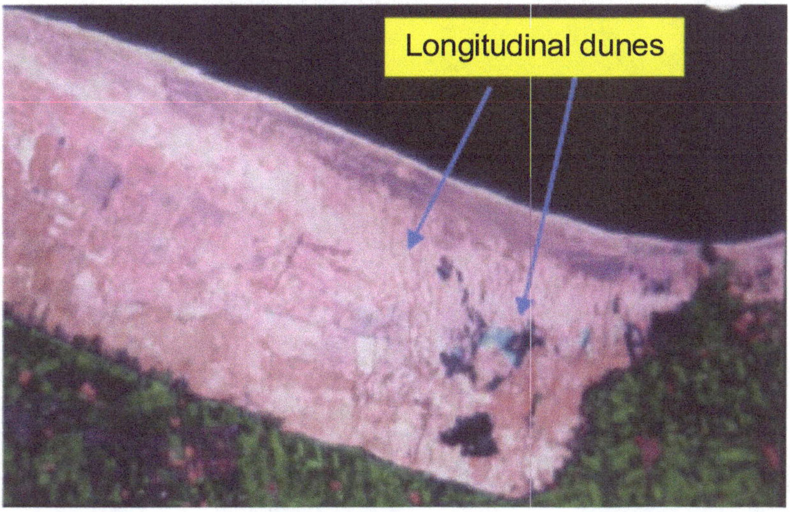

Fig.2 : The longitudinal dunes on TM satellite image of 1990 (After El-Bady, 2001)

Fig.3 : The longitudinal dunes on SPOT satellite image of 1995 (After El-Bady, 2001)

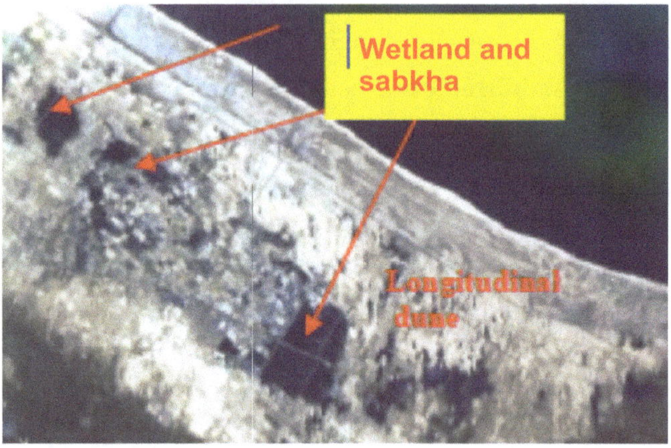

Fig.4 : Satellite image of 2005 showing the longitudinal dunes and wetland and Sabkha near Gamasa

Fig.5 : Satellite image of 2005 showing Barchan dunes from Kitchener to Burg El-Burulus

Fig.6 : The sand dunes belt adjacent to the shoreline in El-Burg

Different methods have been used for measuring the sand dunes movement in the study area such as Howard *et al.* (1978), Fryberger *et al.* (1984) and Livingstone (1989) methods. The dune movement at El-Burg village have reached up to 3 m/yr, while no movement was detected at Baltim lighthouse. El-Asmar (2000) recorded that the movement is related to the wind direction and calculated the movement. It ranges from 3-8 m/y at Gamasa. Now, most Gamasa dunes are removed due to development activities such as reclamations, industrial and tourism activities.

Conclusions

1. There are two type of coastal sand dunes old and recent (longitudenal and Barchan).
2. The coastal sand dune (recent) are very mobile and move toward the many developmental activities.
3. The coastal sand dune encroachment is the main reason for desertification.
4. Sand dune movement effect on the palms in El-Nakhel and El-Hammad areas causing the destruction the palms.
5. The coastal sand dunes attack the tourism and urban areas.
6. The fixation of sand dunes are tar spraying, stone wall and wind fence.
7. The wind fence are the best method for stabilization in the area.
8. Although the coastal sand dune are removed from some sites in the study area, the desertification is advanced by the sands of the other dunes, beach, coastal plain and recently formed dunes.

References

Bartlett, D. 1994. GIS and The Coastal Zone: Past, Present and Future. The Association For Geographic Information.

Carter, R. W. G. 1988. Coastal environments, an introduction to the physical, ecological, , and cultural system of coast, pp1, Harcourt Brace Jovanovich Publishers.

Cooke, R. and Warren, A. 1973. Geomorphology in deserts. London, B.T. Batsford, Ltd.

El-Asmar, H. A. 2000. Geoenvironmental studies on the coastal area between Gamasa and Baltim, North of the Nile Delta. Z. *Geomorph. N .F,* 44 (1): 59-73.

El-Asmar, H. M. 1994. Severe coastal Damage along Ras El Bar shoreline, Nortn of Nile Delta: An effect of the construction of the detached breakwater system. *Journ. Geol.* 38(2), 793-812.

El-Bady, M.S.M. 2001. Geoenvironmental study on the coastal area between Gamasa and Baltim, Northeast of the Nile Delta, Egypt. M.Sc. Thesis, Damietta Fac. of Sci., Mansoura Uni. Egypt.

El-Fishawi, N. and El-Askary, M. 1981. Charachteristic features of coastal sand dunes along Burullus-Gamasa strech, Egypt. *Acta Miner. Petrogr*. 25(1): 63-76.

Fryberger, S.G., El-Sari, A. M., Clisham, T. J., Rizvi, S. A. R. and Al-Hinai, K. G. 1984. Wind sedimentation in the Jafurah sand sea, Saudia Arabia. *Sedimentol.* 31: 413-431.

Howard, A.D., Morton, J.B., Gad ELHak, M. and Price, D.B. 1978. Sand transport model of barchan dune equilibrium. *Sedimentology*, 25, 307-338. http://www.worldbank.org/data/dataquery.html. Accessed January 2, 2003

Livingstone, I. 1989. Monitoring surface changes on a Namib Linear dunes. *Earth Surface Process Landforms* 14:317-332.

Livingstone, I.and Thomas, D.S.G. 1993. Modes of linear dune activity and their significance: an evolution with reference to sothern Africa examples. In: Kpye (ed.) Dynamics and Environment Context of Aeolian Sedimentary System, 91-102, Geological Society of London, Specil publication No. 72, London.

Sestini, G. 1992. Implication of climatic changes for the Nile Delta. In: L.J. Jefftic, J.D. Milliman and G. Sestini (eds) Climatic changes changes and mediterranean. 555-601 Edward Arnold, New York.

UNESCO, 1998. United Nations Educational, Scientific and Cultural Organization. Chapter 2 :Coastal Zone Management (CZM). www.unesco.org

Tetra Tech, 1984. Shoreline master plan for the Nile Delta coast. Progress Report No. 1.

B. RESOURCES, DEVELOPMENT AND MANAGEMENT

12

'ORANS' : The Sacred Groves of Thar Desert, India

Mahesh Kumar Gaur

Introduction

Thar Desert is unique both from floristic and climatic parameters. Biodiversity in this region survives on a very/ fragile ecosysytem under a highly hostile environment. Major area consists of dry undulating mass of loose sand, thus, leading to shifting sand dunes. The sandy plain is full of sand-hills with serve as low depressions where salt and soda are deposited after drying. The whole area is inhospitable and sterile. Trees are rather a few and far between (Bhandari, 2005).

The Thar Desert of western India is known for its rich and ancient culture system and traditions. The communities have long been part of the Thar Desert ecosystem and have evolved specific strategies to live in harmony with its hostile environment. *Orans* have remained mainstay in balancing this arid ecosystem. Man of this region has never been greedy as he is today. *Orans* were repositories of ancient system of gene pool conservation. But, there are unregulated encroachment on the marginal lands and over-exploitation of groundwater resources. *Orans* has been part and parcel of the lifestyle of the Thar Desert inhabitants. *Orans* are used to provide vital grazing land for livestock, water, minor forest produce, medicinal plants and green cover for the villages. They were worshipped as *devbhumi*. *Orans* preserve the endemic, endangered or threatened species, medicinal plants and wild variety of cultivars found.

Status of Common Property Resources in India

Broadly speaking, common property resources include all such resources that are meant for common use by the villagers. They include resources like village pastures and grazing grounds, village forests and woodlots, protected and unclassed government forests, wasteland, common threshing grounds, watershed drainage, ponds and tanks, rivers, rivulets, water reservoirs, canals and irrigation channels. In pre-British India, a very large part of the country's natural resources was freely available to the rural population. They were largely under the control of local communities. With the extension of state control over these resources and the resultant decay of community management systems, CPRs available to villagers declined substantially over the years. Today, in almost all parts of the country, villagers have legal right of access only to specific categories of land and water resources. Nevertheless, it is widely held that CPRs continue to play an important role in the life and economy of the rural population.

A 1999 NSSO study on the role of land, water and forest commons in the life and economy of rural Indians revealed that CPRs provide as much as 58% of fuelwood requirements and up to 25% of fodder requirements. It also provides evidence of large-scale depletion of CPRs, with CPR lands in rural India declining by almost 2% every five years.

Published in 1999, the National Sample Survey Organisation's (NSSO) study is the largest national-level assessment of size, utilisation and contribution of CPRs, with separate estimates for 12 agro-climatic zones. The report is based on a sample survey carried out across 5,242 villages, covering 78,990 rural households, and pertains mainly to the role of three CPRs – land, water and forests – in the life and economy of India's rural population. NSSO has defined common lands as, 'a resource becomes common property only when the group of people who have the right to its collective use is well-defined, and the rules that govern their use of it are clearly and followed universally'. Moreover, common grazing lands and watering points have played a significant role in enabling the inhabitants of Western Rajasthan to pursue an adaptive mix of animal-rearing and crop husbandry.

According to the report, 15% of the country's geographical area is under CPR lands, excluding forests, though the latter in practice may be used as common property. The average landholding size stood at 0.84 hectares, while each household had access to 0.31 hectares of CPR land. The average area of CPR land available to households was highest among the northeastern states followed by Rajasthan and Madhya

Pradesh, states which also have a high proportion of drylands. Most frontline Green Revolution states have a relatively smaller share of per household CPR land availability. Similarly, states with higher population density predictably reported lower per capita CPR land availability.

The report reiterates findings of previous studies on CPR land availability and utilisation across different agro-climatic zones. The percentage of geographical area under CPRs and per household CPR land availability is lowest for the Gangetic plains and highest for the Thar desert. Likewise, the extent of permanent pasture and grazing lands is highest in the western Himalayas, followed by the western coast and central plateau; it is lowest in the Thar desert. The eastern and western Himalayas have the highest proportion of CPR lands under village woodlots and forests, while the Thar desert and the Gujarat and eastern coasts have relatively lower CPR lands under forests.

Origin of *Oran*

In India, one of the earliest documented work on *Oran* dates back to 1897 when the first Inspector General of Forests, D. Brandis, wrote: "Very little has been published regarding sacred groves in India, but they are, or rather were very numerous. I have found them in all provinces------". Most of the traditional societies in India depend on the biodiversity contained in their rural environment for meeting the livelihood concerns. About 4 percent land covers such an area in the country. These serve as sanctuaries for herds of ungulates as well. *Oram* provide vital grazing land for livestock, water, providing minor forest produce, medicinal plants and green cover for the villages they used to serve. About 41% livestock is dependent on *Orans* in Barmer district of Rajasthan. In addition to this livestock economy, *Orans* also benefit the villages through their water bodies and diversity of trees.

Traditional natural resource management in Rajasthan is characterized by community managed lands, consisting of: *Agors* (A) that are areas that traditionally served as catchments for water bodies; *Gouchars* (G) that are areas that served as community grazing lands; and *Orans* (O) that are areas that served as community forests. All three community resources were traditionally managed by the village community. In the Jodhpur, Pali and Barmer districts, AGOs cover 50-60% of the geographic area. But over the years, the total land area under community managed AGO lands has declined and the level of degradation of remaining AGOs has greatly intensified. This is mainly due to encroachment and conversion to agriculture, conversion for settlement, roads, community assets, inappropriate afforestation programmes, and reduction in recharge and over exploitation of ground water due to deep bore-wells, further increasing livestock and human pressure on the remaining AGOs. In recent years, there has been a breakdown of the traditional resource use regime. Grazing lands have effectively become open access resources with no system for controlling and monitoring their use. Though specific data for the Thar Desert region are not available, national data show a decline in the extent of common property lands, ranging from a decline of 26 percent to 52 percent in several states of India during the last four decades, and Rajasthan is among the more severely affected states. Traditionally managed AGO lands are repositories of biodiversity and the source of multiple products such as grass, fodder, fuel wood, timber and non-timber tree products. Protection and management of AGOs are fundamental to the survival of agro-ecosystems in desert areas, as well as maintenance of ecosystem stability, integrity, functions, and services in the face of climate change.

There are about 25,000 *Orans* in Rajasthan (Forest Department, Government of Rajasthan). About 1100 major *Orans* are spread out on an area of more than 1,00,000 hectares. The area under an *Oran* can vary from a few square meters to several hundred hectares, for instance the Bhadriya *Oran* in Jaisalmer district is 15000 hectares, Baankal-Devi-ka *Oran is* spread over 4600 hectares where as Kundla's *Oran* is 7500 hectares in Barmer district. "*ORAN*" means that nobody is allowed to cut trees or shrubs in an area, only animals can graze over the vast areas of "*Oran*" which are full of biomass. "*THAN*" or "*Majisa*", "*Jamboji*", "*Jog Maya*", "*Bhomiaji*", "*Pabuji*", "*Harbuji*", "*Ramdeoji*", "*Mamaji*" are established under the shade of green trees and nobody is allowed to cut trees, twigs or leaves due to the fear of the God/Goddesses. These practices are still followed to maintain a proper ecosystem and reduce the expansion of the process of desertification.

But for the *Bishnoi* community, the artiodactyls, black buck, Indian Gazelle, would have also vanished from the desert landscape. They do not allow anybody to cut Khejari and kill dear, come what may. They offer their lives for conservation and protection of these two.

Such groves are a very ancient and widespread phenomenon in the old world cultures. They are mentioned in Sanskrit classics as *'aranya'* (*अरण्य*) meaning forestland. Many of tree species are found both within and outside the *Orans* and for traditional societies who use it in a variety of livelihood concerns, managing biodiversity is crucial for their sustainable development. Precisely, floral biodiversity in *Orans* is very rich, they not only yield several non-timber forest products, but also harbour multiple-use livelihood goods.

Orans Serve as Gene Pool Conservation System

Orans are repositories of rich biogenetic diversity and venues of local and universal manifestation of aesthetic tradition and societal sacredness. In the management of *Orans*, ecologically valuable species perform key functions in the ecosystem, thereby contributing to the support and enhancement of biodiversity. Generally, the species are selected and valued by the local communities for cultural or religious reasons. But currently, at least 3% of the recorded wild flora and a somewhat larger number of wild fauna are on the threatened category. For the last few decades, this area has lost about 65% of its forests despite the fact that the biological diversity of the area is one of the most significant in India with several thousands species of flora and fauna found in the area.

Species in most of the *Orans* in Thar Desert are *Prosopis cineraria, Zizphus nummularia and Salvadora oleoides, Capparris decidua*. Shrubs include *Calotropis procera* in Jaisalmer, Kair (*Capparis decidua)*, Ber *(Zizyphus nummularia)*, acacia trees, Hingota (*Balanites egyptiaca*), Neem (*A. indica*) in the hills and undulating plains.

Orans preserve the endemic, endangered or threatened species, medicinal plants and wild variety of cultivars are found. '*Orans*' are very useful of the species preserved and are known to be of considerable value for the medicinal plants, while other species such as *ber* and wild fruits and seeds are valuable for home consumption and trade. In addition, the *Orans* aid scientific knowledge in producing value-added products and derivatives, which could be used in the pharmaceutical, chemical and food processing industries. (Fig. 1)

The fairs and festivals have retained religious significance of *orans*. The village community organizes 'fairs' at the *'Oran'* site where they rejuvenate commitment towards their grasslands. In this arrangement, the sanctity of the domain ensures a ready and plentiful availability of an important energy source for the benefit of all. Most of the Fairs are generally organized on specific dates, which vary from *Oran* to *Oran* and are related to some religious or social event of significant importance. Thus the ethnic, cultural, and history also play an important role. In brief, *Oran* serves as :

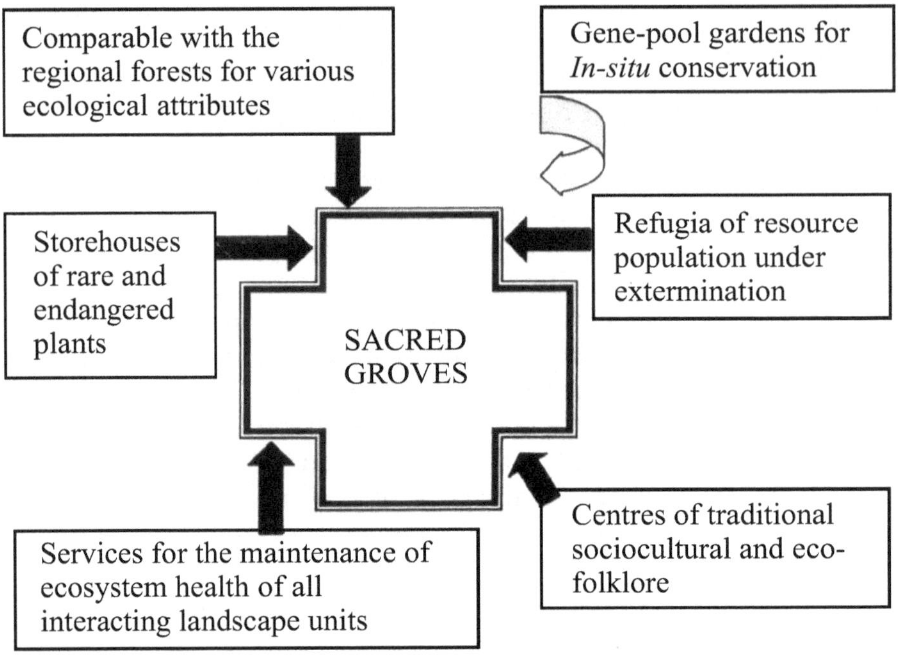

Fig.1 : *Orans* as gene-pool conservation (after Subrahmanyam, 2009)

- Refuges for bio-genetic diversity and gene pool conservation,
- Repositories of ethno-social codes of relation and regulation vis a vis nature,
- Venues of local and universal manifestation of aesthetic tradition and religiosity.

Declining trend in *Orans*

Despite of all the characteristics, *Orans* have undergone decline and shrinkage. *Orans* in western Rajasthan have been neglected and destroyed by modernization and legislation. Also, threatened because of increasing pressure from human and livestock population. Several encroachments have taken place, and worse, the Governments have regularized them. Area and legal status of several *Orans* has not been clearly defined. Forest wealth and water sources/water-harvesting structures, which existed within or adjacent to the *Oran* lands are depleted, dilapidated and facing extinction. This has happened due to abandoning of traditional practices of natural resource conservation and management, and importantly non-people centric government policies and laws play a significant role in this regard. Most of the herbs, which existed in '*Orans*', have either become extinct or threatened to become extinct.

The quality of grazing material has declined due to so much overgrazing that the herbaceous cover is now mostly composed of inedible annuals which have dominated the regeneration cycle of nutritive perennial vegetation. Moreover, weakening of religious faiths and the utilization of woody biomass for human use has further deteriorated the productivity of the arid ecosystem. Table 1 explains tehsilwise distribution of *Orans* in Jodhpur district.

Table 1 : Tehsil-wise distribution of *Orans* in Jodhpur district

Area in Hecare	Tehsils				Total
	Luni	Jodhpur	Bilara	Bhopalgarh	
>320 (2000)	4	0	1	0	5
160–320 (1001 - 2000)	5	5	1	0	11
80-160 (501-1000)	11	6	5	1	23
6-80(101-500)	61	51	39	9	160
8-16 (51-100)	35	73	24	1	133
2-8(12.5-50)	114	88	75	9	286
<2 (<12.5)	40	9	51	14	114
Total	270	232	196	34	732

(Figures in paranthesis indicate area in bigha)

Source : Choudhary, 2002.

Thus, the decline of the *Orans* began once the forests went out of community control and is attributed to the weakening influence of religion with the spread of modern civilization and legislation. Over a million livestock dependent people are struggling hard for survival because of shrinking of *'Orans'* (grazing grounds). These pastoralists have either to undertake large scale migration to adjoining areas or overgraze nearby forest area, which often cause them physical hardship and severe social strain and also lead to conflicts and destruction of vegetation. The decline of forest cover has undoubtedly brought grave consequences. Denudation of tree cover quickly leads towards infertility and erosion. (Table 2)

The Indira Gandhi Nahar Pariyojna (IGNP) in the districts of Ganganagar, Bikaner, Jodhpur and Jaisalmer covers a large tract of sandy plains and hummocks in the northern and north-western parts of Rajasthan. Major part of Shri Ganganagar district is under older canal system since more than 80 years but other areas under IGNP command are under irrigation for varying periods. Colonization of areas falling under Canal Commands has been introduced for (re-)settlement. As a result land (*Oran*) falling close to settlement has been encroached and allotted to inhabitants. As a result, large areas of *Orans* have been lost. Also rich desert biodiversity is lost. Pace of faunal and specially floral biodiversity loss has been augmented with the introduction of changes in landuse patterns. The croplands of this zone do not harbour xeric rodent fauna. Various floral and faunal species are under constant threat due to transformation of sandy grasslands into irrigated croplands.

Table 2 : Approximate Encroachment Percentage of Orans and gauchars (numbers)

Approximate % Encroachment	Tehsils				Total
	Luni	Jodhpur	Bilara	Bhopalgarh	
0	101	93	98	14	306
0-10	157	162	103	94	516
11-20	35	46	59	12	152
21-30	15	15	34	8	72
31-40	8	12	5	4	29
41-50	8	11	14	7	40
51-60	6	5	5	1	17
61-70	2	2	1	0	5
71-80	4	6	5	2	17
81-90	2	0	4	2	8
91-100	4	3	6	0	13
Total	342	355	334	144	1175

Source : Choudhary, 2002.

Orans currently face various threats like mining, quarry, encroachment, clear felling, and other depletive factors. Most of the *orans* are threatened by Vilayati Babool plantation by Forest Department. Eroding community values have made the matter worse. Lack of faith in the younger generation is also a problem. This situation is similar throughout the country.

Status of faunal and floral diversity

The vegetative coverage of the *Ornas* needs special attention for their conservation, regeneration values for the welfare of human kind. Plant diversity of *Orans* has lot of economic, ethical and religious value. Thar Desert of Rajasthan has many endemic plants and vertebrate species. They have been part of the desert biome and needs attention as other 'biodiversity hotspots' are getting. A few studies have been carried out by individuals but these are not sufficient enough. It is difficult to say that they have made any significant impact on the front of conservation. Significance and threats to desert plants and animal species have yet not been assessed completely.

Tal Chhapar Sanctuary is situated in Churu. It is a flat saline depression which is locally known as "*Tal*". In this Tal Chhapar Sanctuary one very special type of grass is found. This grass is locally called MOTHIYA (*Cuperus rotundus*). The word mothiya comes from the word MOTI or from the English word of Pearl. As the shape of the bulb of this grass is like very fine round shaped pearls. This MOTHIYA is very tasty and has sweet taste to eat. Human beings eat it with pleasure and it is found in very less quantity so production is only few kilograms every season. This grass and its bulb Mothiya is also food for Black bucks and other birds who dig it from hooves and beaks. Seeds of this grass were eaten during drought and famines.

Plate 1: Sand dune Biome of Thar Desert

Plate 2 : Sacred groves in Tal Chhapar Blackbuck sanctuary

Ganganagar and Bikaner districts were dense with a small shrub locally called as SAJJI (*Haloxylon recurvum*). Juice extracted from this plant was used in the preparation of *Papar* (Salt cake). But, with the tractorisation of agriculture, such endemic plants are no more found in the area.

Plate 3 : Sultana Natural Grasslands

Desert National Park, Rajasthan: The Desert

National Park is spread over in Jaisalmer-Barmer districts. It is part of Sind and Thar Desert (Desert grassland and open scrub-field). This is one of the largest national parks, covering an area of 3162 km^2. Various endemic plant species are found there. Many of them may be in the threatened category and a few are on the verge of extinction.

Plate 4 : Bio-diversity loss due to farming on marginal lands

Bhadriya Oran: It is located in Jaisalmer district. Area-wise, it is one of the largest oran in the region. This land was donated by the Jaisalmer Maharaja to the Goddess when his wish was fulfilled. It is about 45 km away from Pokaran, towards Jaisalmer. Entire area falling under the oran is cultivated. This area is rich in groundwater resources.

Plate 5 : Threatened Bird: Indian Vultures

Loss of indigenous biodiversity due to introduction of invasive species has yet not been assessed. By the time scientists become aware of it, many a species may be lost forever. Human encroachment has deteriorated the situation further. Like *water hyacinth*, which was not growing in this region as

Plate 6 : Threatened Bird: Great Indian Bustard (Godawan)

there were no perennial water bodies, has now intensively encroached the entire Indira Gandhi Canal system to such an extent that it obstructs the flow of the canal water at certain points. The waterlogged areas are full of tall grasses such as *Arundo, Typha, Phragmitis* and others. Saxena (1991) has reported a large number of newly introduced weeds, which are flourishing owing to high nutrient and moisture availability. About 40 weed species have invaded the desert through transport of soil and water, crop seed impurities and due to migration of birds and animals.

Insects very quickly intrude into new areas and with the presence of integrated crops, a large number of them such as the stem borer *Spenoptera gossypii* of cotton, *Heliothis* spp. of pulses crops, white grub *Adoretus bembinator* from sugarcane have been reported from the transformed regions. Similarly, riverine insect species like *Schizodacty lusmonstrosus* have also been collected from the banks of the canal in the Mohangarh-Ramgarh sector.

It has been found that a number of animals have changed their habitat due to changes in the landuse introduced by the human beings. For example, porcupine *Hystrix indica* is rocky formations, especially in the foothill slopes. But now it inhabits on the slopes of tall sand dunes along the IG Canal. Similarly blue bull, *Boselaphus tragocamelus,* was not found in Jaisalmer and Bikaner districts. But now large herds are encroaching the IG Canal irrigated fields.

Because of the availability of poor quality of food resources, the highly adapted wild animals - invertebrates as well as vertebrates - have suffered. Their population have dwindled (Panther, Jungle Cat, Desert Cat, Desert Fox, Wolf, Jackal, Hyaena, Ratel, Wild Boar, Indian Gazelle, Black Buck, among mammals. (Prakash, 1994); Common Sandgrouse, Grey Partridge, and some have vanished from the desert scene (Lion, Cheetah, Caracal, Wild Ass, etc).

It is thus clearly evident that the changes in the ecological scenario have negatively augmented the species richness in the IG Canal zone of the Thar Desert.

Cause of biodiversity loss in *Ornas*

Vast grasslands of *Lasirus sindicus* (locally called as *Sewan*) used to occur in the north-western desert, almost along the entire length of the IG Canal. Changes incorporated in the land use pattern, after the introduction of IG Canal, have led to delimitation of this grass, and it may ultimately vanish from the region where soil moisture is high and persists all the year round. Already, most part of the grass has been lost to the crop raising. A number of shrubs such as *Calligonus polygonoides, Haloxylon salicornicum, Dipterygium glaucum, Tribulus elatus* and the sedge

Cyperus conglomeatus, which are adapted to survive only in extreme xeric conditions, are also amongst severely threatened one due to irrigation by IG Canal. This author has witnessed only a few hundred trees of *Tecomella undulate* (locally known as *Rohida*, the teak of desert) in the Pokaran Tehsil of Jaisalmer district. In fact this area was densely populated from this plant before 1975. Similarly *Withania somnifera* (Ashwagandha) and Guggul (*Commiphora wightii*) are no more locally found, as was the case before 1970s.

Canal and tubewell irrigation : It is seen that irrigation by the tubewell water in the sandy areas like Churu, Sikar and Jhunjhunun districts has led to depletion of typical deserticolous vegetation and thus replaced by newer weeds and plant species. Jalore district is witnessing typical changes in land use pattern due to availability of water from Sardar Sarovar dam. Like depletion of endemic plants, same impact has occurred in the native animals. A number of reptiles, amphibians and other micro-organism which were significant part of this desert habitat have been lost. This loss of biodiversity is compounded due to ploughing with tractors and spray of insecticides. This area was rich in faunal diversity. Godawan was found in the entire desert tract. The impact of irrigation on the endemic rich avian fauna has not been so far fully studied. Recent study shows that the Indian Bustard (*Godawan*) and Indian Vultures (also known as Scavengers of Nature) are now severely endangered.

Invasion for fuelwood : Burgeoning human population and consequent demand for fuelwood is leading to unprrecedentd pressure on the vegetative coverage of the Orans. Populations are directly dependent upon these resources for meeting their daily requirements. Situation is further compounded because rural as well as urban poverty forces inhabitants to illicitly fell, collect, and sell the bio-resources.

Tremendous changes in the avifaunal structure of the Thar desert are taking place due to the Indira Gandhi Nahar Project (IGNP) and species never seen earlier are now regularly found near the canal. However, this project is playing havoc with the desert ecosystem by changing the crop pattern, traditional grazing regime and because of colonization by new people who do not have the same conservation value system which the desert people had. Due to easy availability of water everywhere, unsustainable livestock grazing is taking place and the famous Sewan grasslands which have survived for hundreds of years with low grazing pressure now under tremendous pressure. These grasslands are the major habitat of the highly endangered Great Indian Bustard *Ardeotis nigriceps*, and the winter migrant *Houbara* or the *Macqueen's Bustard Chlamydotis macqueeni*.

Overgrazing: Excessive pressure of livestock population, traditionally high, has grown to 29.11 million (2007 census). With shrinking vegetative coverage and dwindling productivity of these lands, overgrazing is a self-amplifying problem afflicting orans. The roaming livestock instantly graze the very first flush of plants sprouting with the onset of monsoons from every inch of land. This precludes any regeneration of forests and lead to degradation of ecology of the area with inferior and less productive species replacing the natural flora. Thus, it has caused many fodder grasses to disappear or be reduced to such that the roots or stolons rarely get an opportunity to flower and disperse. The scope of biological invasion due to fast changing ecological scenario in the Thar Desert is enormous. It will be at a great ecological cost due to the irretrievable losses of native biological diversity in the orans. Further, this reverse ecologicl (de)succession is turning large areas of lands into deserts, barren expanse bereft of plants and soils being washed down under monsoon showers.

Marginal agriculture and human settlements : Villagers encroach upon nearby orans for marginal agriculture and settlement. It is difficult to save village orans from intentional encroachment by the inhabitants either for settlement or agriculture. They bring these lands under plough very slowly by the owners of adjoining areas.

Contribution of Commons in Food Security

In a study conducted in 81 villages across 21 dryland districts across seven states in rural India, pioneer of *commons* research in South Asia N S Jodha measured the dependence of poor communities (small farm households and the landless) on CPRs. The commons in these districts included village pastures, community forests, wasteland, common threshing grounds, waste dumps, watershed drainages, village ponds and tanks, rivers, rivulets, riverbeds, etc, and the area under commons ranged from 9 to 28% of total village area.

The data revealed that 84-100% of poor households derived benefits such as food, fuel, fodder and fibre from common lands. By contrast, only 10-28% of large farmers used CPRs, and the benefits they accrued were fewer.

Per household income from CPRs ranged from Rs 530 to Rs 830, depending on the region. Small, poor households benefited more than larger, richer households; common lands also serve as a safety net for the poor during periods of crop failure and uncertainty. Since returns for individual users from degraded lands are not much for the rich, the poor, who have surplus labour and fewer opportunities, accept these low-paying options more readily.

Harvesting of seasonal products and the opportunity to benefit from unskilled labour are additional factors that make the commons more attractive for the poor. In fact, the results indicate that CPRs contribute more to poor households than anti-poverty schemes in some of the areas that were evaluated. An analysis of monetary data from this study also suggests that CPRs play a role in reducing inequalities in income between classes. The results call for their inclusion in poverty alleviation and rural development exercises.

One of the key findings of Jodha's work has been the documentation of the decline of CPRs. This refers not only to physical loss in terms of area, but also recorded declines in productivity and changes in status, ownership and governance. In the three decades following India's independence, the area under common lands declined by 26-63% in the dryland districts that were assessed.

While population growth had a role to play, this decline was largely attributable to privatisation for the benefit of the poor as part of various welfare programmes. However, 49- 86% of privatised lands ended up being allocated to farmers who were better endowed. In a majority of instances where land was given to the poor, they did not have additional resources to develop the land, nor were they provided any support to do so. As a result, much of the land was sold, mortgaged, remained fallow, or was leased on a long-term basis. The remaining common lands were overexploited resulting in declining productivity. These developments also have adverse consequences for drylands which are typically fragile ecosystems and hence prone to greater degradation and poor resilience. The decline in CPRs is often very closely paralleled by a decline in social capital (especially traditional institutions and mechanisms of governance).

According to a study by the ICAR, seeds from grasses, pods from trees, buds from bushes, fruits from herbs, etc. have been traditionally consumed by desert communities to compensate for nutritional deficiencies during periods of acute food shortage. Depletion and loss of CPR lands therefore deprives dependent communities of a vital safety net during periods of food insecurity and agricultural crisis.

Conclusion

In India, concept of scientific pasture management has not been properly planned, despite the fact that India has one of the largest livestock populations in the world, with an estimated 520 million heads. Efforts in India for pasture management have been confined either to improvement of existing grasslands or introduction of suitable exotics. There is no sound management plan for the development of pasture land and protection of existing patches of grassland, some of which are unique and harbour rich fauna. Conservation of orans for floral and faunal diversity is essential for the suvival of local ecosystems and human race, as such. Even though a noteworthy research work has been carried out in the Thar Desert still scientists feel that lot of efforts are required in both conservation and sustainable use efforts. It is therefore difficult to provide a definitive picture of biodiversity status and trends. It may be conjectured that biodiversity in Thar Desert survives in areas not affected by habitat conversion, or in traditional farming systems. But, it can be safely said that habitat degradation is an imminent and immediate problem affecting biodiversity loss in Thar Desert. It is important that biodiversity in Thar Desert should be addressed in two parallel fronts: addressing potential and actual biodiversity loss, through documentation, advocacy, capacity building, and improvement of the enabling environment, and highlighting and encouraging instances where biodiversity is healthy and managed sustainably. It is important to plug holes in the research.

References

Band, L. E. 1986. Topographic partitioning of watersheds with digital elevation models. *Water Resource Research.*, 22: pp 15-24.

Barnes, Peter. 2001. 'Who Owns the Sky? Our Common Assets and the Future of Capitalism'. Island Press.

Bhandari, M M. 1990. Flora of Indian Desert. MPS REPROS, Jodhpur.

Brabyn, L.K. 1997. Classification of macro landforms using GIS. *ITC Journal*, 1997-1, pp 26-40.

Brabyn, L.K. 2002. The use of GIS for biodiversity mapping in New Zealand.

Choudhary, H.C., et.al. 2002. Comprehensive community drought preparedness programme to improve quality of life of women and children in Jodhpur district. Report submitted to UNICEF.

Dikau, R. 1989. The application of a digital relief model to landform analysis. In: Raper, J.F. (ed.). Three dimensional applications in Geographical Information Systems. *Taylor and Francis, London*, pp 51-77.

Dymond, J.R. DeRose, R.C. and Harmsworth, G.R. 1995. Automated mapping of the land components from digital elevation data. *Earth Surface Processes and Landform*, vol. 20, pp 131-137.

Edouard G. Bonkoungou. Biodiversity in drylands: challenges and opportunities for conservation and sustainable use. *The Global Drylands Partnership. (IUCN).*

Francis, H. D. 1985. Soil landscape analysis, London : *Routledge & Kogan Paul.*

Gahegan, M., 1999. Four barriers to the development of effective exploratory visualisation tools for the geosciences. *Int. Jrnl. Geographical Information Science* 13(4): 289-309.

Gaur, Mahesh, Pathak, S. and Sharma, J. R. 2004. Mapping of potential rangelands for optimizing biomass production in Thar Desert of India: A Remote Sensing and GIS Approach. Presented at 24th ISRS Annual Convention and National Symposium "*Converging Space Technologies for National Development*" November 3-5, 2004 held at Jaipur.

Gaur, M. and Gaur, H. 2004. Combating Desertification: Building on Traditional Knowledge System of the Thar Desert Communities. *International Journal of Environmental Monitoring and Assessment, Kluwer Academic Publishers, New York.* pp. 89-103.

Gaur, M. & Gaur, H, 2003. Impact of Human Activities on Land degradation & Desertification in the Thar desert- Using Remote Sensing Technique. In V.C. Jha (Ed.), *Land Degradation and Desertification,* Rawat Publications, Jaipur. pp. 386-402.

Gaur, M. 2002. Indian Desert of Western Rajasthan: A Potential Area for Conflict Management or An Opportunity for Sustainable Development, paper presented at the International Course "*Environmental Conflict Management*" on 05-12-2002 at the UN University of Peace, San Jose', Costa Rica.

Geiger, R. 1971. The climate near the ground. *Harvard University Press, Cambridge.*

Goswami, Bhasker 2008. 'The Nuts and Bolts of Appropriating Agriculture Land', in *Agenda,* April 2008 (http://infochangeindia.org/200804057058/Agenda/Battles-Over-Land/The-nuts-and-bolts-of-appropriating-agricultural-land.html)

Hardin, Garrett : 'The Tragedy of the Commons'. *Science*. December 13, 1968, pp 1243-1248.

IEMSD, 1997. Biodiversity Related survey of indigenous knowledge of Thar Desert.

IPCC. 2008. Climate Change and Water. IPCC Technical Paper VI, June 2008.

Jodha, N.S., 2005. Paper presented at the UNDP workshop on "Sustainable land use in Dry Lands: Global and National Perspectives", 2nd February 2005, New Delhi.

Mazhar, Farhad, Daniel Buckles, P V Satheesh, Farida Akhter. 2007. 'Food Sovereignty and Uncultivated Biodiversity in South Asia : Essays on the Poverty of Food Policy and the Wealth of the Social Landscape'. Academic Foundation : New Delhi.

National Sample Survey Organization (NSSO). 1999. Study on "Common'. NSSO-54 Round, January, 1998-June, 1998.

Olson, Mancur. 1965. The Logic of Collective Action : Public Goods and the Theory of Groups. Cambridge, Massachusetts. Harvard University Press, 1964, p 4

Prakash, I. 2001. Biological invasion and loss of endemic biodiversity in the Thar Desert. *Resonance,* pp. 76-85.

Prakash, I. 1988. Changing patterns of rodent populations in India. In *Rodent Pest Management* (Ed. Ishwar Prakash). CRC Press, Boca Raton, pp. 179-190.

Rahmani, A.R. 1997. *Wildlife in the Thar,* WWF, New Delhi.

Saxena, S.K. 1991. Impact of canal irrigation on the ecology of arid tract of Rajasthan. *Prospects of IG Canal Projects* (Ed. Abrol & Venkateswarlu). ICAR, New Delhi, pp. 65-73.

Singh, G.R. *et al.* 1992. Soil Erosion Rates in India, Journ. of Soil & Water Conservation 47 (1) : 97-99.

Singhi, Manju and Ramesh Joshi. 2010. 'Famine Food in Arid Rajasthan : Utilisation, Preceptions and Need to Integrate Social Practices by Bio-Resolutions'. [I] Studies on Ethno-Medicine, Vol 4 (2), August 2010, p121-124

Tivy, J. 1993. Biogeography - A study of plants in the ecosphere. *Longman, Harlow.*

Wadsworth, R. and Treweek, J. 1999 Geographic Information Systems for ecology. *Longman, London.*

Websites :

1. planning commission.nic.in/aboutus/committee/..../tfll_grass.pdf

❑❑❑

13

Sustainable Management of Small Scale Irrigation Projects, Case Study : Um Jawaseer Desert Farm Project, Sudan

Tarig EI Gamri and Fadul Basheer

Abstract

This chapter highlights the significance of irrigated desert agriculture as a prime cause of the settlement and welfare of nomads and the consequent reduction of conflicts over limited land resources. Um Jawaseer Desert Farm Project is located in Bayouda Desert, the Northern State about 200 km north of the capital Khartoum. The planning and management approaches of such irrigated projects were discussed with special emphasis on the participatory approach. In this respect the roles of the donor (NGO), traditional tribal institutions and the project beneficiaries were illustrated. The project beneficiaries were organized into the farmers, well and women committees. The environmental and socioeconomic impacts of the project were evaluated. The project maintained food security, improved nutritional status, hygiene, housing environment and assisted in the provision of primary education and basic health facilities. Cost/benefit analysis showed the high profitability of the project. Other positive impacts include conservation of the fragile environment and biodiversity and improvement of animal productivity. Consequently, the area which witnessed an out-migration, during the seventies and the eighties, due to drought and desertification is now experiencing incoming migration.

However, adaptation to the sedentary mode of life was a real challenge. The most striking benefits of the project was that of women training for the, involvement in the decision-making process and empowerment, since 20% of the land is allocated for women. In addition to that, problems like build up of salinity, sand encroachment, low organic matter content and increased susceptibility to some water-related diseases due to irrigation were tackled successfully.

Introduction

Desertification is land degradation in arid, semi-arid, and dry sub-humid areas resulting from various factors, including climatic variations and human activities.

It was revealed by the UN Secretary General during the World Summit on Sustainable Development (WSSD) which was held in Johannesburg in the year 2002 that about one billion people are directly affected by the process of desertification (www.unccd.int). However, the impacts of desertification were found to be escalating (El Gamri and El Nasikh, 2003).

According to FAO, (1995) the increasing human population in developing countries is putting pressure on their finite land resources and causing land degradation. About 3,600 ha or 70% of the world's dry lands (excluding hyper-arid deserts) are degraded. Land degradation in marginal dry lands covers about 30% of the total area of the globe. Poor irrigation practices were put among the five major desertification causes that include also deforestation, climate variability, overgrazing, and over-exploitation of land resources (UNCCD, 1998).

It is estimated that the population of Sudan is between 35 and 40 million of which about 70% live in villages, hamlets or lead semi-nomadic existence meanwhile 30% are town and city dwellers or live in displaced persons settlements (UNEP, 2007). Recent studies revealed the fact that the country is under water stress with an amount of water of less 1500 m^3 per capita per annum. It is worth mentioning that the amount of water below 1000 m^3 per capita per annum is considered as the water poverty zone.

Rainfall has a major impact on the socio-economic conditions of the country. In terms of distribution, it varies from almost nil at the Sudanese-Egyptian borders to about 1500 mm at the extreme Southwest. (Fig. 1).

El Mustafa, (2005) revealed the fact that 1.78 million km^2 of the area of the country (total area 2.5 million km^2) is located in climatic zones that are vulnerable to desertification. Thirteen states in the northern part and two states in the southern part of the country are affected.

Ayoub (1998) showed that 64% of the soils of the Sudan are degraded. Table 1 shows the extent and severity of soil degradation in the Sudan and Table 2 exhibits the different forms of land degradation in the country. Soil erosion is a real threat to natural resources, infrastructure and housing especially in northern, western and eastern Sudan. It is evident that in some locations in the northern state, sand dunes have invaded the banks of the Nile River. Plate 1 shows animal houses invaded by wind-blown sands.

Table 1 : The extent and severity of soil degradation in the Sudan (million ha)

Degradation severity	Hyper-arid	Arid	Semiarid	Dry sub-humid	Moist sub-humid	Total
Very high	8.0	24.7	8.8	2.1	2.3	45.9
High	0	4.7	5.4	0	0	10.1
Medium	0.2	0.5	2.1	2.2	1.7	6.7
Low	0	0	0.9	0.2	0.2	1.3
Total	8.2	29.9	17.2	4.5	4.2	64.0

Source : Ayoub, 1998

Table 2 : Soil degradation types in the Sudan (million ha)

Aridity zone	Wind erosion	Water erosion	Chemical deterioration	Physical deterioration	Total degraded
Hyper arid	5.8	2.4	0	0	8.2
Arid	20.0	6.9	3.0	0	29.9
Semi-arid	1.2	7.7	5.3	3.0	17.2
Dry sub-humid	0	0.7	3.8	0	4.5
Moist sub-humid	0	0.5	3.7	0	4.2
Total	27.0	18.2	15.8	3.0	64.0

Source : Ayoub, 1998

Um Jawaseer Project is located about 200 km northwest of the capital Khartoum in a desert climate with a precipitation of less than 100 mm per annum. The project harnessed groundwater from an aquifer of about 40 m deep to irrigate 800 feddans (Feddan is a unit of land area in Egypt and Sudan and One Feddan is about 1.038 acres) divided among 210 households. However, each household is estimated to host about three people of the extended family in addition to that the project also supported some casual labour specially during land preparation and harvesting.

Plate 1: Animal houses invaded by wind blown sands in the State of Khartoum

Irrigation in Desert Lands

Dry lands occupy about 64%, 39% and 81% of the total areas of Africa, Asia and Australia respectively (WMO, 1990). In such areas there is no season during which a crop can be grown without irrigation. Yet it is well recognized that such lands has got an underexploited agricultural potential (NAS, 1974). On the other hand, agriculture represents some 38% of the total gross domestic production of the semi arid tropical countries of Africa and a similar proportion to that of Asia (Gibbon, 1988).

Exploitation of land and water resources in the dry and semi-dry region began at slow rate after Second World War but gained full momentum during 1960s. Irrigation plays a pivotal role in meeting the basic food needs for the people of the globe especially in Asia. It provides more than 50% of the most important two basic staples and about 30% of all food crops. However, sub-Saharan Africa has the smallest regional area under irrigation that produces about 10% of its total food production (Wolff, 1997).

According to Sinha (1996) and Madeley (1997), diverting riparian waters to desert lands has improved agricultural productivity in both India and Egypt. In India the practice assisted in protecting the fragile environment of the Thar Desert.

In this respect El Gamri (2004) enumerated the general prospects and constraints as follows:

General constraints

- The difficulty and complexity of dry land management have led to the disregard for their problems (Agnew and Warren, 1996).
- Scarcity of data: Data are inevitably scarce due to the sparse population, poverty and the harshness of the environment that can cause instruments to fail.
- Confusing definitions of drought and desertification. Fortunately as mentioned before the UNCED has adopted a comprehensive definition, which was later, adopted by the Convention.
- Lack of solidarity, peace and setting of the appropriate priorities at the global level
- Poverty is a major retardant on development including that of agriculture.
- Global economic situation: World recession in the 1980's has severely hindered agricultural development (Barrow, 1987).
- The price policies for export trades, which are determined by industrialized countries, are putting at risk commercial exchanges even between developing countries. Soares (2001) pointed out "since developing countries have no access to more developed technologies, they cannot compete with the prices of industrialized countries which can achieve high scales of productivity".
- Inequitable terms of domestic trade: It is not unusual for farm prices to be held artificially low in order to ensure cheap food supplies so as to avoid unrest in cities (Barrow, 1987).
- Moisture variability and other unfavorable climatic conditions.
- Arid lands are usually characterized by having limited soil resources. Soils are usually very shallow, unbalanced in texture, with high amounts of salts and hardpans may exist WMO (1990).
- In arid and semiarid regions the soil is subjected to severe water erosion. This is attributed to high rainfall intensity (daily rainfall of more than 100 mm has been recorded in such areas) and consequently frequent runoff. According to WMO (1990) the rate of soil loss is directly proportional to rainfall intensity, slope, vegetation cover and soil type. In addition to that wind erosion is also a serious problem e.g. northern Sudan where wind-blown sands denuded agricultural lands and threaten the River Nile.

- Lack of efficient and appropriate extension services.
- Land tenure system: Few farmers in such areas own their lands, and consequently they have little incentive effort or money in improving production.

General prospects

- Development of appropriate technologies: As mentioned before arid lands have under-exploited agricultural potential, according to NAS (1974) this potential can be best developed adopting methods specifically suited to dry regions.
- The temperature regimes and the abundance of sunlight that prevail in such areas create favorable conditions for the growth and production of some important crops. In this respect Sinha (1996) discussed the potential of some crops (pearl millet, sorghum, wheat etc), horticultural crops (Date-palm, pomegranate, guava etc) and some pastoral crops (*Cenchrus ciliaris*) in the Thar Desert of India. According to him some crops also provide industrial raw materials for explosives, soap, drug and pharmaceutical, perfumes and petroleum industry. Vogt (1995) made a comprehensive study on the identification, propagation and utilization of the trees and shrubs of arid lands of the Sudan. A huge number of such plants were found to be of medicinal (*Cassia senna*), pastoral, nutritive (date palm) and industrial (*Gum Arabic*) uses. Other can be used for fuel wood (*Acacia* spp.) and to prevent land degradation viz. sand dune stabilization (*Salvadora persica*) and shelterbelt (*Acacia melifera*).
- Desert pasture: According to WMO (1990), the positive features of desert pasture include the following: diversity of fodder, the relatively high nutritional value, the duration of the grazing period and the possibility of using cheap fodder all the year around.
- Water conservation: In some arid lands the largest opportunity for increasing water supplies is to improve existing water systems especially in agriculture (NAS, 1974). In this respect, Saeed and El Gamri (2000) reviewed the appropriate technologies for the efficient utilization and management of water resources with special emphasis to arid and semiarid regions.

- Indigenous knowledge: In this respect Gibbon (1988) and Reij et al (1996) called for combining indigenous knowledge and modern knowledge.

In the Sudan, agricultural projects constructed for the resettlement of the people who are displaced due to the construction of Merwoe Dam were irrigated by constructing channels from the River Nile. Two of these projects were located in the vicinity of Um Jawaseer Project

Irrigation in such areas is usually associated with serious problems like excessive seepage and deep percolation, high crop water demands, bad soil chemical conditions like salinity and water logging. However, the agricultural potential can be best developed adopting methods specifically suited to dry regions (El Gamri, 2004). For instance, the use of canal lining, modern irrigation systems, biofertilization and land drainage were found to improve the performance of the irrigation systems tremendously. (Plate 2)

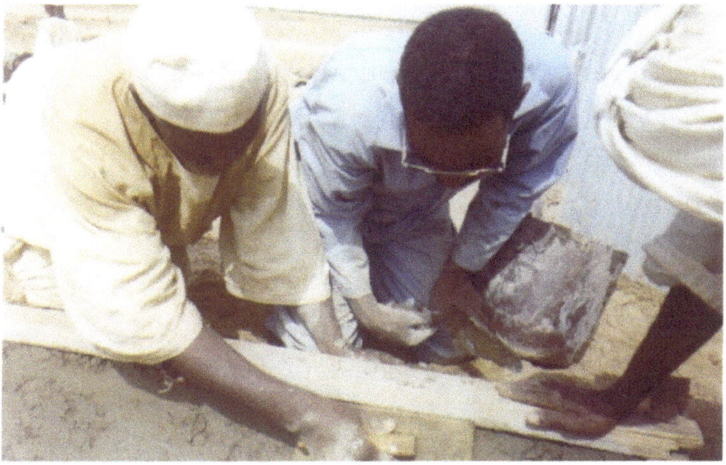

Plate 2 : Canal lining in Um Jawaseer

Problems like excessive seepage rates and undulating topography can be solved by using modern irrigation systems. Now a days, there is considerable expansion in the use of centre pivot systems around the capital and in the northern dry parts of the Sudan.

Planning and Management of Irrigation Projects

Planning

This analysis concentrates mainly on smallholders i.e. a farmer who normally derives his/ her livelihood from an irrigated holding of less than 2-5 ha and is often less than 0.5 ha (Cornish, 1998).

Based on the concepts of participation, FAO formulated guidelines for planning development projects for rural poor including that of irrigation (FAO, 1986). These guidelines can be summarised as follows :

1. Identifying the target groups.
2. Function and sequence of project design:

   A project design must provide the basis for a workable project and must supply the information needed for project approval.
3. Project design decision making :
   a. Specifying project objectives.
   b. Specifying project components.
4. Finalizing the project design :
   a. Structuring the project design.
   b. Phasing project interventions.
   c. Relating project analysis to key feasibility issues.
   d. Preparation of a realistic implementation.
   e. Designing a monitoring and evaluation system.

The above-mentioned guidelines evolve around what is called the "*Process Approach*" rather than the more structured "*Blue print Approach*" which was suggested by UNDP during 1970s.

The following are the advantages of the *Process Approach* over the *Blue Print Approach*:

1. It depends on the participation of the intended target group and, hence it is more responsive to their needs and potentials.
2. It allows for variations in bureaucratic structures and thus is adaptable to political, economic, and social changes.
3. It is based on learning by doing and on building institutional capacity, so it is well directed towards the promotion of self-sustaining development benefits.

Management

The traditional approach to natural resources development and management is basically anthropocentric and strongly influenced by conditions in water rich countries. This approach takes its starting point trends in birth rates and mortality. The resources and systems needed to support these populations are generally discussed in fragmented fashion (SRFAC, 2001).

IWRM was defined as: *"The process which promotes the coordinated development and management of water, land and related resources in order to maximize the resultant economic and social welfare in an equitable manner without compromising the sustainability of vital ecosystem"*.

In the past irrigation management approaches were either structure-oriented or process-oriented. Recently the participatory approach has been introduced (SRFAC, 2001).

The success of rural development projects depends mainly on the voluntary commitment and participation of the beneficiaries in the decision making process. However, developing local organizations was found to be the most effective mean of promoting participation (FAO, 1986). On the other hand, participation is also one of the Dubin Principles that governs integrated water resources management (GWP, 2000). In Um Jawaseer the community is organized in the farmers', well and women development committees. Since 2005, the donor concentrate in capacity building specially the development and support of community based organizations.

The recent approach of Irrigation Management Transfer (IMT) aims at the reduction of the government's role in irrigation management and the corresponding expansion of the role of water users and other local institutions (Vermillion and Johnson, 1995).

IMT include approaches like Joint agency- users' management, privatization and others. IMT will fail unless it is clearly seen as a benefit to both governments and the farming communities. However, post transfer guidance is needed to help protect the integrity of the irrigation structures and protect the environment. It is advisable to link such support to financial and technical performance.

Although there are global evidences on the success of the IMT approach, some authors argued that the process has some negative impacts on women. Zwarteveen, (1995) pointed out that IMT will increase the time women directly or indirectly have to contribute to operation and maintenance of irrigation systems, without fully benefitting from the benefits generated.

Role of local institutions

Sudan, like most African countries, is basically organized on the tribal system especially in the rural areas, and the country is estimated to consist of about 570 tribe. Each tribe inhabit and own specified piece of land. During the Condominium (Anglo- Egyptian) (1898- 1965) era this land ownership has been legalized and delineated on official maps

(MAF, 2002) Each tribe practice a different land tenure system as stemmed from its culture and modified by the modern Condominium administration system. In the year 1927, El Sheikh Ordinance Law was issued. Although the system has been weakened due to different factors like instability of policies, MAF (2002) concluded that the system still plays an important role in protecting natural resources. At each level land ownership and natural resources utilization are governed by certain laws and institutions. It is worth mentioning that such institutions play an important role in the resolution of conflicts especially within the same tribe.

Actually Um Jawaseer Project utilized the traditional tribal system in the planning and management of its functions. On the other hand, the system proved to be very influential in conflict resolution. Table 3 shows the five basic methods for resolving conflicts as suggested by SRFAC (2001).

Mobilization of indigenous knowledge is anticipated to promote desert agriculture and the development enterprise in general (El Gamri, 2004). Although the inhabitants of the project have practiced flood cultivation in the wadis as part of the nomadic system, it is not clear that such knowledge has affected their productivity substantially. However, they showed high skills in animal production activities.

Planning and management of Um Jawaseer project

The project adopted the participatory approach and the beneficiaries were strongly involved in all stages of the project. The planning and management structure created, was made use of the traditional tribal institutions with some modifications to suit the project objectives. The beneficiaries were selected so as to represent the different clans of Hawaweer Tribe in the vicinity of the project equitably. In addition to that, the project has allocated 20% of the area for women.

The project management was carried at two levels:

1. Technical and supervisory role that was carried out by ADRA. This role usually will phase out in about 5 year after the project commences.
2. Field operations which were carried out by the community who will lead the whole process after donors phased out.

Table 3 : Five Basic Methods for Resolving Conflict

Method	What happens when used	Appropriate to use when	Inappropriate to use when
Denial or withdrawal	Person tries to solve problem by denying its existence. Results in win/ lose.	Issue is relatively unimportant; timing is wrong; cooling off period is needed; short-term use.	Issue is important; when issue will not disappear, but build.
Suppression or smoothing-over	Differences are played down; surface harmony exists. Results in win/lose in forms of resentment's defensiveness and possible sabotage if issue remains suppressed.	Same as above, also when preservation of relationship is more important at the moment.	Reluctance to deal with conflict leads to evasion of an important issue; when others are ready and willing to deal with the issue.
Power and dominance	One's authority, position, majority settles the conflict. Results in win/ lose if the dominant party sees no hope for self.	When power comes with position of authority; when this method has been agreed upon.	Losers have no way to express needs; could result in future disruptions.
Compromise and negotiation	Each party gives up something in order to meet midway. Results in win/lose if "middle of the road position ignores the real diversity of the issue".	Both parties have enough leeway to give; resources are limited; when win/ lose stance is undesirable.	Original inflated position is unrealistic; solution is watered down to be effective, commitment is doubted by parties involved.
Collaboration	Abilities, values, and expertise all are recognized; each person's position is clear, but emphasis is on group solution. Results in win/win for all.	Time is available to complete the process; parties are committed and trained in use of process.	The conditions of time, abilities and commitment are not present.

Before phasing out of the donor the beneficiaries were trained in:

- Training of women in cooking, food processing, soap production, and sewing and tailoring at the women development centre,
- Farm management under participatory conditions,
- Revolving fund management, and
- Training of youth in pump operation and tractor driving.

Impact of the Irrigation Project

Environmental degradation in developing countries is clearly linked with agricultural production so that many writers refer to the "South's environmental crisis" (Barrow, 1987). Significant part of such degradation could be rehabilitated with better water use; this is partly because bad irrigation practices that is one of the five major desertification causes.

Environmental impact assessment plays a central role in acquiring information on the social and environmental implication - including water resources implications- of development programmes and projects; identifying the measures necessary to protect the resources and related ecosystem and then ensuring that such measures are implemented. On the other hand, the IWRM approach implies that sectoral developments are evaluated for possible impacts on water resources and that such evaluations are considered when designing as well as giving priority to development projects.

Social impacts

1. Settlement: Irrigation paves the way for the settlement of the nomadic communities, and hence essential services (health care, water supply, sanitation and education) can be provided.
2. Awareness raising.
3. Minimization of educational and functional illiteracy.
4. Improved nutrition.
5. Improved housing environment.
6. Food security: During 1920s China's population was about 400 million people and famine occurrence was common. Nowadays China has achieved food security mainly because of improved water supply and management.

Social impacts of Um Jawaseer project

1. Food security: Due to recurrent drought, the area was stricken repeatedly by famine (ADRA was among the relief providing agencies to the area). According to Abdalla *et al.* (1998) the annual average grain (sorghum and wheat) production of a farmer is about 2000 kg. On the other hand, it is estimated by the World Standards that the annual per capita requirement of grain is 140 kg. Hence, an average family of 5 members consumes 700 kg annually and the surplus could be sold.
2. The inhabitants of the area were the most marginalized group in the Northern State. The Project assisted them to restrict to their land as indicated by the incoming migration from towns and villages along the banks of the River Nile.
3. One of the most striking benefits of the Project is that of women training, enlightment and involvement in the decision-making process.
4. Improved nutrition both in terms of quantity and quality.
5. Improved hygiene and housing environment where mud houses replace the huts of the nomads.
6. Provision of primary education and basic healthcare facilities.

Economic impacts

Action programmes for combating desertification, which are not integrated with poverty alleviation, are doomed to failure (El Wakeel, 2005). Economic impacts of irrigated agriculture include the following:

1. Income from agriculture.
2. Income from improved animal production due to the availability of feedstuffs and the introduction of improved animal breeds as practiced in Um Jawaseer.
3. Wages from off-farm work.

Economic impacts of Um Jawaseer project

1. The project sustains the livelihood of 210 farmers (households) with a very high production potential since they are provided with the appropriate technical packages and extension services. Break Even Point is defined as the minimum yield above which a positive net profit is guaranteed. As shown in Table 4, the different crops were found to achieve high profitability.

2. Women are able to generate income from food processing, soap production and sewing and tailoring.
3. Improved animal production : Due to the availability of feedstuffs and the introduction of an improved goat breed (Sa'anin). And consequently milk production in the project site increased from 1 pound/ goat/ day to about 4 pound/ goat/ day. On the other hand, the number of animals increased from 6 per household in the year 1990 to about 8 per household in 1999.

Table 4 : Crop Yield and Break Even Point in 1998

Crop	Break Even Yield (sack/feddan)	Farmers' Yield (sack/feddan)
Wheat	5.4	7.0
Sorghum	3.72	5.0
Broad Beans	3.47	5.0
Onion	8.94	10
Alfalfa	1046 (bundles)	2000 (bundles)
Okra	14.5	20

Source : Mohamed *et al.* (1998)

Plate 3 : Alfa alfa field in Um Jawaseer

Environmental impacts

The positive environmental impacts include the following:

1. Conservation and promotion of biodiversity.
2. Creation of a better microclimate.

3. Assist in mitigating global warming through carbon sequestration.
4. Creation of a healthy environment through reducing suspended dust particles and the combat of the water washed diseases.

On the other hand, the negative impacts include the following:

1. Build up of both soil and groundwater salinity.
2. Harbour of water related diseases.

According to McDonalds and Kays (1988) water related diseases can be classified as follows:

1. Faecal-oral diseases: (e.g. cholera, typhoid, diarrhoea, dysenteries, hepatitis, ascariasis). These are all infection transmitted by the faecal-oral, and also faecal-nasal and faeco-ocular) route. They may be water-borne or water washed.
2. Water-washed diseases: (e.g. infections of the skin and infections of the eyes). These are all infections related to poor hygiene (therefore water-washed) but which are not faecal-oral and not water-borne.
3. Water-based diseases: (e.g. guinea worm, schistosomiasis, clonorchiasis). These are all the helminthes that have an aquatic intermediate host.
4. Water-related insect vector: (e.g. malaria, filariasis, yellow fever, trypanosomiasis). These are all infections transmitted by insects which either breed in water or bite near water.

Environmental impacts of Um Jawaseer project

1. About 35% of the area of the first project was invaded by sand encroachment. Consequently the project adopted a combination of earth embankment and shelterbelt (3 rows of Eucalyptus trees) in its subsequent phases to control the phenomenon with very encouraging results. However, introduction of bushy tree species is recommended. Shelterbelts in the project area have resulted in the reduction of wind speed, sand movement and dust storms markedly.
2. Temperatures drop of about 3° C was recorded inside the project area.
3. The availability of feedstuffs has alleviated pressure on the natural rangelands.

4. The project maintained biodiversity since some migratory species were observed in the project area. On the other hand, the relative availability of organic matter due to agricultural activities supports the existence of organic decomposers and their predators as well (A. Latif, 2003).
5. Environmental awareness raising that is reflected in the commitment to protect the environment and the enthusiasm to plant shelterbelts.
6. Reclamation of about 800 feddans (acres) and utilization of ground water in a sustainable way.

Strategies for Economic Sustainability

FAO (1986) suggested that during the planning phase the following economic sustainability strategies are to be considered:

1. The technological packages, organizational arrangements and service delivery system should be acceptable to the beneficiaries and maintained with minimum of external inputs.
2. The economic and financial returns of the project should be sufficiently high.
3. The relevant institutions should be capable and supportive.
4. Consistency with the microeconomic policies.

Others may include the following:

1. Provision of credit: Governments usually prefer to finance large enterprises because they are not confident that small farmer will pay their debts.
2. Environmental protection:
   - Land degradation lowers agricultural productivity.
   - Environmental pollution will harbour water related diseases and hence agricultural productivity will be reduced due to the farmer's absence and the expenses spent in medication which otherwise can be directed to farm inputs.
3. The use of water conserving technologies like canal lining and modern irrigation systems. That is because water availability is the most governing factor for agricultural expansion in such areas.

References

Abdel-Latif, M.A. 2003. Impact of Agricultural Practices on Selected Soil Decomposer Fauna. Ph. D. Thesis, Department of Zoology, Faculty of Science, University of Khartoum.

Agnew,G. and Warren A. 1996. "A Framework for tackling drought and land degradation", *Journal of Arid Environments*, No. 33, PP 309-320.

Ayoub, A.T. 1998. "Extent, severity and causative factors of land degradation in the Sudan". *Journal of Arid Environment* 38(3) 397 - 409.

Barrow, C. 1987. "Water Resources and Agricultural Development in the Tropics". Longman Development Studies.

Cornish, G. 1998. "Modern Irrigation Technologies for Smallholders in developing Countries". ITDG,UK.

El Gamri, T. and El Nasikh, M.H. 2004. "Desertification Studies in the Sudan": *Annals of Arid Zone*. Vol. 42, No. 3, 2003. India.

El Gamri, T. 2004. Prospects and Constraints of Desert Agriculture: Lessons from West Omduman. Pp 57-73 : Environmental Monitoring and Assessment Journal. Vol 99/2004. The Netherlands.

El Mustafa M. A. 2005. "National Action Programme for Combating Desertification" Presented at the Workshop on the implementation of the UNCCD, Organized by Sudan National Assembly in collaboration with the Fedral Ministry of Agriculture and Forestry and the Dryland Coordination Group.

El Wakeel, A. S. 2005. "Desertification: the Number One Environmental Problem in Sudan". Presented at the Workshop on the implementation of the UNCCD. Organized by Sudan National Assembly in collaboration with the Fedral Ministry of Agriculture and Forestry and the Dryland Coordination Group.

FAO 1986. Guidelines for Designing Development Projects to Benefit the Rural Poor. FAO, Rome.

FAO 1995. Planning for Sustainable Use of Land Resources: Towards a new approach. FAO Land and Water Bulletin 2. FAO, Rome.

Gibbon, D. and Pain, A. 1988. "Crops of the Drier Regions of the Tropics". ELBS/ Longman.

GWP, 2000. "Integrated Water Resources Management". TAC Background paper, Global Water Partnership, Sweden.

National Academy of Sciences (NAS), 1974. "More Water for Arid lands, Promising Technologies and Research Opportunities", Report of Ad hoc Panel of the Advisory Committee on Technology Innovation Board on Science and Technology Innovation for International Development Commission on International Relations, Washington, D.C., USA.

Madeley, J. 1997. "Water for the Desert: How Irrigation Can Make The Difference. Development and Cooperation (D+C) *Journal, No.* 2/1997.

McDonald, A. T. and Kay, D. 1988. "Water Resources: Issues and Strategies". Longman Group Ltd., UK.

Ministry of Agriculture anf Forestry (MAF) 2002. "Sudan National Action Plan to Combat Desertification". In Arabic. MAF and Arab Organisation for Agricultural Development, Khartoum, Sudan.

Mohamed, Y.A., Mohamed, Y.Y. Mustafa, M.M. Abdalla, M.S. Musa, F.S.and Khalifa, A. 1998. Evaluation of Um Jawasir Irrigated Project. ADRA-Sudan and Land Use and Desertification Control, Ministry of Agriculture and Forestry.

Reij, C., Scoones, I. and Toulmin, C. 1996. "Sustaining the Soil: Indogenous Soil and Water Conservation in Africa". Earthscan Publications Ltd, London.

Sinha, R. K. 1996. "Making Aridlands Productive: Conservation and Utilization of Thar Desert Biodiversity for Sustainable Development in India". Desertification Control Bulletin. No. 26, UNEP.

SRFAC, 2001. "Integrated Water Resources Management". Scientific Research and External Relations Centre, Sudan University for Science and Technology.

Soares, H. S. 2001. "Poverty Eradication From Dream to Reality: Hope in the Sahel". Spore Bulletin, 96, CTA. Wageningen, The Netherlands.

UNCCD 1998. "An Introduction to the United Nations Convention to Combat Desertification (Fact Sheets)" CCD/98/4, Paris.

UNEP, 2007. Sudan Post-Conflict Environmental Assessment. UNEP, Nairobi, Kenya. http://en.wikipedia.org/wiki/Activated_sludge (visited January 2008).

Vogt, K. 1995. "A Field Worker's Guide to the Identification Propagation and Uses of Common Trees and Shrubs of Dryland Sudan". SOS Sahel International (UK).

WMO, 1990. "Final Report: Working Group on Monitoring, Assessment and Combat of Desertification".

Wolff, P. 1997. "Solving Conflicts Over Water Uses". *Development and Coperation Journal.* No 2/ 1997.

14

Indigenous Water Saving Technologies in Arid and Semi-Arid Regions : Examples from Sudan

Mohamed A. M. Abd Elbasit, Abdelbagi M. Ali, and Hiroshi Yasuda

Abstract

Indigenous water saving technologies in arid and semi-arid environment is a local traditional practice for effective use of the limited water supply. These technologies can be used as a base for future development that can be made on these technologies through incorporating the current industry tools. The importance of studying the indigenous knowledge and practices is that it can help to develop enhanced versions of these technologies, and to understand the ability of local communities to adopt the newly developed technology. In Sudan, several indigenous water saving approaches has been used by local societies at various scales. An insight for these technologies, their strength and weaknesses are given here. On "Haffir" (water ponds), Wadi control, and water diversion methods were used for a long time of history for water collection and conservation. At farm scale, terraces, mulching, intercropping, low tillage, fallow, low-water consumption crops and fast maturing crops are common indigenous water saving technologies. Traditional water storage in trees such as Baobab and water-melon is an indigenous proactive along the history in western Sudan to store rain water for long dry season.

Introduction

Due to rapid expansion of world population, the pressure on limited natural resources is increasing day by day. Arid and semi-arid regions face the higher rate of population growth around the globe. This emphasizes that the competition on natural resources will increase with time. The common water shortage in the arid regions suggests that the development track of nations' life in such areas should be different from that of the humid areas. The agricultural and industrial rapid expansion will be faced by water shortage obstacles. Most of the countries in arid areas rely either on ground/desalinized water for their water supply with consuming rate higher than the recharge rate (IPCC, 1977), or surface water with catchment areas outside their territories. This situation may widen the term of water security, which emphasizes the protection of water sources in terms of quantity and quality. However, the existence of the country water supplies outside its territory creates several future uncertainties. Scarcity of good water quality is one of the characteristics of the arid and semi-arid regions. For long time of history, the people living in dry areas try to adapt with these conditions. Searching for water and fodder by the ancient societies in arid and semi-arid societies was a common practice called nomadism. The search for water and good pastoral lands was not apart from the natural cycles of rain and heat. For example, in Sudan the trading of "Baggara" tribes between south and north following the inter-tropical convergence zone (ITCZ) is one of the great indigenous practices for saving water. Thus, studying the traditional water saving practices in arid and semi-arid zones can be the starting point for development of efficient water saving technologies. Here, the water saving technologies that have been practiced in Sudan will be overviewed and their prospective also highlighted.

Sudan : Geography and Environment

Sudan is located in the north-east of Africa extended approximately, as of July 9th 2011, from latitude 10.0 to 22.0 degree north and longitude 22 to 38 degree east, covering an area of 1.89 million km^2. The major climatic zones in Sudan are (from north to South): desert, hyper-arid, arid, semi-arid, and sub-humid. The rainfall ranges from 0 mm in the north to 700 mm in the south (Figure 1). The rainy season is only three months in average, starting from July until September.

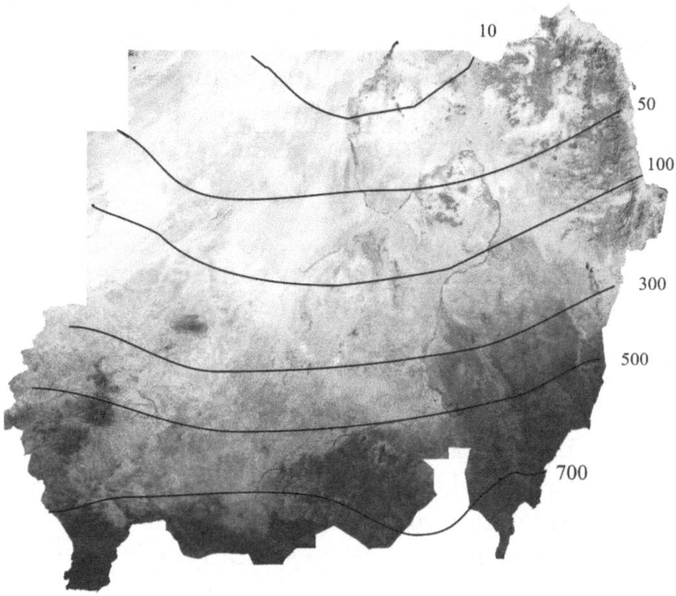

Fig. 1 : Sudan rainfall and climatic zones

Drought and Aridity

The term "drought" is most controversial term describing the situation of water shortage. However, the general concept of drought is a negative change in water balance between the supply and demand. Although aridity and drought are not synonyms to each other, however, arid areas have a higher susceptibility to drought spells. Drought can be classified in groups: meteorological, hydrological, agricultural, ecological, and social droughts (Sen, 2008). This classification is based on the tools that have been used in order to assess the drought conditions. Along all of these categories, the most common type of drought assessment tool is the meteorological drought which is mainly described by the lack of rainfall. By this point of view, the drought as a phenomenon can occur in arid, semi-arid, sub-humid, and humid environment. However, the scarcity of rainfall in arid regions made this confusion between the aridity and drought. The aridity describes stagnant status of ratio between the rainfall, as a major environmental moisture source, and potential evaporation, as a major environmental moisture loss. Wilhite and Buchanan-Smith (2005) defined the drought as insidious event results from deficiency of precipitation from expected and becomes insufficient to cover the demand of human activities and environment. Coping with drought condition has been called as drought mitigation and adaptation procedures. The drought mitigation is a procedure or group of procedures and rules work to prevent societies and environment from the impact of drought. These procedures are perhaps terminated under no-drought conditions. On the other hand, aridity is a permanent status and the societies in such areas develope their knowledge to live normally under such status.

Water Saving Categories

Generally, water saving technology can be defined as practices for improvement of water uses and reduction of water losses. From view point of development, the water saving technologies can be divided into major classes: traditional (indigenous) and standard (modern) technologies. The indigenous water saving technologies are the water saving practices that has been developed and adopted within the local societies. These traditional technologies helped the local communities in arid and semi-arid regions to live in such condition of water shortage. The modern water saving technologies could be developed originally from a traditional practice and then improved and got the advantage of current industrial facilities. The modern water saving technologies include for example: concrete dams, roof rain harvesting, pressurized irrigation systems, soil tillage implements, water reuse, irrigation scheduling, and desalinization. However, the modern water saving technologies is out of the scope of present discussion.

The indigenous knowledge is defined as a local information base for certain society (Warren, 1991; Flavier, 1995). The indigenous knowledge in arid and semi-arid societies was used to develop technologies in order to improve the livelihood in various aspects including: water, agriculture, livestock, education, etc. The water saving indigenous knowledge in arid environment, generally, and specifically in Sudan arid and semi-arid zones has been discussed.

Importance of Indigenous Technologies

One of the most difficult stages of technology transfer process is the stage of technology adoption by local societies. In this regard, development of new technologies based on the local knowledge may help the local communities to adopt these technologies. Local technologies are usually cost and effort effective practices, which can be tackled by local communities without external subsidy. Studying the indigenous water technologies will help to understand the level of communities' development.

Types of Indigenous Water Saving Technologies

Several indigenous water saving technologies (IWST) have been practiced in Sudan on various scales, ranging from watershed to domestic scales. Although the general concept of IWST is well known, there is no internationally unified definition. Various definitions are used implying the same or the partial meaning of IWST. The UNEP (1983) defined water saving technology (WST) as a collection of rainwater from surface and its storage to provide water supply. Oweis *et al.* (2001) considered the WST as the concentration of precipitation through runoff and storage for beneficial uses. Both definitions conform on the efficient use of available water supplies through capturing, diversion, and storage (Ahmed and Naggar, 2003). The WST is technical and practical procedure done by human beings in order to collect, spread, divert and store the rainwater for direct or indirect uses. The major purpose of WST is to reduce the losses of rainwater and runoff using flow regulators to ensure the maximum water infiltration into the soil profile. In Sudan arid regions, the rainy months range from May to September which means surface storage can only support the people during the long dry periods. Several types of IWST are practiced in Sudan, which can be classified based on the available water source, topography, soil type, rainfall depth/intensity/distribution, scale, water use and level of knowledge held by the local people (Farah *et al.*, 2003). Generally, these can be classified into following categories:

1. Rainwater collection,
2. Agricultural water saving techniques, and
3. Seasonal water courses (*Wadi*) management.

Examples of IWST in Sudan

i. Baobab tree

The direct collection of rainfall for drinking is the simplest method for water saving in arid and semi-arid regions. In western parts of Sudan, the use of Baobab tree (*Adansonia digitata L.*; locally called Tabaldi tree) hollow trunk for water collection has been used since long time back and still in use (Figure 2). The Tabaldi tree branches work as catchment area, and direct rainfall and stem flow are collected in a hole made in the main stem of the tree. The stem can be filled naturally or artificially by local people. By this method, the villagers can save reasonable amount of rainwater with good quality to be used later during the dry season, mainly for drinking. The stored water in a single tree is usually between 3-5 m^3, which enough for one family of 5 to 10 members to drink for 2 to 3 months (Farah *et al.*, 2003).

Fig. 2 : Baobab tree in western Sudan

ii. Water ponds (*Haffirs*)

Collection of water in large natural or man-made depression, called locally as *Haffir*, is one of the most important water saving techniques in Sudan (Figure 3). Haffir is land pit with different dimensions naturally existing or constructed in areas with low infiltration rate such as clay soil or bedrocks. Haffir method is common IWST in the central, eastern and western Sudan. Haffir provides water for domestic and livestock water use (Figure 4). However, due to the high evaporation rate, Haffir can sustain only for few months after the rainy season. The Haffir size depends on its catchment area, topography and base material. Several improvements have been developed in order to reduce the water losses due evaporation and seepage. Moreover, siltation pools have been recently used in order to protect the Haffir from the sediment deposition and to improve the water quality in the Haffir. In some areas where underground water is very deep, very big Haffir is constructed to collect water which can then be pumped up through pipe system to supply water to big villages during dry season (Figure 5).

Fig. 3 : Traditional Haffir in Abu-Dulayq area east Khartoum

Fig. 4 : Local people using Haffir for their domestic purpose and livestock

Fig. 5 : Haffir system for municipal water use

iii. Residual soil moisture

This method is common on the alluvium beds of seasonal water courses (Wadi) and on the flood plain of perennial streams (Giroof). In this method, the farmers start to grow their crops and vegetables on the wadi bed or on Girrof after the recession (Figure 6). The residual moisture will be the only source of crop and usually farmers use fast maturity cultivars. The production from this type of farming systems can be used by farmers in case of small area and can be delivered to local and central markets in case of large areas. Little water diversions are made from stones, earth levees, and tree branches are also used by framers to drive more water to large areas.

Fig. 6 : Wadi bed cultivation in Wadi, North Sudan

iv. Home-yard gardening (Gobraka)

Cultivation inside and around the house is common practice even in small cities. However, the term Gobraka is used in the western part of Sudan (Kordofan and Darfur). In this method, they get benefits from the high elevation of sheep and goats fences to interrupt the water flow and enhance the infiltration (Farah *et al.*, 2003). The existence of animal residuals improves the quality of the produced crops and can be used locally by families.

v. Soil tillage

Soil tillage using manual and animal forced tools is an important water saving technology for agricultural purpose. Various types of tools are used; however, the most common tool is called

"Sallooka" which is known in different parts of Sudan. The Sallooka is bended stick, with sharp end used to make small ground pits used for sowing seeds and to generate depression to collect the rain and flood water around the seedlings (Figure 7). Using animals for soil ploughing, also generate these kinds of pits or furrow on the ground for water collection and to reduce the soil compaction due to heavy machineries and enhance the water infiltration. Usually farmers grow their crops along the bottom of these furrows. Another type of tillage practiced for water conservation is cutting of the soil capillaries before planting or during dry spells to increase water conservation (Figure 8). In some case farmers enhance moisture through un-weeding of small pushy weeds known for their less water use to maintain ground cover and to enhance water conservation (Figure 9). In some other cases inter-cropping is widely used between prostrate like water melon and legumes and erect plants like sorghum in order to efficiently utilize scarce water in the soils and diversify crop yield for family consumption and for local market (Figure 10)

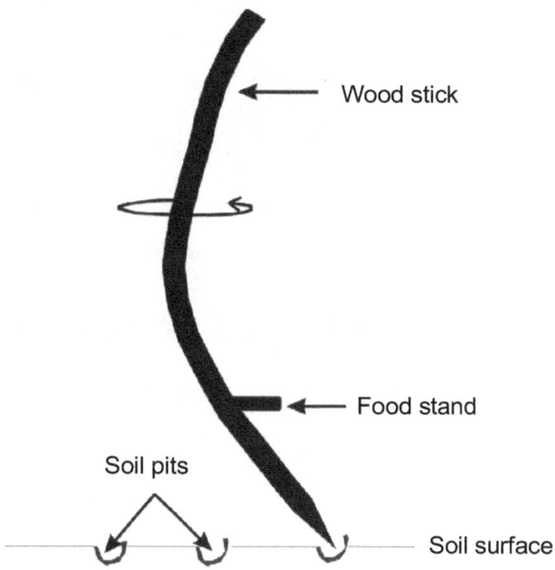

Fig. 7 : Traditional Sudanese tool for seeding (Sallooka)

Fig. 8 : Tillage practices for water conservation, North Sudan

Fig. 9 : Ground cover using grass in Wadi bed cultivation system, North Sudan

Fig. 10 : Inter-cropping in Wadi bed cultivation system, North Sudan

vi. Earth and stone terraces

Terraces are common water saving method all around the globe and it has been used for long time back. In Sudan, terraces have been used in all Sudan and specifically in rain-fed agriculture in central, east and western Sudan and Wadi bed cultivation in arid regions (Figure 11). These kinds of terraces are usually having low elevation which range from 0.6 to 0.9 m, and it can be made manually or by animal-drawn simple equipments. The terraces have also been used to improve the Haffir water yield by diverting the runoff water (Figure 13). These type of terraces are relatively with high elevation and done by the help of heavy machines.

Fig. 11 : Terraces use in rain-fed area, Gadarif, East Sudan

vii. Wadi control

Developments of Wadi control systems mainly depend on the size of Wadi catchment. Wadi control systems range from small check dams for water diversion and increasing water moisture content to large dams with capacity to store water that can be used for agriculture and livestock. Small check dams are commonly used by villagers in order to guide the flood water to their lands and to enhance the soil water content. The Wadi control method, terraces method and water residuals are integrated methods which may exist on the same location. However, generally big dams constructed on Wadi with large catchment areas in order to ensure large water inflow to the dam pond. Irrigation practices can be used around the dam area using small canals or water pumps.

viii. Farmers traditional varieties

Over time and in the drought prone areas, farmers practice their knowledge and techniques to select the most adapted varieties of the local crop that can grow in their limited water environment and provide sufficient yield for their food supply. These varieties have different mechanism for survival on limited water or fast growing to escape dry terminal periods. Some other varieties have special morpho-physiological characteristics such as less transpiration, deep roots, stay-greenness, etc which enables them to efficiently use the limited water resources. In Sudan many lsorhum land races known by the locality are know for their drought tolerance e.g., *Arfa gadamak, Fetarita managil,* etc.

Research and Development

Studying the indigenous water saving techniques and understanding their scientific background represents an important source of knowledge for arid and semi-arid livelihood. The natural feedback from the local communities will help in developing efficient water management systems for arid areas. Several practices have been widely used and proved to be efficient for cropping systems without having a scientific knowledge about the underlining reasons. Development of research project focusing on understanding the moisture distribution mechanism under Wadi cultivation, and role of inter-cropping and conventional tillage on water availability can provide crucial base for future development. Haffir water saving system can be improved by introducing the remote sensing and geographical information system and link them with hydrological models for selection of suitable Haffir site. The high resolution digital elevation models (DEM) can be utilized for depression, water streams, and catchment areas identification. Also, the water quality conservation is quite important issue, especially for sediment and pollutant control. The bioprospecting of the traditional farmers' landraces will provide enormous information for the development of the most suitable cultivars that efficiently utilize the limited water resources and provide most adapted varieties to the arid and semiarid regions.

References

Ahmed, A. A., Naggar, O. M., 2003. Rainwater harvesting: concepts and techniques. Proceeding of the Conference on Water Harvesting and Future development in Sudan, Khartoum 19-20, 2003. Pp. 3- 14.

Farah, S. M., Fadull, H., Omer, M. A., Salih, A. 2003. The Sudanese experience on water harvesting. Proceeding of the Conference on Water Harvesting and Future development in Sudan, Khartoum 19-20, 2003. Pp. 255-276.

Flavier, J.M. *et al.* 1995. "The regional program for the promotion of indigenous knowledge in Asia", pp. 479-487 in Warren, D.M., L.J. Slikkerveer and D. Brokensha (eds) *The cultural dimension of development: Indigenous knowledge systems.* London: Intermediate Technology Publications.

IPCC, 1997. An introduction to simple climate models used in the IPCC Second Assessment Report. In Houghton Jt, Filho LGM, Griggs GJ, Maskell K, IPCC Group I.

Oweis, T., D. Prinz and A. Hachum, 2001. Water Harvesting Indigenous Knowledge for the Future of Drier Environments. International Center for Agricultural Research in the Dry Areas (ICARDA), Aleppo, Syria. pp. 36.

Sen Z., 2008. Wadi hydrology. CRC Press, Taylor & Francis Group: Florida. pp.347.

UNEP. 1983. Rain and storm water harvesting in rural areas, Technical Report, Nairobi. pp. 92-113.

Warren, D. M. 1991."The Role of Indigenous Knowledge in Facilitating the Agricultural Extension Process". Paper presented at International Workshop on Agricultural Knowledge Systems and the Role of Extension. Bad Boll, Germany, May 21-24, 1991.

Wilhite DA, Buchanan-Smith M. 2005. Drought as hazard: understanding the natural and social context. In Drought and Water Crises: Science, Technology, and Management Issues, Wilhite DA (ed.). CRC Press, Taylor & Francis Group: Florida; 3-29.

15

Water Budgeting in Arid Lands

Shil Niyogi

Why Water Budget?

The reliability of water supply in desert regions is at risk due to a multitude of issues including low rainfall, low snow runoff, court-ordered water transfer restrictions and hence higher price of wholesale water. Agencies in desert regions understand the importance of improving water supply reliability, facilitating better water storage, adapting to changing climate patterns and reduced precipitation, improving Environment and Water Quality.

The Desert regions have always fought back the perpetual water scarcity with infrastructure designed around saving water and utilizing alternative water resources. Off late, not only new construction but also new policies and programs have added to their water conservation activities.

One such program attends to the fair distribution of water by tending to consumers who are price sensitive and hence use water without wasting it and by extra-charging those who do not know where to put a stop to their wasteful water use.

There are two types of water rate structures that water agencies have traditionally used to price water for its customers – uniform rates and inclining block rates. Uniform rates, as the name suggests, uses one rate for any amount of consumption. On the other hand, inclining block rates or tiered rates, are those where the water price/gallon increases with more consumption. However, how do these tiers get created and where does one get bumped from one tier to the other? Imagine Mr.

Singh, with 5 children in 2 acre land that his father passed on to him. And his neighbor, Mr. Verma, with 2 children and 1 acre land. According to, inclining block rates, Mr. Sinha will always get penalized. So, should households with many members or a big house now pay more (per the uniform rate) or extra more (per the inclining block)?

This is where Water Budget Based tiered rates come in. According to the water budget program, water gets allocated to consumers based on the number of household members and also the land they irrigate. In this scenario, Mr. Singh receives an allocation of water - his water budget. Similarly, Mr. Verma also gets his allocated water budget. Both are different. Mr. Singh, due to the amount of land he owns and the number of family members in his household, receives more water in his budget (remember he is already taxed on his land and by his family members!). So, now he pays a higher rate only when he goes over his allocated budget.

There are formulae from AWWA and other leading water management agencies, who have studied various ways to budget water. In all calculations, how much water is fair and equitable gets decided by a number of factors - such as the soil type, the type of grass, vegetation or crop, the area of landscape, the evapo-transpiration factors, among others.

After the budget for individual household is worked out, the agency conducts public outreach to educate the ramifications of wasteful water use. The agency also provides instructions for saving water and using water efficient fittings.

Once the water budget based tiered rate structure is set into practice, the agency gives a chance to report any variance to its customers. The agency then starts tracking water consumptions, from water bills. Subsequently, it compares water consumed with the corresponding water budgets. If the consumption is more, the consumers are placed in a higher tier. Now, how much higher, depends on how much more than the budgeted water have they consumed. If they have consumed 50% more, they can be termed excessive, and now they pay almost twice the rate. If they consumed 100% more, they can be termed as wasteful and forced to pay three times the tier 1 rate. Again, these tiers and terms vary from agency to agency, but the central idea is to be efficient, equitable and fair.

How to go About Planning Water Conservation, With Water Budget in Mind?

The agency should review its current and past conservation projects and assess all the water conservation goals alongside a water budget program and see how they help achieve the conservation goals of the agency. The first step in reviewing the agency's existing water conservation programs will be to identify:

- The water users (customers) that are being targeted for water efficiency measures (e.g. commercial, residential, agricultural). And identify the water conservation stakeholders.
- The legal requirements for conservation and estimated water volumes that could be saved from conservation and associated program costs.
- Where will these savings be achieved (e.g., indoor and outdoor) and which ones are a priority (i.e., top users to target)?
- When can these water savings be realized (short-term, long-term)?
- How will these demand reductions be realized—from what specific conservation incentives and programs?
- The agency's water conservation program's activities and results to date, including previously and currently implemented Best Management Practices (BMPs) and other water conservation programs (e.g., residential audits and retrofits, toilet rebates) and incentives (e.g., public education, watering restrictions) and water rate structure.

The key drivers of this entire process are:

1. Better Water Demand Management Practices
2. Better Staff skills to educate residents and manage resources
3. Reduction of Financial, Institutional and Political Constraints
4. Customer communication and Water Conservation Program Marketing

Better Water Demand Management Practices

Under water scarcity conditions, water demand management becomes imperative. Reducing avoidable water waste, inefficiencies and unnecessary losses, by both customers and the water infrastructure, becomes important so that the agency continues to meet the water needs for the foreseeable future. It is vital to assess and assure that annual average day demand does not exceed the reliable capacity of existing and planned supply sources. It is also important to manage peak day water demands in ways that will allow the deferral of major capital projects, and associated finished water pumping and storage facilities.

Another important driving factor is implementation of conservation measures that are cost-effective and acceptable to both the agency and its customers, with special consideration given to economic, environmental, and social costs and benefits. And the most important driver is to implement the demand management program (such as a water budget program) in a timely manner that provides cost benefits to customers and minimizes the near-term revenue and cash flow effects caused by reductions in customer demand.

- Water Efficient Fixtures and Appliances - the agency should see that all new constructions comply with the plumbing codes and use water saving fixtures. The use of water saving fixtures reduces the rate of water consumption primarily for residential customers. As growth within the agency continues and natural replacement of old plumbing fixtures occurs, increased efficiency will be achieved through the use of modern fixtures.
- Identify Large Water users and conduct customer Water Use Audits - Agency should look at its water auditing practices for large use residential customers. Those surveys can be used to make immediate simple repairs and retrofits such as changing water faucet heads. The agency should also perform water audits for large volume commercial, institutional and industrial users. Audits of buildings and sites by qualified personnel would identify water savings that are highly cost effective for facility owners.
- Water Reuse - the agency should look into areas where water recycling (refers to a direct application of water to an additional use after an initial use, with no or minimal treatment) can be done.

Better Staff Skills to Educate Residents and Manage Resources

Management of water resources is a process which depends mainly on the skills of the people who are in charge of its implementation. Development of personnel is a long term and expensive process with a long term return as it takes time to train them and then retain them. The adoption of water management and conservation policies requires high degree of political maturity and an educated public supporting the cause. It is therefore, necessary to raise public awareness before the adoption of demand management policies. It is also important to support and implement a comprehensive set of water conservation measures that will result in measurable and significant water savings, thereby demonstrating to the public the Agency's commitment to an efficient and environmentally sustainable water system.

Reduction of Financial, Institutional and Political Constraints

With water conservation comes an unavoidable stress on the finances of the agency. Institutional weaknesses enhance the bureaucratic behavior that further influences the efficiency of any reform action. Water Budget programs have a cost recovery element that can fund and support the entire Water Conservation Program of the agency. Institutional and political issues may result in water not being available which could otherwise have relieved water scarcity. The agency needs to review measures taken for reducing other institutional and political challenges such as:

- Emphasizing the growing situation of water scarcity and the occurrence of droughts and water stress and the establishment of the appropriate water management strategy.
- Institutional reforms that include changes in functions, responsibilities, work flows and day to day operations.
- Funding levels for public information programs at the agency, to educate and market the program to its customers, even prior to implementing the Water Budget program.

Customer and Stakeholder Communication and Water Conservation Program Marketing

The agency should come up with a marketing program and use of new technologies such as GIS to enable access to the end user and explain the benefits of water savings.

Educating the public is important but getting the active support of the stakeholders is imperative. As the success of water conservation programs has a huge equity built in customer participation, it is very important that the agency should engage in one. The question that the agency should ask is: How can it induce customers to change water-using behaviors? These fundamental implementation challenges require effective customer communication, skillful marketing, and proper incentives.

- Incentives and Rebates - the agency can set aside a budget to provide citizens with a rebate for water efficient fixtures.
- Education/Information Dissemination- Building awareness of the need for water conservation, though a necessary first step, may not easily result in water savings. The agency needs to distribute conservation information in news releases, utility bills, and other places.

A water budget-based tiered conservation rate structure is fair and equitable in rewarding customers who use water efficiently and discouraging wasteful users. It also enables all the water conservation drivers that we discussed in this section:

- Water budget program improves Water Demand Management Practices, and introduces water use efficiency, effectively.
- Water budget program enhances staff skills and helps them educate residents and manage resources.
- Water budget program funds Water Conservation Programs and builds a stronger business case against institutional and political challenges during implementation.
- A successful Water budget program also improves customer satisfaction and helps the agency develop closer communications with the community.

How to go About Designing a Water Budget for an Agency?

In recent times, a lot of attention has been given to meeting unmet needs and removing unnecessary needs. So, every retail company has done a lot to make their solution more customer facing and customer centric. Very similarly, the early adopter agencies of a water budget structure have tried identifying customers - they have created buckets, based on their water use.

We all do that – don't we? We put a tag to our customers – people who buy more from us become loyal customers and those who rarely buy are stragglers. However, in the water conservation market, traditionally, a customer who used much more water was not necessarily considered loyal – it was actually just the contrary – they were considered wasteful.

An examination of the additional benefits of an activity compared to the additional costs of that activity (or Marginal Analysis) led some of these agencies to believe that there was a point of water sales beyond which an extra gallon of water sales cost more (in overheads and capacity requirements) than the benefits from the sale. They used marginal analysis as a decision-making tool not necessarily to help them maximize their profits, but to save them from losses.

However, the scientists, who developed and refined water budgeting were of the opinion that something else led to their overuse. That is when they tried to find out customer specific characteristics and defined water budgeting as "Volumetric allotment of water to customers based on customer specific characteristics & conservative resource standards to promote water use efficiency" (Mayer *et al.*, 2008)

Outcomes of Water Budget

Studies done on this have also put forward the fact that agencies successfully implementing water budgets realize:

- Sustained revenues.
- Revenue sources to fund conservation programs.
- Reduction in residential water use and landscape water use.
- Fairness to all customers.
- Increased customer satisfaction.
- Decreased infrastructure maintenance.
- More efficiency in responding to drought conditions.
- Identification of cases where excessive consumption is natural (due to the size of the household or the landscape).
- Better relationship between the wholesalers, retailers and customers.
- Tracking of all water production and consumption.
- Tracing of infrastructure leaks and non-revenue water sources.
- Utilization of internet and other outreach method to make the customers knowledgeable.
- Use of information technology integration to have the key performance indicators in place.
- When tied to water rates, the agency can tweak the rates to get to revenues (and profits) without affecting the costs.
- Use of a billing system to enable comparison of water allocated and water consumed.
- Use of an asset management system to track all agency owned assets (pipes, valves, etc).
- Use of a hydraulic model system to manage power and supply needed for distribution more effectively.
- Use of a hydraulic model, along with a work order system, to identify distribution assets that need more maintenance.
- Use of Geographic Information System, to identify landscape areas for outdoor consumption analysis.

- Use of a document management system to document all variances requested by customers.
- Better water harvesting and water storage methods by agencies and even customers, as an indirect benefit.

When to Introduce a Water Budget System

So, now the more critical question that an agency should ask itself is when is a good time to introduce such a system. That is when the agency should examine, if:

- There is commitment within the agency.
- There is coordination between the departments at the agency to have a successful implementation.
- The stakeholders at the agency are educated about water budgets and its benefits.
- Customers accept that water budget is an equitable and fair method.
- Customers agree with the allocations.
- The agency has a fairly good GIS - having information on lot sizes of the customer households also would help if available in a table format
- The agency has an electronic customer information system or billing system (as that would help in comparing between actuals versus budgeted).
- The agency has a hydraulic model of some kind (as a simple hydraulic model can help in analysis of water demands).
- The agency has an asset management system (as that would help track faulty demands due to leaks and not necessarily due to over-consumption by the customer).
- The agency has sufficient staff to deal with the introduction of the new water allocation method.
- The agency has trained departments to understand the technology.
- The agency has customer service people, who could handle request for information, request for variance and challenges for allocated budget by customers.
- The agency chooses an off-peak season when it can focus on the water budget allocation much better.

- The agency knows the soil and plant types to understand crop coefficient.
- The agency knows the evapotranspiration factors that determine the evaporation and transpiration rates.
- The agency knows the groundwater recharge and surface runoff of different areas as that helps assess the irrigation efficiency.

How to Calculate Water Budget

When water budgets are built, they consider two types of consumption: indoor consumption and outdoor consumption.

(a) Indoor Consumption : Water budgets for indoor consumption consider the amount of water consumed in drinking, cooking, showers and other domestic uses. Through studies, agencies now allocate either a flat rate (6,000-7,000 gallons per month) or a per-capita rate (60-70 gallons per person per day).

With efficient bathroom fixtures in the market, controlling wasteful users has become much easier these days. Also, indoor consumption accounts for a smaller percentage of overall water consumption in households in drought-prone areas. The bigger concern is the outdoor consumption.

(b) Outdoor Consumption : Water budgets for outdoor consumption has been calculated by different methods. One that is primarily being used these days considers four factors.

- Water Consumption is directly proportional to Evapo-Transpiration : Evapotranspiration (ET) is a term used to describe the water losses due of evaporation and plant transpiration from the Earth's land surface to atmosphere. Evaporation is the loss of water to the air from sources such as the soil, in our case. Transpiration accounts for the movement of water within a plant and the subsequent loss of water as vapor its leaves. Types of vegetation and land use significantly affect evapotranspiration. Consider this - as water transpired through leaves comes from the roots, plants with deep reaching roots can more constantly transpire water. Factors that affect evapotranspiration include the type of plant, its maturity, percentage of soil cover, sunlight, humidity, temperature, and wind. Since there are so many factors affecting ET, it is extremely difficult to formulate an equation that can produce estimates of ET. Therefore, the idea of reference crop evapotranspiration was developed by researchers. ET from a standardized grass surface is commonly denoted as ET_0.

- Water consumption is directly proportional to crop coefficient :

  The crop, as discussed in the last paragraph plays an important role. Plants with deeper roots and large leaves lose more water than smaller plants with shorter roots, smaller leaves. Crop coefficient (K_c), is used to calculate the actual evapotranspiration (ET_c) for a specific crop, when multiplied by ET_0, the standard evapotranspiration for grass.

- Water consumption is directly proportional to Landscaped Area:

  Similar to crop, the more is the landscaped area of a house, the more the evaporation that takes place there. The landscaped area can be calculated using GIS - by drawing polygons on aerial/satellite images and measuring the area. As this process can be quite time consuming, some agencies assign an average landscape area (as a percentage of the lot size) to all homes in particular vicinity.

- Water consumption is inversely proportional to irrigation efficiency :

  Irrigation efficiency (Howell, 2003) is a factor that determines the amount of water needed for irrigation because of the spatial uniformity of the crop, the amount of water that percolates beneath the crop root zone, the amount of water that return to surface sources for downstream uses or to groundwater aquifers that might supply other water uses, and the amount of water lost to unrecoverable sources. Irrigation efficiency depends on the conveyance efficiency, application efficiency, storage efficiency among others.

Hence the water budget is an equation that is derived from these four factors.

A Step-by-Step List to Implementing Water Budget

(a) Identifying mandates or unmet need – the initial justification

The first thing one has to do is identify the losses the supply agency is incurring. You should ask yourself if the agency is incurring any cost due to the excessive water use of wasteful users. Some states have mandates that ask the local governments to lower their water use by a certain percentage and by a certain date (For example, in California,

SBX 7-7 Water Conservation Act of 2009 requires the state to achieve a 20% reduction in urban per capita water use by 2020). If that is the case in your state or even in any other group or consortium your agency belongs to, that might be a good starting point for you. Even if you agency does not have any mandate that it has to fulfill, it might be a good idea to meet with the operation supervisors, the finance department and the governing board and see if water conservation using water budget, makes sense. You have to remember with a change in water allocation, there will be some resources needed, hours spent and changes made. So, everyone in the agency should be committed to make this successful.

(b) Strategic Plan

Now you have some interest in implementing a water budget from some departments. So, you have to start putting the pieces together. Identify stakeholders (including representatives from the public), objectives, resources, budget, timelines, and the team. Meet with the team and refine the work breakdown structure, the activities, the critical path, the roles and responsibilities, the risks and the communication methods - to the public as well as to the department staff and board members. A lot of attention will be needed to determine the IT systems (GIS, asset management, Billing System, Hydraulic data, among others) and personnel with skills to manage the change. With the preliminary strategic plan in place, the next step is to identify a pilot area to conduct a proof of concept.

(c) Proof of Concept

The proof of concept will take into consideration a small area representative of the whole agency. Surveys to identify indoor and outdoor consumption need to be done. The other parameters (Irrigation efficiency, crop coefficient, among others) should be determined through different formulae as well as from these surveys. Inputs from other departments, customers and staff and previous research are very important to arrive at reasonable values for these parameters. Through iterations these values can be improved. Integration of billing and GIS (among other systems) in the pilot area gives the actual versus the budgeted consumptions.

(d) Presentation and Approval

Presentation of the proof of concept is important at this stage as the board and other stakeholders have to be informed of the standards used. The pilot group of customers has to be in consensus with the team on the outcome of the pilot water budget, along with other departments and staff.

Offline meetings and focus groups are also very helpful to make all stakeholders understand. After iterations, the implementing team should get an approval from all the stakeholders. An approval gives a go ahead to the team for the next step - the whole agency area implementation.

(e) Whole Agency Water Budget Implementation

After the approval, the implementation team should divide the whole project into phases. The team should ensure that everyone has an understanding of the objectives and the methods to be achieved for accomplishment of goals. The phases have to be divided among teams and action has to be taken to track and monitor progress of the projects. During the project, the outdoor consumption factors have to be determined and extrapolated from the pilot results. Integration of billing system and development of a full blown GIS have to happen to enable a solid water budget system. Asset Management System and Hydraulic models have to be established as well - for ensuring water losses are minimal.

The result of all this will be a database with allocations of water (in cubic feet or gallons)to all residents (or customers) of the agency.

(f) Public Outreach

Alongside the implementation, the agency should also train its staff in IT, GIS and customer management. The Public Relations department should send out fliers educating customers about the fairness of water budget system. The communication should also contain plans with funds gained from water budget and how that will help the agency further its water conservation and asset maintenance goals. It is imperative that the public agrees with the plans of the agency, as they are the ones who are going to benefit the most.

(g) Track Using GIS and Budget Tool

Once the implementation is done, the IT, GIS, Water model, Customer Service staff will be engaged in overseeing the consumptions and integration of IT systems, as well as implementing variances requested by customers (variances are requests by customers in the event some information such as number of household members or the area of landscape was wrongly input in the agency water budget system). The staff should also gear itself for other requests and challenges by the customers.

A water budget system developed from GIS and integrated with billing system can work wonders in such cases and be easily implementable across different agencies. There are still a lot of parameters that need to be standardized as more and more agencies are implementing water budgets.

References

Howell, Terry. 2003. "Irrigation Efficiency" in Encyclopaedia of Water Science.

Mayer, Peter; De Oreo, Willian; Chesnutt, Thomas; Summers, Lyle 2008, Water Budget and Rate Structure : Innovative Management Tool", AWWA Journal vol. 100 (5), 117-131.

16

Issues for Future Urban Planning in Rajasthan

Amal Kar

Introduction

Definition of 'urban' areas varies from country to country. Consequently, it is difficult to provide a universally acceptable meaning of the term. The difficulty has been summarised by UN (1998) as follows: "because of national differences in the characteristics that distinguish urban from rural areas, the distinction between urban and rural population is not amenable to a single definition that would be applicable to all countries" (UN, 1998). In many cases the divide between a rural and an urban area is based on the minimum population, but the number varies widely between the countries, e.g., 200 in Iceland, 2000 in some countries of West Asia, 5000 in India, and 10000 in Italy! Despite this challenge, most towns and cities are historical phenomena and have their defined administrative boundaries for services and other purposes, including municipalities. These places straight away qualify as the urban centres.

In India, the word has been defined by the Census of India (2001) as: "(a) All statutory places with a municipality, corporation, cantonment board or notified town area committee, etc.; (b) A place satisfying the following three criteria simultaneously: (i) a minimum population of 5,000; (ii) at least 75 per cent of male working population engaged in non-agricultural pursuits; and (iii) a density of population of at least 400 per sq. km (1,000 per sq. mile)." This definition was adopted for the 2011 Census.

According to Census of India (2001), "an Urban Agglomeration is a continuous urban spread constituting a town and its adjoining urban outgrowths (OGs) or two or more physically contiguous towns together and any adjoining urban outgrowths of such towns. Examples of OGs are railway colonies, university campuses, port areas, etc., that may come up near a city or statutory town outside its statutory limits but within the revenue limits of a village or villages contiguous to the town or city. Each such individual area by itself may not satisfy the minimum population limit to qualify it to be treated as an independent urban unit but may deserve to be clubbed with the town as a continuous urban spread".

The above definition of urban agglomeration has been further qualified for the 2011 Census in the following way: "an Urban Agglomeration must consist of at least a statutory town and its total population (i.e. all the constituents put together) should not be less than 20,000 as per the 2001 Census". For determining the outgrowth of a town, "it has been ensured that it possesses the urban features in terms of infrastructure and amenities such as pucca roads, electricity, taps, drainage system for disposal of waste water, etc., educational institutes, post offices, medical facilities, banks etc. and physically contiguous with the core town of the UA" (Census of India, 2011).

A related term for urban planning is "slum". Census of India (2001) has defined the term as "a compact area of at least 300 population or about 60-70 households of poorly built congested tenements, in unhygienic environment usually with inadequate infrastructure and lacking in proper sanitary and drinking water facilities".

Keeping in view the above definitions, the status of urbanization in Rajasthan, including its growth, future prospects and the need for planning has been discussed.

Urban Growth in Rajasthan

Rajasthan is basically an agriculture-and-service-dominated state, with more than 75% of its population living in rural settings. About 25% of the population lives in urban centres, which is less than the national average of urban population (i.e., ~28%). The 2011 census determined that out of 68.62 million people in the state, 17.08 million lived in the urban areas, which was 3.87 million more than during 2001 census. Yet, some of the urban centres in the state have registered a sharp increase during the last 100 years, especially after the independence and more specifically during the last three decades, while some others have failed to register growth.

At the dawn of the Twentieth Century the erstwhile Rajputana State had 104 towns with a total urban population of 1.55 million (15.06% of the total population). By 1941, the number of towns increased to 124, and the total population to 2.12 million. The first major increase in urban population during this pre-independence era was witnessed during the 1931 census when the total population was enumerated as 1.73 million (a rise of 17.18% over the previous decade), and the towns numbered 119. The urban population, however, declined to 14.71% of the total population (from 15.06% during 1901 census).

The 1951 census recorded 147 towns and cities in the state, having a population of 2.96 million (18.50% of the state's total population). Although new hopes might have been a cause of the sudden increase, this rise also includes an influx of population from Pakistan after the partition. At least four moderate to large droughts in the western part of the state between 1946 and 1951, which forced people to take shelter in the towns for food and livelihood, might also have partly influenced the enumeration. Since then, the urban centres and urban population started increasing steadily. By 1981, the urban population was enumerated as 7.21 million in 201 towns and cities, which formed 21.5% of the state population.

Based on statistical analysis of the town population, town density and spacing in a district, literacy of the people, etc., an analysis was made by Sati (1988) to identify the level of urbanization in the state between 1901 and 1981. According to him, only Jaipur and Kota qualified for 'very high level of urbanization', followed by 10 towns under 'high' category, 8 under 'moderate' category, 4 under 'low' category and 2 under 'very low' category. Mean growth rate of urban population in the arid western part of Rajasthan was 280% during 1941-1981. In the semi-arid eastern Rajasthan the growth rate during the same period was 314%. Surprisingly, during the previous four decades (1901-1941), arid western Rajasthan had a better urban population growth rate (52%), as compared to the eastern Rajasthan (35%), which was attributed by Sati (1988) to the relative political stability in the western part during the period.

Using statistical methods, Sharma and Mishra (2007) found the distribution pattern of urban settlements in the north-western and south-eastern parts of the state 'uniform', but in the south-western part it was random. Most urban settlements in the state were within 80 kms of each other. Jaipur and Kota districts were found to have the highest percentage of urban population (42.5%), followed by Bikaner, Churu, Jodhpur and Ajmer (based on 2001 census). However, when a hierarchical analysis was done using data on occupation structure, amenities and facilities, it was found that the arid western part of the state had only two order-I cities, Jodhpur and Bikaner. Most of the towns in this part were of order-V. In the state as a whole, Jaipur and Kota, Jodhpur, Ajmer and Bikaner qualify for hierarchical order-I urban centres, while Ganganagar, Sikar, Alwar, Bharatpur, Bundi, Bhilwara, Udaipur and Pali were found to be in order-II. There were 70 order-III towns, 105 order-IV towns and 34 order-V towns. Looking from another angle, the studies suggested that the urban facilities in Rajasthan picked up considerably during the period 1981-2001, putting more number of towns and cities in a higher order as compared to 1941-1981 period.

The above ordering system is different from the ordering by Census of India, which usually puts the urban centres in hierarchical orders based on population size only, and calls them 'class'. According to the 2001 Census, out of 222 urban centres in Rajasthan, 20 were in class-I (population >100,000), accounting for 57% of the total urban population. Jaipur, Jodhpur, Bikaner, Kota, Ajmer and Udaipur together had 70% of the total population in class-I category. Among the other towns and cities, 26 were in class-II (50,000-99,999), 90 in class-III (20,000-49,999), 65 in class-IV (10,000-19,999), 17 in class-V (5,000-9,999) and 4 in class-VI (<5000).

The class-I and II urban centres of Rajasthan have significant slum population. As per 2001 census, out of 13.2 million people in the 34 class-I and II urban areas, 1.21 million (9.13% of the total) were slum dwellers (Raj Bala, 2007). The highest slum population was in Jaipur, where 15% of the city's population lived in slums. A total of nine urban centres in the above two categories reported absence of any slum. These were Jhunjhunun, Churu, Bhilwara, Bundi, Dholpur, Ganganagar, Gangapur, Hindaun and Tonk.

A recent study by Joshi (2011) reveals a rapid growth in the population of the class-I cities after the 1971 census, as well as in the class-II and class-III towns, but the class IV, V and VI towns registered a decline. The major cause of this shift is migration from the smaller towns to the larger ones, as well as from the rural areas to the large urban centres.

It has been found that Jaipur recorded the fastest growth, but Udaipur, Bikaner and Jodhpur had slower growth, possibly due to poorer infrastructures that failed to attract investment and opportunities. Surprisingly, among the new class-I cities, Hanumangarh and Ganganagar experienced high growth rates. Located as these cities are in the granary of the state, amidst highly productive agricultural lands, the fast rise of the two cities needs to be noted for its implications. As a recent report of Govt. of Rajasthan, prepared by the Price-Waterhouse-Coopers (PWC) suggests, "this is in line with empirical research that suggests that areas with higher agricultural productivity experience faster urbanization as they can support a larger population in non-farm employment" (Anon., 2003).

By 2011 census, Rajasthan had 29 class-I urban centres, five having municipal corporation. The population of the 29 class-I urban centres is shown in Table 1.

Table 1 : Provisional population of class-I cities/towns in Rajasthan during 2011 census

	City/Town	Civic status	Population
1.	Jaipur	Municipal Corp.	3073350
2.	Jodhpur	Municipal Corp.	1033918
3.	Kota	Municipal Corp.	1001365
4.	Bikaner	Municipal Corp.	647804
5.	Ajmer	Municipal Corp.	542580
6.	Udaipur	Municipal CL	451735
7.	Bhilwara	Municipal CL	360009
8.	Alwar	Municipal CL	315310
9.	Bharatpur	Municipal CL	252109
10.	Ganganagar	Municipal CL	224773
11.	Sikar	Municipal CL	237579
12.	Pali	Municipal CL	229956
13.	Tonk	Municipal CL	165363
14.	Kishangarh	Municipal CL	155019
15.	Beawar	Municipal CL	145809
16.	Hanumangarh	Municipal CL	151104
17.	Dhaulpur	Municipality	126142
18.	Gangapur City	Municipality	119045
19.	Sawai Madhopur	Municipality	120998
20.	Churu	Municipal CL	119846
21.	Jhunjhunun	Municipal CL	118966
22.	Baran	Municipality	118157
23.	Chittaurgarh	Municipality	116409
24.	Nagaur	Municipality	100618
25.	Hindaun	Municipality	105690
26.	Bhiwadi	Municipality	104883
27.	Bundi	Municipality	102823
28.	Sujangarh	Municipality	101528
29.	Banswara	Municipality	100128

Changing Structure of the Urban Settlements

Traditionally, most urban centres in Rajasthan evolved either as small business hubs (e.g., Hanumangarh), or as pilgrimage places (e.g. Pushkar), or as capitals of kingdoms (e.g. Jodhpur, Nagaur). Consequently, the towns were compact and well-knit, with an almost circular to semi-circular shape. A Central Business District (CBD) near a palace or a fort or a temple was the most densely populated part of the town, with heterogeneous land uses (including residences, business houses and markets). Residential areas along the radial roads around the CBDs did not spread much due to the limited scope for livelihood at the CBDs and little variability and choices of opportunities from the towns. The arterial roads to the neighbouring towns and large villages, therefore, did not attract fast expansion of the towns outward, and the peri-urban fringe changed very slowly. Travel time and security demanded that the citizens reside as closely as possible near the CBD, and the void spaces within the town limits was more in demand for new residences than any outward expansion along the arterial roads. Consequently, the towns mostly had a compact and semi-circular shape, which had very high relevance for functional efficiency. Many localities in a town were multi-functional, having both residential and commercial spaces, and different kinds of amenities and utilities. The compactness of the settlements allowed residents to maintain a better social contact in the neighbourhood, share the common resources amicably and judiciously, and maintain the resources as a responsible member of the well-knit society. It also helped them to commute to their places of work in a short period of time, without spending much energy and resources.

The most notable exception was Jaipur, which was the best planned city of the country during the pre-colonial period. Other exceptions were the towns like Nasirabad, Ajmer, Beawar, etc., where a strong presence of the British administration provided for a space for experimenting with segregation of the CBD from the residential areas, which led to sprawl. However, the towns were developed in a planned manner, especially Nasirabad and Ajmer. As railway lines started opening up the state to neighbouring areas from the last quarter of the Nineteenth Century, and livelihood opportunities started increasing, a number of towns sprang up or grew along the railway lines (e.g., Abu Road, Phulera, Pali-Marwar, Bandikui, Mavli and Reengus).

Topographic barriers played a crucial role in the expansion of the towns, especially in the Aravalli hills region where the hill ranges compelled many towns to spread linearly (e.g., Beawar).

The overwhelmingly compact pattern of urban settlement started changing with the opening up of the opportunities for livelihood in the state, as has been enumerated above. This was especially after the independence when new concepts of urban planning also began to be experimented in a democratic set up. Influx of population from the rural areas in search of livelihood and setting up of new industries and other institutions demanded a planned development of towns, especially those attracting large influxes.

Planning and Administration for Structural Changes

The impetus for a major shift from the traditional compact structure of the urban centres was provided to a large measure by the administrative mechanism for city and town governance when the Urban Improvement Trusts and other related government departments took up the challenge of providing amenities and settling the increased population in the fringe areas of the existing towns in 'modern townships'. Urban master plans were developed and rules were framed for appropriate use of the urban land.

The first major change in this direction was to disperse the CBDs in a manner that the dense core of the old towns did not have to suffer from still higher population pressure. This was effected through re-locating the major food grain and vegetable markets as well as the major public bus stations along the fringes of the towns. It created inertia of settlement growth around the new edifices, and the towns spread automatically thereafter. New industries were invariably located in the outer fringes. Unfortunately, lack of proper control on the urban land uses soon led to unbridled growth of many towns to unmanageable sizes. This resulted in problems of waste management, management of crime, etc., and of providing of proper infrastructures, public goods and services. The institutional framework that is supposed to deliver on these issues has lagged behind in addressing them, especially because of the confusion created by a single issue being addressed by multiple bodies. Presently, the institutional framework for urban services in the state includes:

(a) Town and Country Planning Organization that prepares development master plan

(b) Municipal Corporations, which address the problems of sewerage O&M, city roads and solid waste management

(c) Other municipal bodies like Municipal Councils/Municipalities, which look into land development, housing and duties of Municipal Corporations

(d) Urban Development Authorities, which are mandated for land development

(e) Urban Improvement Trusts that are also mandated for land development

(f) Public Health & Engineering Departments, which look into water supply and sewerage disposal, and

(g) Rajasthan Housing Board that is tasked with water supply, sewerage, solid waste management and housing.

Thus, the urban services are being addressed by different institutions, and in some cases a single issue is under the domain of more than one institution, which creates confusion in addressing the issues and leads to compounding of the problems with consequent displeasure of the residents. Therefore, the vision for a brighter urban scenario should address the issues related to urban governance and services.

The problems of urban service delivery need to be solved adequately and early to make glitch-free new urban development possible. As in every service sector, investment and capital are the key requirements to maintain the services in top order. An assessment by PWC (Anon., 2003) suggesed that a minimum of Rs. 9851 Crore (1 Crore is 10 million) might be required as investment for meeting the future demands (in 2025) of water supply, sewerage, solid waste management, drainage, street lighting and city roads in the top 20 cities of the state (the maximum being Rs.16509 Crore).

PWC assessed that urbanization in Rajasthan might pick up soon and the mean rate of urbanization could be 40%, which is comparable to the current rates in Maharashtra (42%) and Gujarat (38%). This assumed faster growth rate could be mainly due to the impetus given by the expected higher investments in the state by the mining and industries (including the petroleum exploration) sectors, as well as by the service sector, and the estimated high agricultural outputs. Obviously such a growth rate would mean spilling out of the population into newer adjoining areas, especially from the mega-cities.

Ministry of Urban Development, Govt. of India, estimated that the cost of setting up a new township for 5 lakh population is about Rs. 20375 Crore, including the cost of land, transport linkage, water supply, sewerage and drainage, and power (Anon., 2003). The cost estimate for at least one major township development to relieve the pressure on Jaipur may have to be included in the total cost liability of urban services by the designated local urban bodies. A growing urban centre may be able to finance the services in large measure, especially through the high-value industries and services that come up to aid the urban growth, provided the management mechanism is efficient.

Compared to the growth of the class-I cities the other towns in the state are either stagnating or are showing signs of negative growth. In fact, many of the large cities are developing partly at the cost of the neighbouring small towns due to the dismal livelihood opportunities there. Livelihood opportunities are the prime drivers for the health of any urban centre.

Livelihood Opportunities as Determinant of Urbanization

Migration in search of livelihood is a major trait of human settlements in Rajasthan, especially due to the high vulnerability of agriculture in the state imposed by erratic behaviour of monsoon in an arid and semi-arid landscape. Another major factor of migration is the increasing livelihood opportunities provided by the mining and industries sectors that have flourished and have started proliferating since the 1970s, especially as a result of infrastructural growth and capital investments in the state. The major influxes have taken place in the class-I cities like Jaipur, Kota, Jodhpur and Ajmer, but several other smaller towns have also become hubs of urban growth. Opening up of mining areas and of small and medium industries that are based mostly on mining or processing and assembly of light industrial products, have become the key attractants. Broadly, the towns and cities to the east of the Aravallis in the semi-arid tract have become the hubs for industrial growth (e.g., Bhilwara, Dausa), while those to the west of the Aravallis in the arid tract have mostly experienced mining-related growth (e.g., Jalor, Barmer). Some new small centres have also come up in the wake of mining activities (e.g., Gotan, Balesar). Only few towns have registered growth due to agricultural marketing or agro-industries, especially in the northern irrigated alluvial plains (e.g., Ganganagar, Sangariya, Hanumangarh, Suratgarh). Barmer town may soon experience a phenomenal growth due to the near-by oilfields and coal mines. Sanchor town may also witness fast growth due to the commissioning of the Narmada Canal.

With the commissioning of the Delhi-Mumbai Industrial Corridor (DMIC), it is expected that urbanization, as driven by industries and related sectors, will pick up fast in the state. About 58% of the state falls within the DMIC. It is hoped that class-IV to class-VI towns that have registered negative growth rate of 25-30%, will eventually benefit from DMIC. If these small towns with less than <20,000 population each are able to attract industrial and services investments, they will possibly restrict the migration to the large cities (>50,000 population), and thus help in facilitating a well-spread and better-served urban environment.

Restricting the migration from the rural areas and from the small towns to the large cities needs a major policy and management intervention for a healthy urban growth in the state. A perusal of census data shows that between 1971 and 1981 about 16.4% of the total migrants in the state lived in the urban areas. The figure rose to 22.4% during 1981-1990, and to 25.4% during 1991-2001.

It was gradually realized from the 1990s that the creation of livelihood opportunities in the cities and towns, especially in the class I, II and III categories, was faster than expected, which led to faster spread of the townships, and development of slums in the larger cities as demand for different kinds of services started increasing. Amenities and services became poorer, commuting to places of work and to market places more difficult, time-consuming, energy-intensive and costly, and security very uncertain. Urban sprawl became somewhat chaotic and unmanageable, and therefore, needed a re-look and re-designing through a better planning process.

This brings us to the question of whether the present sprawling township concept is a viable proposition, and whether the traditional compact type of settlement with multi-functional arrangement in-built within it can be re-worked with some modern planning and designing for a more energy-efficient and less clumsy living experience. As we shall see in a future section, the majority of global experiences now favour the compact type of settlement with multi-functional localities, in place of sprawling towns with segregated land uses.

Models of Urban Forms

Conceptual models for spatial planning of urban forms and designs are undergoing a slow transformation the world over from their early- and mid-Twentieth Century vestiges, as the needs of the inhabitants of city-region are changing, the cost of energy, land and basic amenities are escalating, and awareness of environmental goods and services are increasing. The traditional British plan is of a sprawling town with dispersed habitations, a defined market place and a central business district (CBD), interspersed with wide patches of green, and rimmed by a green hinterland of agricultural fields and small villages. This plan is replicated in numerous small colonial towns of British India, but such a townscape is difficult to reproduce in the present era. The typical American plan of the early and the middle Twentieth Century believed more in segregation of the habitation space from the CBD, as well as from the market places (e.g., Chandigarh). Building of such a town not only encourages urban sprawl at the cost of green hinterland, but more importantly it also encourages private transport vehicles and thus increases the time and cost of travel by the inhabitants. With the fast escalating fuel prices, this segregated urban design becomes more unsustainable and less popular with the inhabitants. In fact, sustainable transport is one of the key factors that govern the plan of a successful urban form.

In today's world, urban form often antagonizes the natural environment, especially through usurpation of the green cover and the productive agricultural land, and through air, water and soil pollution due to the growth of industries, vehicular traffic and unmanaged waste disposal. As the interference of the natural environment increases, the built-environment itself gradually becomes a victim of its own waste generation, especially in the form of smog, sludge, overflowing sewage, as well as depleted water level and its quality, urban heat belt, etc.

Despite all the problems posed by the growth of an urban centre, one must admit that growth is fundamental to the well-being of a town or a city. As a town becomes a centre of attraction for certain kinds of economic activities it pulls not only the humans from its hinterland, but also the other physical resources and capitals that help to sustain its growth and creation and maintenance of the amenities. It is quite natural that with time a vibrant city will encroach upon its hinterland and generate large quantity of wastes that must find an outlet for disposal. The issues of management of the resources, infrastructures and the wastes of a grown up city, as well as the planning of future expansion of the city, keeping the inflow of the resources and the capitals, then assume more importance and a challenging task than ever before. Although this challenge leads us to the emerging concept of 'sustainable' urban form, one wonders if any of the modern-era urban centers have ever become 'unsustainable', and by implication set for a collapsing or collapsed urban space with no sign of growth. Stagnation of growth signifies a slow death of an urban centre.

Designs of Sustainable Urban Forms

Growth may be in-built in the functionality of an urban centre, although the pace of growth might vary with the attention it receives from the investors and the state administration. The concept of 'sustainability' could, in a sense be utopian, yet a workable concept for an ideal city with desirable traits. What becomes more realistic for planning is a dynamic scheme for management with the available resource that is commensurate with the city-growth, as well as a visionary plan to manage the future growth through proper allocation of resources, spatial planning, infrastructure development and creation of an enabling environment for capital inflow.

The human and environmental challenges posed by the seemingly unplanned growth of the towns and cities have compelled the urban planners to think of new model frameworks of design and spatial form of the urban centers. The consensus opinion for a robust and so-called 'sustainable' urban plan, as gleaned from the literature, may include the following (Jabareen, 2006):

(a) Plan and design for **compactness** of the urban form that helps to reach places of work and amenities easy and quick, but at the same time is large enough to hold a variety of work opportunities, services and ingredients of a rich socio-cultural urban life, while maintaining the privacy of a hassle-free residence;

(b) Provision of cost-effective, fast and rapid public **transportation services** that are accessible to all, keeping emissions and waste levels within permissible limits, so that the need for daily commuting in individual private vehicles is drastically reduced;

(c) Maintenance of an appropriate **urban density** that helps in efficient utilization of the urban space for different services and residence, and at the same time does not give the feel of crowding and cramming;

(d) Provisioning of a **mixed land use** of residential, commercial, business and services, and institutional within a city zone, which reduces the commuting time of a resident to work and shops;

(e) Allowing **diversity of activities** that includes the mixed land uses mentioned above, as well as a mix of residential plots that include different income groups, cultures, housing types, etc.;

Since energy consumption and environmental concerns are now major issues, the robust urban plan also requires considering the following two:

(f) Introduction of passive solar designs and improving the **urban micro-climate**; and

(g) **Greening** of the city.

The purposes of the above seven design concepts are not only to stride for a energy-efficient, cool and less-polluted city dwelling, but also to make the living experience functionally more integrated that reduces the commuting time, provides amenities and services within reachable distance and in adequate measures, and overall helps to provide an enjoyable and hassle-free living condition.

To achieve the desirable traits urban planners look for some workable models (Jabareen, 2006), the more common of which are the concepts of:

(a) **Neo-traditional development** that puts emphasis on compactness, diversity of housing, mixed land uses with wisely crafted green spaces within the city limits, and short and fast mass transit system;

(b) **Urban containment** that believes in compactness of the existing city, and finds solutions towards checking the unrestricted urban sprawl, although it is difficult to implement;

(c) **Compact city** that emphasizes on maintaining a high density within the city limits with efficient utilization of the inner-city space for mixed land uses; and

(d) **Eco-city**, which seeks to provide greenness to the urban areas, as also alternate energy sources and environmental services. Contrary to the concept of compact city, the eco-city concept allows dispersal of the urban settlements to accommodate green spaces, and thus encourages urban sprawl.

All the above models have their own merits and limitations. No singular model is ideally suited to address all the seven traits that have been described above. Yet, looking from the perspectives of Rajasthan's present socio-economic development and future aspirations it appears that the neo-traditional development model and the compact city model are better suited for adoption.

The two models on neo-traditional development and compact city, advocating "New Urbanism" and "Smart Growth" concepts, are antagonistic to those that so far have suggested urban segregation and sprawl, exposing the city-region to significant environmental damage, especially through uncontrolled expansion into the green hinterlands and daily long-distance commuting. The urban sprawl also segregates people and decreases the communications between different socio-cultural groups.

Suggestions from UN-Habitat

In detail, the future urban growth models in the state may like to incorporate what the UN-Habitat Draft Working Paper on urban patterns for sustainable development suggests for adoption (UN, 2011):

(a) **Embrace land mosaic patterns that provide for large green patches and more sustainable urban development**. Urban models having "*compact concentric zones*" and "*satellite cities*" are considered to preserve a greater number of open spaces that provide a natural ambience, and allow for growth within, while restricting the sprawl. In the process, this model of urban structure satisfies both human and ecological systems.

(b) **Promote compact cities and planned extension of urban areas**. A denser growth within the city limit allows people to take advantage of agglomeration economics, while protecting the natural environment. This also reduces the fraction of space for commuters. In a developing country like ours, the compact city helps people to spend less on commuting to work and market facilities, which is a major gain for the city dwellers.

(c) **Balance strategic facilities with diversified local economic opportunities**. As the city grows in a healthy way, it attracts many different functions and services that make the city more preferable to the investors and residents alike. The diversified economies and opportunities within the city limits make the residents feel at home, as it reduces the hazards of uncertainty and search for opportunities elsewhere.

(d) **Expand network infrastructure while getting the most out of existing networks**. In this era of fast communication for timely delivery of goods and services, efficient e-connectivity and public transportation are now basic to economic well-being of any city. Better connectivity with other major cities in the country, as well as within the city limits, are crucial factors in determining a city's worth. Frequent rail and bus services for commuting and business not only help in reducing the atmospheric pollution load, but also in increasing the actual work hours and better output through a less fatigued mind and body.

(e) **Construct greener built environments that use water and energy efficiently**. In view of the global warming scenarios, controlling of the Green House Gas (GHG) emission is a major requirement at all levels. Towns and cities usually emit more GHGs due to the industries, transport and buildings, Apart from having a long-term effect on the atmosphere this also artificially heats up the city environment. Therefore, creation of green spaces within the cities, and construction of green buildings make good senses.

(f) **Protect valuable ecosystems services and biodiversity hotspots while increasing resilience to some natural disasters.** A related environmental issue is protection to the valuable ecosystems in the surroundings and the natural biodiversity. Towns and cities usually transgress upon the natural space and create an artificial environment of its own. This often becomes counter-productive as the services provided by a natural system gradually become less and less available. In case of Rajasthan, the two major valuable natural resources that are crucial to a healthy city environment are (1) the lakes and ponds (with their unhindered catchment areas), and (2) a green belt of trees and shrubs around the cities. Both are necessary for survival and a cooler environment, although these also have aesthetic components and attract faunal diversity (including birds) that encourages eco-tourism.

(g) **Promote clusters of green industries and jobs.** Most new cities need new industries for their proper growth. Unfortunately, the industries are also the major pollutants through smoke, dust, wastewater, etc. Consequently, the surroundings of the new cities are becoming more polluted and unworthy of their ecosystem services. The large farming communities in the peripheries of the cities become a major victim of the pollution when their soils and groundwater begin to get polluted. This results in unnecessary conflicts with the urban dwellers and the urban authorities. As the hinterlands become affected, the impact also starts affecting the city, especially as contamination of water, milk, vegetables, etc., lead to numerous health hazards. It is, therefore, necessary to promote green concepts in the establishment of the industries, and enforce the laws so that there is adequate restriction to and safe guard from environmental pollution. Part of the solution will lie in new research, but a major part is implementation of the research results already available as technologies, and enforcement of the numerous legislations that are already in place. Proper dialogue between the research establishments developing the technologies and monitoring the environment, the concerned industries and the local urban authorities should also be held on a continual basis to improve the situation.

Need for Policy and Investment Support

Despite the best efforts of the urban planners to regulate the growth of urban areas, the cities and towns grow in pulses, largely in response to industrial and service-sector capital inflow, and job opportunities, as well as to infrastructural build-up. This is where the government schemes and policies can help to regulate the growth and management of the towns and cities of Rajasthan in 'sustainable' manner, especially based on the best utilization of the available local resources.

Majority of the towns in Rajasthan are of small and medium sizes, and are located in an overwhelming rural situation. Most of these are endowed with small industries/trade facilities that thrive on local resources. Therefore, it may be sensible to let these small towns develop as compact cities that can thrive on the local resources, be that as hubs of agro-processing, agri-engineering, tourist/pilgrimage destination, or as new centres of higher education, IT-based and other industries being opened by the State (especially along the National Highways and the Delhi-Mumbai Industrial Corridor, DMIC). In the cases where the growth regulator for the towns would be the production and marketing of agricultural produces, an enabling policy intervention will make the hinterland of the concerned town develop as sustained agricultural producer. Thus, a healthy symbiotic relationship will develop between the town and its hinterland, making it more vibrant and livable. As Planning Commission has noted in its Eleventh Plan document, one of the major challenges would be to bring about planned urbanization for the country and creation of new growth centres around existing small and medium towns (Anon., 2008).

For the mega-cities, a string of satellite towns develop and grow simultaneously with the main city, gradually making a near-continuous strip of urban households and industries along main highways. This continuous urban space is called *conurbation*. Jaipur is fast developing into a conurbation, and is expected to engulf soon the rural spaces from Shahpura to Dudu as the city turns into a megalopolis. Managing the indiscriminate growth and land uses of such a huge cluster of linked towns and cities needs special attention.

For other large cities (e.g., Jodhpur, Ajmer, Kota), the indiscriminate growth at the cost of prime agricultural lands needs to be regulated properly, although the task is becoming increasingly difficult. Although industrial development helps in the growth of an urban centre through capital inflow and livelihood generation, improper planning for the industrial location, transportation network for routing of its supplies, waste management, etc., can become a source of huge pollution of water, soil, dust and smog, as well as of other forms discomfort for the urban dwellers (e.g. urban heat belt development, noise pollution, slum development, huge dumps of slow-to-non-degradable wastes and transport congestion). We discuss here briefly the cases of pollution due to textile printing and urban heat belt development in the arid western part of Rajasthan where the problems are assuming unmanageable proportion due to lack of proper planning and a sense of apathy.

Pollution due to textile printing : Cultivable good soils along ephemeral streams and potable groundwater along those streams are considered to be the scarce agricultural resources and a major source of drinking water in arid Rajasthan. Unfortunately, these two resources have gradually become highly polluted in the vicinity of some towns and cities like Jodhpur, Pali, Balotra and their surroundings, especially due to the release of untreated effluents of textile dying and printing industry into the ephemeral channels. The textile printing factories flourished in these towns because of the good access to potable water. According to an assessment by CAZRI in the late 1990s, effluent discharge into the Jojri River near Jodhpur had affected at least 5500 ha of land, while that along the Luni River at Balotra contaminated more than 7000 ha, and along the Bandi River downstream of Pali more than 22,000 ha. The potable shallow groundwater wells along these ephemeral streams were traditionally used for drinking and irrigation purposes. As the textile printing factories mushroomed along their banks and huge amount of untreated toxic waste water began to be discharged from the factories into their beds, the potable groundwater became highly contaminated and both the wells and the croplands had to be abandoned. The ephemeral streams began to be treated by the factories as wastewater drains! On an average the effluent waters have pH >9, RSC of ~35 me L^{-1} and SAR exceeding 150.

When the problem at Pali became acute and the farmers along the streams found their soil and water spoilt beyond any chance of recovery, their protest compelled the state to start common effluent treatment plants (CETP) at Pali and to construct a reservoir on the Bandi River at Nehda, about 50 km downstream of Pali, to conserve fresh water for drinking and minor irrigation.

Unfortunately, the CETPs have been found to be inadequate to take the calculated load from the clusters of textile dying and printing mills for treatment. The CETPs are also reported to be operating at less than half of their designed capacity, as many industries simply bypass the plants to discharge the effluents directly into the rivers. It was found out by some independent studies by NGOs that all the harmful elements in the effluent water could not be sensed by the CETPs. Recent studies by the Centre for Science and Education, New Delhi revealed that despite the establishment of CETP at Pali and Balotra, there were problems of lower capacity of the plants than the discharge, bypassing of the CETP to channelize untreated effluents to the river, inability of the CETP to treat all the harmful elements, etc. The concentrations of arsenic, copper, lead and zinc in the groundwater in village wells downstream of Pali are alarmingly high and pose health hazard (Kar *et al.*, 2009). With time and with continued discharge from the factories into the Bandi, the reservoir at Nehda became a collector of polluted water and its transmission further downstream. The problem has now become a rallying point for different water users. Several litigation cases have been filed in different courts during the last one decade and some judgments have been received by the state for corrective measures, but a solution is still to be found.

Problem has also become acute in the vicinity of Jodhpur and downstream of Balotra. At Jodhpur, the textile printing factories were initially located along the small Jojri River flowing NE to SW along the eastern fringe of the city. The major concentration of the factories was between Banar and Kuri where the wells along the channel banks were used for the printing process, and the channel bed was used as a waste water drain of the factories. Gradually the factories spread southwestward, beyond Kuri and towards Mogra and Dhawa, and more untreated wastewater began to be discharged into the channel. As a result the groundwater along the Jojri became highly polluted. Irrigation with this polluted water, especially for vegetables, soon spoiled the soils in the crop fields. A recently constructed drain to carry the effluents out of the city limits southwestward into a pond became a flash point for confrontation with the surrounding villagers as the discharge has increased manifold, breaching the pond itself and flooding parts of Dhawa village, including a National Highway to Barmer town. The otherwise dry channel now looks like a perennial stream, although containing dark polluted water that can be traced beyond Kalyanpur village, more than 50 kms to the southwest of Jodhpur.

Near Balotra the effluent water from more than 700 textile printing factories at Balotra, Jasol, Bithuja, etc., now flows more than 50 kms along the Luni River to Sindri village and beyond, making the Luni look like a perennial stream, though with highly contaminated water. This has prompted the judiciary to ask the state very recently to ensure immediate stoppage of effluent flow directly into the Luni River, even if the water is treated.

Urban heat belt formation : Another problem faced in the major desert towns and cities is the growing oppressive heat, even at night. Earlier, the desert towns, including Jodhpur, used to be cooler at night due to adequate space for air circulation and faster out-radiation. As the towns started growing without much of planning and surely without any thought for environmental development, the congestion led to shutting of the natural air passage, and release of more heat from the larger concrete entities. The heat now gets trapped within the narrow spaces between the crowded buildings. This has started to develop stifling heat in not only the core of the towns, but also along its fringes, creating large heat islands, when compared with the cooler non-urban areas. In the case of Jodhpur, a cool breeze from south and southwest during the nights of March to September and a rapid out-radiation used to be the characteristic features of the city, at least since the mid-1950s. This was especially because of a large farm of the Central Arid Zone Research Institute (CAZRI) in the southwest of the city, which had numerous planted trees and shrubs. The dusty southwest wind, blowing in from across wide open crop fields, used to get filtered in the large CAZRI farm area and then get slowly released as a cool breeze north and northeastward through Shastri Nagar, Sardarpura, Jalori Gate and Sojati Gate area. Because of this, the city residents used to enjoy a much cooler and comfortable night, despite the city growing to the north of CAZRI farm. The efficiency of this natural cooling system has gone down significantly in recent decades due to faulty city planning, especially as new buildings and industries have been allowed to come up just to the south of CAZRI farm (and partly on the erstwhile CAZRI farmland). Quite a large area has been cleared of the tree and shrub cover after the land has been acquired for other land uses, and has been replaced with concrete jungle. Gradually significant part of the tree-shrub cover in the farm was destroyed to make space for other edifices. Apart from the density of tree and shrub cover, the long fetch of the air stream through open fields before it entered the CAZRI farm, and the typical elongated shape of the farm in the direction of dominant wind, were the other critical factors that helped to process the warm air in the farm and release it for circulation as a cool breeze into the city. As parts of CAZRI farm were gradually taken away for other purposes (although the Master Plan for the city marked this whole area for developing a green belt), and new buildings and factories came up to the south and southwest of CAZRI farm, right on the flow path of the air stream, the fetch became not only shortened but also severely blocked and fragmented. The heat generated by the buildings and factories clustering in an almost unbroken stretch from Basni to Pal to the south of CAZRI Farm, save and except a small gap of farmland, not only make the air much warmer around CAZRI Farm, but also interrupts the airflow. The small opening left between the Basni Industrial Area and the Purbi Pal Road scheme is now the only passage left to allow the natural airflow into the farm.

We suspect that this small passage will soon be shut by the unplanned city growth (or by a differently 'planned' growth), which will cut off the unobstructed wind flow into the farm, and severely restrict the cooling efficiency. In other words, a natural cooling system for this desert city, which worked so efficiently to the satisfaction of the city dwellers, will end due to the unplanned/wrongly planned growth of the city. Although it is argued that small gardens within a city help to keep the area cool, the heat load of a desert city is much larger to be tackled by those small green patches within. To reduce the large heat load of a desert town one needs a longer fetch from the right direction and an elongated but sufficiently wide green space with adequate tree-shrub cover to process the incoming dusty warm air and funnel it into the city as a cool breeze. CAZRI Farm is still very appropriately located to cool a part of the city. To increase the efficiency of this natural system, city planners should take urgent steps to restore the long fetch and allow the air passage into CAZRI through as much wide a green corridor as possible, especially as global warming and urban growth are increasing the heat load too much in the city.

One of the ways to tackle the above two problems of pollution and heat control could be to shift the new industries and other growth regulators to some of the satellite towns that are well connected with the main city and are easily approachable. At the same time issues of infrastructure management, environment development, airflow maintenance and green cover restoration need to be given adequate attention, including financial and legislative support. The satellite towns would gradually become the new hubs of activities, and will relieve the pressure from the main city.

Encouraging Rurbanization : An Alternative to Urban Sprawl

Another concept is of developing amenities and infrastructures of towns and cities in the prosperous rural areas, so that the villagers do not feel an urge to migrate to the cities in search of a seemingly different lifestyle. Called "**rurbanization**" this is a slow and almost imperceptible process of developing selected urban lifestyle goods and services in the more affluent parts of the rural landscape. One notable example is the irrigated plains of Punjab where as per the demands of the prosperous villagers, several urban lifestyle goods and services have trickled in, along with road networks for fast mobility. Another example is Kerala where the rurbanization is taking shape due to the regular cash flow from the financially secured Non-Residential Indians (NRIs) in the Middle East. Considering the above two successful examples of rural areas getting urban luxuries and oddities and thereby foregoing the option for migrating to the cities, it may not be difficult to conceive that parts of the Indira Gandhi Nahar Paryojana (IGNP) area will gradually transform into a rurban space with affordable city-amenities.

Realizing the importance of providing urban amenities in rural areas in de-congestion of the urban space and in bridging the rural-urban divide, Government of India has recently embarked upon a scheme called "PURA" (Provision of Urban Amenities in Rural Areas), through the Ministry of Rural Development. The primary objective of PURA scheme is provisioning of livelihood opportunities and urban amenities in the rural areas so that the rural-urban divide is adequately bridged. It envisages a holistic and accelerated development of areas around a potential growth centre in a Gram Panchayat (or a group of Gram Panchayats) in a public-private partnership framework. Details of the scheme are available at http://pura.net.in/.

PURA was originally proposed by Dr. APJ Abdul Kalam, former President of India. Simply stated, it proposes that urban infrastructure and services be provided in rural hubs to create economic opportunities away from the urban areas. Based on an elaboration of the concept several pilot projects were undertaken in the country since 2004. Another alternate model, RISC (Rural Infrastructure and Services Commons), proposed by Atanu Dey, speaks of building rural centres around some existing infrastructures like the railway stations. Both the models, however, aim at increasing the livelihood opportunities in rural areas for checking rural to urban migration (Dhar and Sen, 2006).

Epilogue

As a recent report from McKinsey Global Institute finds, India will witness over the next 20 years an urban transformation the scale and speed of which has not happened anywhere in the world except in China. Urban India will create 70% of all new jobs in the country, which will be twice as productive as equivalent jobs in the rural sector (Sankhe *et al.*, 2010). By 2030, about 33% of the total population in Rajasthan is expected to reside in urban centres, as compared to ~24% at present. It is time that we take proactive steps to facilitate a planned growth of the towns and cities so that the future of our urban centers is more secured and bright.

Acknowledgements: I thank Director, CAZRI, for asking me to represent the institute as a Member of the State Commission on Urbanization, Govt. of Rajasthan, for preparation of a 'Horizon Year 2030' document, especially for Future Settlement Pattern, Urban Form and Urbanism. I also thank profusely Shri P.K. Pandey, Secretary, State Commission on Urbanization, who encouraged me to reflect back on my several decades-old learning of town and country planning, and allowed me the indulgence of deliberating on current and upcoming issues for urban planning in Rajasthan.

References

Anonymous. 2003. Rajasthan Infrastructure Agenda: Vision 2025. Key Deliverable 2 (Phase 1), Volume-II: Urban Infrastructure. Govt. of Rajasthan and PriceWaterhouse Coopers.

Anonymous. 2008. Urban infrastructure, housing, basic services and poverty alleviation. In, XIth Five Year Plan (2007-2012), Volume-III: Agriculture, Rural Development, Industry, Services and Physical Infrastructure. Planning Commission, Govt. of India, pp. 394-422.

Anonymous. 2011. Census Data – Provisional results. http://censusindia.gov.in/2011-prov-results/paper2/data_files/India2/1.%20Data%20Highlight.pdf

Anonymous. 2011. Census Data 2001- Metadata. http://censusindia.gov.in/Metadata/Metada.htm#2b

Anonymous. 2011. Urban patterns for sustainable development: Towards a green economy. UN-Habitat Draft Working Paper, January 2011. 12p. www.unhabitat.org/downloads/docs/9539_39812_3077_alt.pdf

Dhar, V.K. and Sen, R. 2006. Planning for human settlements in India: Spatial perspective. Working Paper, National Institute of Urban Affairs. New Delhi, 31p.

Jabareen, Y.R. 2006. Sustainable urban forms: Their typologies, models and concepts. Journal of Planning Education and Research, 26:38-52.

Joshi, K.N. 2011. State level background paper on Rajasthan. Institute of Development Studies, Jaipur.

Kar, A., Moharana, P.C., Raina, P., Kumar, M., Soni, M.L., Santra, P., Ajai, Arya, A.S. and Dhinwa, P.S. 2009. Desertification and its control measures. In, Trends in Arid Zone Research in India (eds., Amal Kar, B.K. Garg, S. Kathju and M.P. Singh). Central Arid Zone Research Institute, Jodhpur, pp. 1-47.

Rajbala. 2007. Urbanization in Rajasthan at the dawn of the 21st Century. In, City, Society and Planning, volume-1 (eds., B. Thakur, G. Pomeroy, C. Cusack and S.K. Thakur). Concept Publishing Co., Delhi, pp. 213-222.

Salvatore, M., Pozzi, F., Ataman, E., Huddleston, B. and Bloise, M. 2005. Mapping Global Urban and Rural Population Distributions. Environment and Natural Resources Series, No. 24. FAO. 88 p.

Sankhe, S., Vittal, I., Dobbs, R., Mojan, A., Gulati, A., Ablett, J., Gupta, S., Kim, A., Paul, S., Sanghvi, A. and Sethy, G. 2010. India's Urban Awakening: Building Inclusive Cities, Sustaining Economic Growth. McKinsey Global Institute, 232p.

Sati, D.N. 1988. Process and levels of urbanization in Rajasthan. In, Geographical Facets of Rajasthan (eds., H.S. Sharma and M.L. Sharma), Kuldeep Publications, Ajmer, pp.219-231.

Sharma, H.S. and Mishra, R.N. 2007. Hierarchy and spatial distribution pattern of urban centres in Rajasthan. In, City, Society and Planning, volume-1 (eds., B. Thakur, G. Pomeroy, C. Cusack and S.K. Thakur). Concept Publishing Co., Delhi, pp.247-258.

United Nations. 1998. Principles and Recommendations for Population and Housing Censuses. Revision 1. Series M, No. 67, Rev. 1 (United Nations publication, Sales No. E.98.XVII.8).

17

UAE Space Reconnaissance Center as a Valuable Imagery Resource for Monitoring the Environment : An Investigation on Sand Movement

Major Ali M. Alshehhi and Ghassan A. Abdelhamid

Abstract

The UAE Space Reconnaissance Center (SRC) is accounted as a major national and regional source of up-to-date geospatial data, with direct reception of IKONOS, IRS and KOMPSAT satellite imagery as well as access to some other satellite and aerial imagery resources; an activity which covers the whole Middle East region. The main concern in monitoring the environment of UAE and surrounding countries is the desertification process, the monitoring of which requires careful watching of the sand movement and some other relevant features.

This paper throws light on the role of various types of satellite images in monitoring desertification, by utilization of IKONOS, IRS Pan and Landsat TM images covering part of Liwa and Alain areas, representing the two main dune regimes; Liwa and Manadir, which constitutes the northeastern edge of the Empty Quarter sand sea.

The availability of high resolution images makes it possible to detect the yearly or even seasonal sand movements, being controlled by the prevailing wind conditions and distance from the main sand source. The results clarify that movements of up to 25m was encountered at some small sand dunes and about 10m for larger sand dunes, towards the SSE and SSW in Liwa area. In Rimah area a movement of up to 30m was noticed at a small sand dune towards the ENE and about 12m shift in the hinge of a larger sand dune towards the NW. Temporary or permanent? movement of thin sand sheets can reach 100's of meters, was easily detected on Landsat images; ~ 25km NE of Alain city. The 5m panchromatic images, on the other hand, helped in tracing sand movements measured in few 10's of meters, and interpretability of such images was enhanced by merging them with Landsat imagery.

In addition to providing accurate images and image maps to the concern units in the UAE Armed Forces, the SRC is enhancing mutual cooperation with the civilian sector, where satellite images can benefit in applications such as monitoring of environment and agriculture activity, exploration of natural resources, urban planning and mapping.

Introduction

UAE nature

The largest active sand sea on Earth (560,000 km^2) is the Rub'al Khali or Empty Quarter, located in the southern Arabian Peninsula. About 80 percent of this area is in the Kingdom of Saudi Arabia, and covers about a quarter of the Kingdom's land surface. Its eastern edges extend into the United Arab Emirates, and its southern margin is in the Sultanate of Oman. Parts of its southern and southwestern limits are in the Republic of Yemen.

The UAE lies at the northeastern edge of this giant sand sea, and bounded to the north by the southern coast of the Arabian Gulf, where the UAE stretches for more than 650 km along the southern shore. Arabian peninsula generally characterized by a hot dry climate caused by air masses moving from the eastern Mediterranean towards the Arabian Gulf. The rainfall in Arabia peninsula ranges from 300 mm in the eastern Red Sea Mountains to 10 mm in the Rub'al Khali in the southeast.

Most of the surface of the present day UAE is a sand desert, stretching from the Arabian Gulf coast south to the unbroken and uninhabited sands of the Empty Quarter, and east to the gravel plains bordering the Hajar Mountains. The desert is a geologically recent feature, the result of prolonged sub-aerial erosion and deposition in an arid environment. The sands overlie the thick, oil-rich sedimentary strata of the Arabian Platform which constitutes the bedrock of most of the UAE topography.

Dune patterns and effects of climatic change

Sand dune formation is controlled by a combination of wind strength and direction, and sediment supply. However, the formation of dune patterns is complex and remains poorly understood. Within a given area the dune pattern may be quite regular, but also very intricate. Physical features are typically created on several different scales: giant sand ridges on a scale of hundreds of meters to a few kilometers, sand dunes measured in meters to tens of meters, and ripples on a scale of centimeters to a meter or more. This hierarchy can be readily observed in the deserts of the UAE.

Since dune patterns vary with wind direction, seasonal or occasional variations in wind direction introduce new elements into the overall pattern. These elements may reinforce or cancel each other, in the same manner as ocean waves. In addition, because sand dunes cannot move or change as quickly as ocean waves, past history may play a significant part in what we see today.

Despite relatively consistent prevailing wind directions in the present day UAE, dune patterns and alignment vary considerably from area to area (UAE Internet). The largest dune features of the present day UAE, including the major transverse dunes of Liwa and Manadir, the smaller eroded dune ridges of the Northern Emirates, and the longitudinal dunes of the south-west, are believed to date from the most recent glacial period, more than 10,000 years ago. A glacial origin for these major features is consistent with the fact that they do not seem to be aligned with today's prevailing winds and according to Andrew et., al. 2000, the Awafi dune accumulation rate is ~ 3.3m per 1000 years.

The present day wind regime appears to be transporting material from the coast inland and reworking the surface of the major earlier structures without, so far, removing or reorienting them. For example, one may observe between Abu Dhabi and the Liwa oasis that extensive tongues of pale sand resembling a choppy sea of dunes (aligned NE-SW) are filling in broad troughs between higher, flatter ridges of red sand (aligned WNW-ESE). The latter are known as *suruq* , or easy travel zones, and are interpreted as the eroded cores of older, larger ridges.

Further inland, however, the major transverse dune ridges of Liwa and Manadir are neither in motion nor are they being eroded at the present time, although the smaller dunes on their surfaces conform to present day winds. In addition to changes in wind regime, the UAE deserts have experienced changes in rainfall at various times in the past. This is indicated by the widespread occurrence of outcrops of lightly cemented, cross-bedded dune sands.

Dune accumulations in the Arabian Peninsula occur within or near topographic depressions (El-Baz, 1979). This must be explained in any theory regarding the origin of the sand and the evolution of the dune forms in space and time. Two other important observations must also be taken into account. The first is that wind in this desert moves from the north toward the south during most of the year. As in the Eastern Sahara, such wind patterns are measured by meteorological stations (Manent and El-Baz, 1980), deduced from dune forms, or estimated by measuring the direction and rate of motion of sand dunes (Wolfe and El-Baz, 1982). The second observation is that sands in these dune fields are composed mostly of well rounded quartz grains.

The exposed rocks to the north of the sand seas are mostly limestones, which could not have been the source of the vast amounts of quartz sand (El-Baz, 1992a). The two observations discount the possibility of the origin of the majority of the sand by wind erosion and transportation from the north. Therefore, it is more likely that the areas covered by sand dune were relatively low areas that received sediments from northward and eastward flowing stream channels in the geological past. When the conditions of climate changed and dry conditions prevailed, the wind sculptured these sand accumulations into various dune forms and sand ridges. The patterns of dunes in Rub'al Khali in particular support this theory (El-Baz, 1982).

Thus, areas that encompass large sand dune accumulations may host vast groundwater resources.

Environmental implications of sand movement

Sand dunes are moving in the main wind direction and converting some agricultural area to wasteland. Based on United Nations records, the dry areas of the world are particularly prone to desertification. An estimated 80 million people are affected annually (Albaltagy, 2004). The deterioration of vegetative cover, wind and water erosion, salinization, and the degradation of soil fertility and structure are all manifestations of desertification. Nearly 50% of the arid regions globally, including one-quarter of the irrigated land, one-half of the rainfed cropland and threequarters of the rangeland are estimated to be degraded.

In northern China, the development of sandy desertified land has been accelerated during the past 50 years, with annual rates from 1,560 km^2 during 1950-1972, to 2,100 km^2 during 1972-1987 and 3,600 km^2 during 1987 to 2000 (Wang, 2004).

Bofah and Owusu (2004), highlighted the fact that aeolian (wind blown) sand constitutes a very serious problem to development in sandy desert lands and causes equally serious problems in lands that are undergoing desertification. Vegetation is emphasized and recommended as the ultimate viable solution to combat desertification and aeolian sand problems. In UAE, the urbanized areas are heavily vegetated and continuously maintained to reduce the effect of moving sands on the public health and environment. In addition, the government greatly supports agriculture activity around the country, where vast areas of farm land can be seen in the vicinity of Liwa, Alain-Abu Dhabi highway, Diba, Ras Alkhaimah, etc.

Role of Remote Sensing in Monitoring Sand Movement and Desertification

Remote sensing of sand dunes can help in the understanding of aeolian process and desertification. Remote sensing data combined with field studies are valuable in studying sand dunes and regional aeolian depositional history. The main advantage of satellite remote sensing is providing various kinds of images that have regional and even worldwide coverage's and collected regularly each 14days or 24days or even less, making them a reliable tool for monitoring environmental changes, including sand movement and other features related to desertification.

Both optical and radar imaging satellites proved to be useful in monitoring of environmental changes and exploration of natural and cultural resources. Landsat, SPOT, IRS, Radarsat, etc. and very recently the high resolution IKONOS, Quick-Bird and Kompsat images are useful in the above applications.

Because of their large spatial extent, distinctive color and morphology, the study of the classification, dynamics and evolution of sand dunes is greatly enhanced by the application of satellite images, through mapping their distribution and monitoring their movements by repeated imaging capability. The optical images can assess mapping and classification through detecting the color, morphology and texture whereas radar images can assess mapping in all weather conditions, and detect morphological changes and moisture variations, and further more have the ability to map subsurface features such as buried water channels or ruins. Longer radar wavelengths provide more descriptive images than shorter wavelengths, and are better at detecting dune formations (Blumberg, 1998). AVHRR may be used in conjunction with the Normalized Differential Vegetation Index (NDVI) to detect desertification over large swaths of land (Symeonakis & Drake, 2004).

Howari *et al.* (2004), used Landsat and SPOT images to study the geomorphology and mineralogy of sand dunes in the eastern part of Abu Dhabi, where the dunes in 3-types based on their composition has been classified. The first class is low dunes composed of a mixture of quartz and carbonate, the second class of dunes covered with desert grass and shrubs and is stable, and the third class of dunes are without vegetation, and believed to be mobile. Morphologically, the dunes in the investigated area belong to three types. The first is linear and is trending NE-SW dunes, mostly reflects eastward wind direction. The second type is barachonid dunes consisting mainly of chains of crescent and indicates eastward wind direction. The third type is star dunes and appears in images as radially symmetrical to mound shape that characterize multidirectional wind systems.

It is quite evident that Landsat satellite imagery is probably the most practical source for mapping dunes movement and tracking desertification at country-wise scale, but details can be depicted from higher resolution images (figure 1). Here, IRS LISS-III images can help, but they have only a limited historical coverage of ~ 10 years, and the absence of true color is sometimes a disadvantage. Landsat ETM+ images can be used to track dunes and aerial photographs can be combined with satellite images to lengthen time for change detection. High resolution satellite imagery, which became available about 7 years ago, can permit mapping of individual dune details, and tracking small scale movements.

Dune movement rates can be calculated and brightness values gathered from Landsat TM images and aerial photos may indicate aeolian activity and desertification. Actively moving sand dunes have values ranging from 150-250, and inactive from 60-70 (grayscale) (Bunch *et al.* 2005).

False color analysis may be used to identify dunes, and supervised classification of Landsat TM bands (7,4 and 2) may be useful in identifying the distribution of various types of dunes.

Method of Investigation and Imagery Data Utilized

Imagery data collection and registration

Multi-date and multi-source satellite images were utilized. Landsat TM, IRS (1C,1D and P6), IKONOS and Kompsat images of different dates were investigated to check their ability to record and highlight sand movement in two test sites; one west of Alain city (Rimah) and one in Liwa area. Landsat images dated 1984 through 2004, IKONOS images dated 2002 and 2006 and IRS 1C-1D Pan and P6 (LISS-IV and LISS-III) images were studied. Apart from Landsat TM images, all other images were collected and processed at SRC facility. All above images were geometrically corrected and co-registered using IKONOS imagery samples to ensure the best level of location accuracy. In addition, relevant reference information was compiled from various sources.

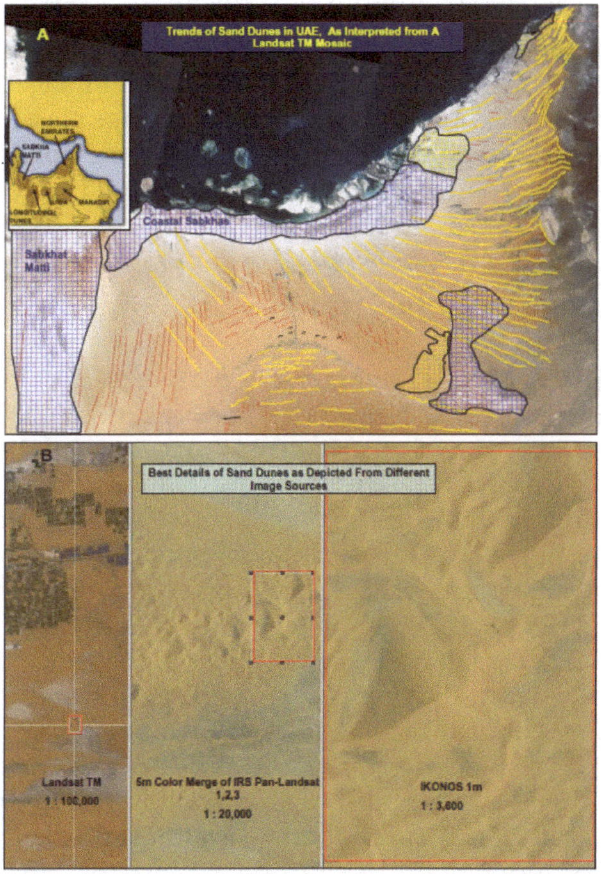

Fig. 1 : Main trends of sand dunes in UAE as interpreted from a Landsat TM mosaic (A), and best details of sand dunes as depicted from different image sources (B).

Image processing manipulations

Image enhancements and proper selection of suitable spectral bands were performed, for all multi-spectral images. This investigation was mainly based on direct correlation of sand dunes or sheets extent, using multi-date images and the absolute movement within a certain period of time was measured, mainly by using the measurement tool in the ERDAS Imagine, image processing software. It was tried to achieve the best level of geometric accuracy and fitting of correlated images, by re-doing geometric correction and using the set drop point capability in ERDAS Imagine. In addition to utilization of the original Landsat TM, IRS 1D-Pan, IRS-P6 (LISS-III and LISS-IV) IKONOS and Kompsat images, subsets from IRS 5m pan and Landsat images were merged to produce a 5m color image product , and similarly one trail of merging the 15m Landsat band 8 with the RGB (28.5m) bands was performed.

High pass and summary filters, were partly employed and resulted in enhancement of the sand dunes crest lines and sheet boundaries. Trails of principal components analysis and image classification were also made in order to investigate their ability to enhance the mapping and correlation processes of sand dunes and related features.

Discussion of Results

Landsat TM and IRS LISS-III images

A number of Landsat images, dated 1984, 1995 & 2004 covering Alain test site, "Scene 160-43"and a Landsat mosaic of UAE, were utilized, whereas in Liwa area Landsat images dated 1989 and 2000 "Scene 160-44" were investigated, with the main aim being to track sand movements. In addition IRS (1C, 1D and P6 LISS-III), 23m resolution images (figure 4A), were also used to verify their capability of mapping and tracking sand dunes. They have almost the same appearance of Landsat TM FCC images; i.e. it can help only in detecting medium scale sand movements measured in 10's of meters. However they are only available for the past ~ 10 years and they did not provide true color products. Using Landsat TM images, an ESE movement of a sand sheet for about 650m was recorded (55° 55′ 52.50″ E, 24° 32′ 31.50″N) over 20 years time (figure 2A). It is quite possible that this movement may partly be due to local storms, because there was almost no change, or even a reverse WNW movement, in the first 10 year of 1984-1995. Some movements of sand dunes and sheets in the order of ~50m were recorded in some parts N of Alain. These movements are consistent with the prevailing westerly winds in this area. The large sand dunes in both Alain and Liwa areas, did not show any evidence on movement over the past 20 years, using the Landsat or LISS-III images.

It was possible to map the main trends and patterns of sand dunes, in the UAE, using the Landsat TM mosaic of 2004 collection, (Figure 1-A). This figure shows the distribution, trend and patterns of the large to medium sand dunes, which is consistent with the known general classification of san dunes; Northern Emirates dunes with NNE and NE trends, Manadir dunes of E-W to ESE trends, Liwa dunes of ~ E-W trends and southwest UAE longitudinal dunes of ~N-S trends (UAE Internet).

Merging of the 15m Pan image of Landsat ETM with the 28.5m true color bands of, can enhance the interpretability of images, but this band (if available since 1999), can't help for long term monitoring of sand movement.

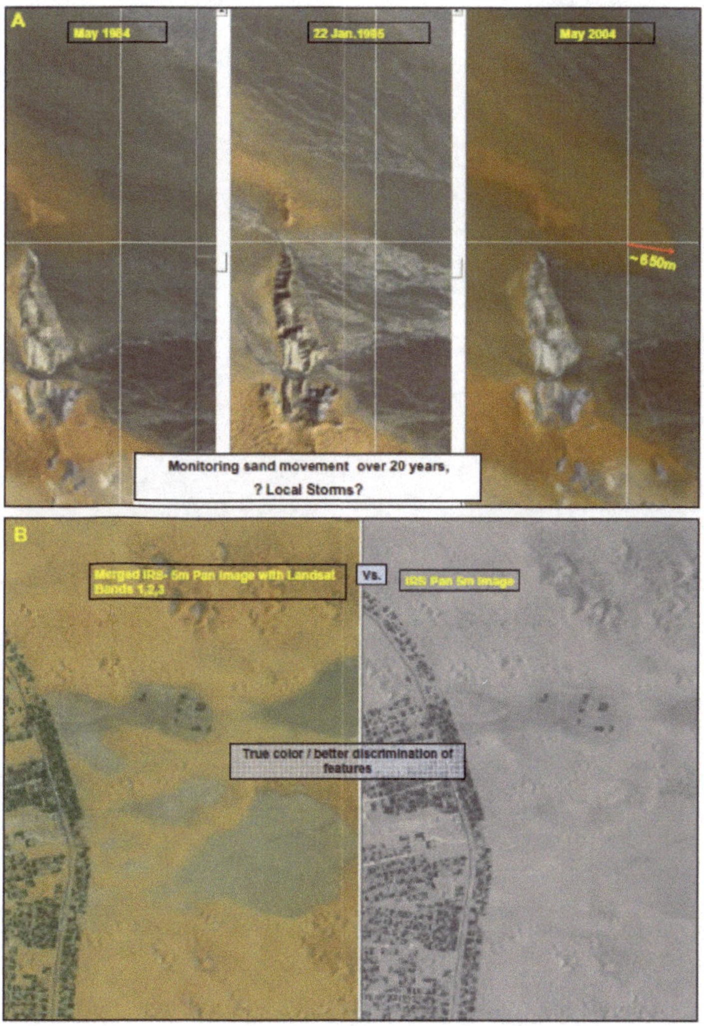

Fig. 2 : Recording sand 'sheet' movement over 20 years, using Landsat images, N of Alain (A), and a comparison color merge and IRS panchromatic 5m images (B).

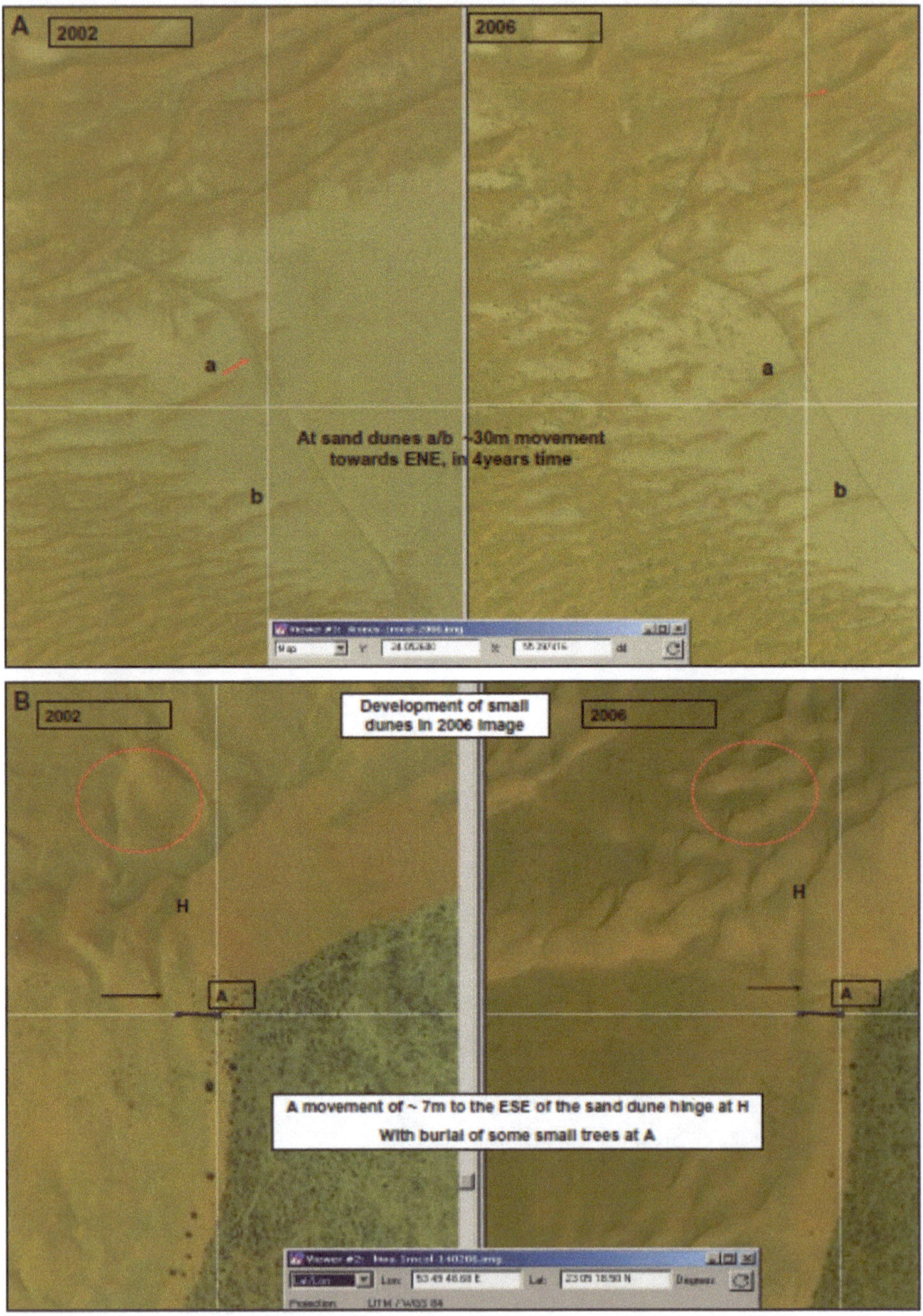

Fig. 3 : IKONOS 1m color images clearly showing movement and development of small sand dunes, (A) Alain area and (B) Liwa area.

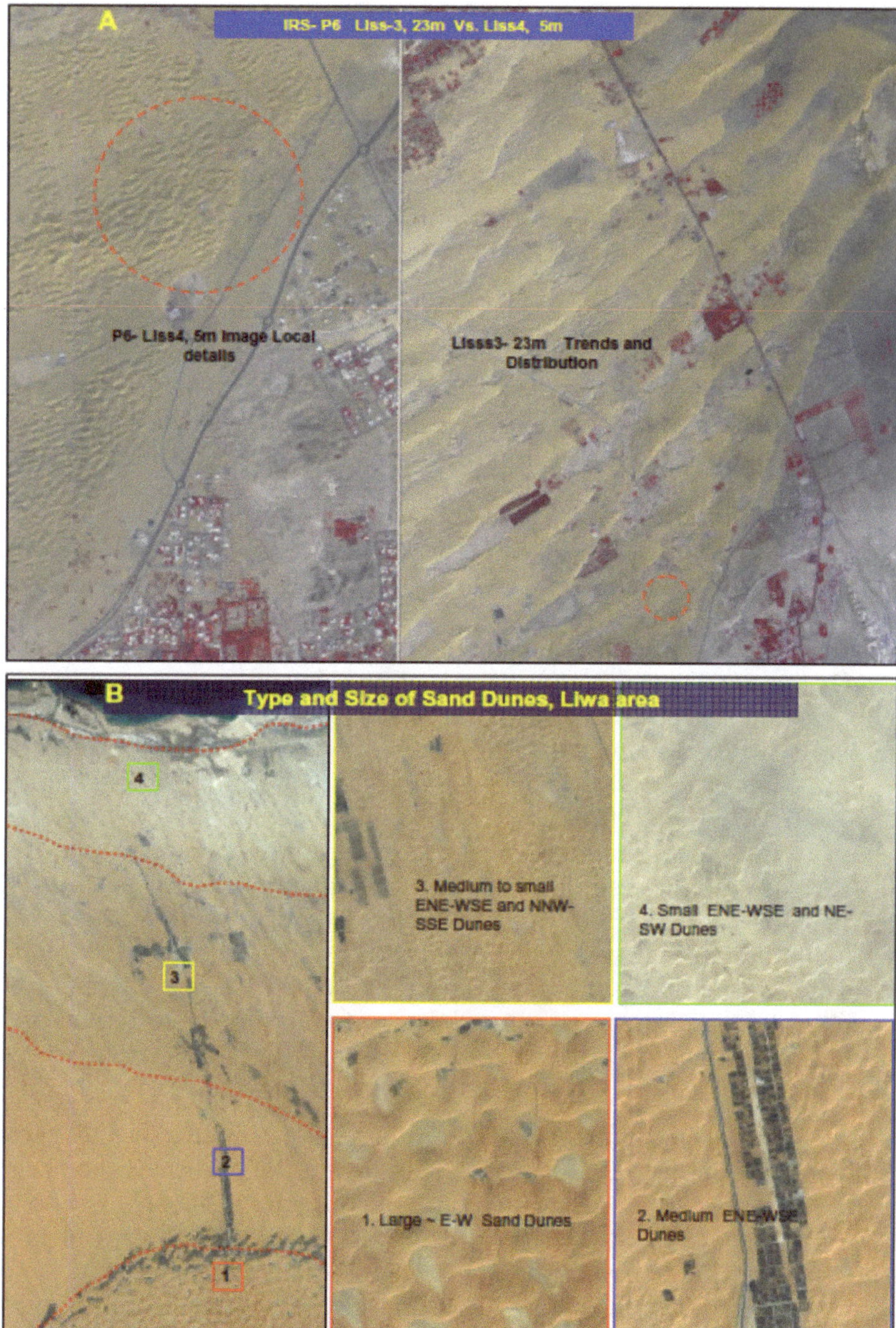

Fig. 4 : A comparison between IRS P6 LISS-IV and LISS-III images, Alain area (A), and type and size of sand dunes in Liwa Area, as interpreted from IKONOS 1m color images (B).

IRS P6 (LISS-IV) and merged IRS 5m Pan images with Landsat TM images

The 5m FCC color images of IRS P6 (LISS-IV) images and the 5m true color images generated from merging IRS Pan 5m images with Landsat TM (figure 2 B), show the distribution and trends of sand dunes (Figure 1B and Figure 4A), and can help in detecting sand movements in the order of 10's of meters. However, the IRS Pan 5m images are only available since 1995 and the P6 images are only available since 2004. Kompsat-1 Pan 6.5m images can be helpful for recording of movements in the past 7 years, but their small coverage and unavailability of color image products make them less practical. The Kompsat-2 which provides 1m Pan images, and scheduled to be collected soon by SRC, can help in detecting small movements and picking details of relevant features.

IKONOS 1m color images

Multi-date images collected in 2002 and 2006, for the two test areas, were used to verify the changes in sand dunes and measure their absolute movements. These true color high resolution images clearly shows the details of sand dunes and other relevant features, and from which small scale movements measured in few to a few 10's of meters were easily detected. Movements of small sand dunes (Figure 3) are so obvious, where ~ 30m of shift was recorded to the south of Rimah (55° 53′ 20.50″ E, 24° 03′ 08.50″N). In Liwa area, an easterly movement of about 7m was recorded at the crest of a medium to large sand dune and the formation of new small sand dune was also noticed (53° 49′ 49.50″ E, 23° 05′ 19.50″N). Here, the small trees, roads or tracks and other permanent features are helpful in measuring the absolute movement. However, it should be emphasized here, that when dealing with features such as high sand dunes, it is much better to use orthorectified images to reduce the effects of differences in angles of acquisition and seasonal changes of sun angle which produce shift in real location of same feature.

When dealing with small sand dunes, they can vary dramatically in shape and pattern due to occasional or seasonal variations in wind direction, hence, the average absolute movement over a few or 10's of years, can be misevaluated. A number of farms and residential complexes, in Liwa area, shows a rapid southerly encroachment of sand and, hence, requires implementing a strategy to overcome or minimize such effects on the various human activities. The UAE government is doing its best in this regard, but the use of images can help a lot. Using a strip of IKONOS images extending right from the Gulf in the north to Liwa sand dunes to the south, it was possible to generally predict the types and trends of sand dunes and subdivide them in four inter-fingering types (figure 4 B). Going northwards of Liwa, the dunes became smaller and scattered and tend to vanish near the coastal Sabkhas. The color of sand is also changing to a bit whitish northwards, which might reflect the carbonate composition. This requires a more detailed investigation. The longitudinal ENE to N-S dunes to the north of Liwa, are clearly filling troughs between or developed over the relatively older ~ E-W larger dunes; zone 3 and partly zone 2 (figure 4B).

Conclusions and Recommendations

a. High resolution IKONOS satellite images, or similar imagery data, are good at mapping local details of sand dunes; pattern, composition, trend, etc., and measuring movements in the range of 5 -10 meters and above.

b. Landsat TM, IRS LISS-III and IRS LISS-IV images, which has large area coverage, and quite cheap source, can successfully monitor considerable sand movements measures in 50m or above, and map the general trends of sand dunes on a country-wise scale.

c. It is recommended to fit all images or "redo-geometric correction, if necessary", of all other image sources, using IKONOS 1m images, if available, to ensure reliable position accuracy. Using one unified projection, spheroid and datum is necessary.

d. One should be careful when evaluating the average absolute movement over a certain time period. It is true sometimes, that the movement of small sand dunes or sheets in particular, is only due to seasonal or occasional variations in wind direction.

e. It is recommended to utilize IKONOS images to make a more detailed classification of sand dunes and inter-dune finer sediments.

f. Orthorectified images need to be used for accurate measurements of large "high' sand dunes, the crest line real position may be missed by few to 10 meters on IKONOS images.

g. Aeolian sand encroachment on highways, farms and residential/industrial communities are evident from IKONOS images thus can help in implementing a strategy to overcome such problems, wherever it affects.

Acknowledgments

Sincere thanks are due to Colonel Mahash S. Alhameli, Director of the UAE Space Reconnaissance Center, for his support and continuous encouragement of technical activities, and many thanks go to the SRC-Production team for help in procuring satellite images and image processing.

References

Andrew S. Goudie, Alison Colls, Stephen Stokes, Adrian Parker, Kevin White and Asma Al-Farraj. 2000. Latest Pleistocene and Holocene dune construction at the north-eastern edge of the Rub Al Khali, *United Arab Emirates Sedimentology* 47 (5), 1011–1021.

Blumberg, D. G. 1998. Remote Sensing of Desert Dune Forms by Polarimetric Synthetic Aperture Radar (SAR). *Remote Sensing of Environment* 65: 204-216.

Bofah .K. K. and Yaw A. Owusu. 1986. The aeolian sand problems arising from desertification, *Environmental Monitoring and Assessment*, Volume 6, Number 3 / May, 1986, pages 283-292.

Bunch F., *et al.* 2005. Twentieth century dune migration at the Great Sand Dunes National Park and Preserve, Colorado, relation to drought variability. *Geomorphology* 70: 163-183.

El-Baz, F. 1982. Genesis of the Great Sand Sea, Western Desert of Egypt, p. 68. Abstracts of papers, International Association of Sedimentologists. Eleventh International Congress on Sedimentology, McMaster University, Hamilton, Ontario, Canada.

El-Baz, F. 1992a. Origin and evolution of sand seas in the Great Sahara and implications to petroleum and groudwater exploration. Geology of the Arab World, Cairo University Press, Cairo, Egypt vol. II: 3-17.

Howari, F. M., Ghrefat, H. A., and Goodell, Ph. C., 2004. Geomorphological and mineralogical characterization of sand dunes using multispectral Landsat TM ETM+ and SPOT, Eastern part of Abu Dhabi, United Arab Emirates. In: Apollo-Soyuz Test Project, summary science report, vol. II: 285-299, NASA, Washington D.C.

Manent, L.S. and F. El-Baz. 1980. Effects of topography on dune orientation in the Farafra region, Western Desert of Egypt, and implications to Mars. Reports of Planetary Geology Program, Tech. Memo. 82385. NASA, Washington D.C.

Symeonakis, E., and Drake, N. 2004. Monitoring desertification and land degradation over sub-Saharan Africa. International Journal of Remote Sensing 10: 573-592.

UAEinternet, 2006 . uaeinteract.com/education/geography.asp.

Wang Tao, 2004. Quantified Evaluation on Development of Sandy Desertified Land for Last 50 Years in the North China. Director, Cold and Arid regions Research Institute, China.

Wolfe, R.W. and F. El-Baz. 1982. Wind patterns in the Western Desert. In: Desert landforms of southeast Egypt: A basis for comparison with Mars. F. El-Baz and T.A. Maxwell (eds.), NASA CR-3611, 119-139. NASA, Washington D.C.

18

The Impact of Land and Water Management on Soil Degradation in Case of Siwa Oasis, Egypt

Rafat Ramadan Ali and Adel Shalaby

Abstract

The Western Desert covers approximately 700,000 km², which is more than two-thirds of the total area of Egypt. This immense desert, to the west of the Nile, spans the area from the Mediterranean Sea to the Sudanese border. The Great Sand Sea lies within the desert's plain and extends from Siwa Oasis to Jilf al Kabir plateau. Scarps and deep depressions exist in several parts of the Western Desert. Limited agricultural production, the presence of some natural resources, and permanent settlements are found in the western desert oases (Siwa, Bahariya, Farafra, Dakhla and Kharga), all of which have fresh groundwater. Siwa Oasis, close to the Libyan border, is isolated from the rest of Egypt but has sustained life since ancient times. The Siwa's cliff-hung Temple of Amun was renowned for its oracles for more than 1,000 years. Herodotus and Alexander the Great visited the temple in the pre-Christian era. The Oasis is a natural depression about 23 meters below sea level. It covers an area of about 1050 km² (2,50,000 acres); of which only 87.95 km² (20,940 acres) are currently cultivated. The total numbers of inhabitants in Siwa were 21,482 in 2006. The main activity in Siwa oasis is agriculture which depends on the groundwater that outflows from about 1199 wells and springs giving a total annual discharge of about 255 million cubic meters from which, about 222 million cubic meters are lost due to evaporation and evapotranspiration, while the remainder goes to the natural lakes of Siwa Oasis. This has led to a continuous rise in the water table level and soil salinity.

Present research paper on land degradation processes relates to the improper management of water and land resources in Siwa Oasis, Egypt.

Introduction

Siwa Oasis is a closed depression located some 65 km east of the Egyptian-Libyan borders and some 300 km south of the Mediterranean Sea. The depression floor lies at 18 m below sea level between latitudes 29° 06′ and 29° 24′ N , and longitudes 25° 16′ and 26° 12′ E (Figure 1). The Oasis is bounded on the north by a cliff rising about 100 m above the depression floor. The great sand sea limits it from the west and the south with an elevation of 25 to 50 m above sea level.

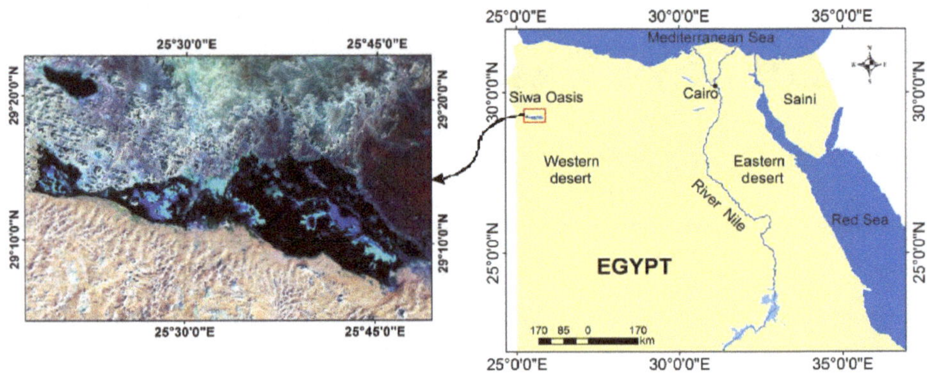

Fig. 1 : Location of Siwa Oasis on Egypt map (right) and Landsat ETM+ image of Siwa (Bands 7, 4, 2) (left)

The desert climate is recognizable in Oases, the difference in day and night temperature degrees are vast especially in winter. According to the Egyptian Meteorological Authority (1996), the Siwa Oasis is characterized by an average temperature of 21.7 °C. The maximum monthly temperature is 38 °C in July, while the minimum is 3.8 °C in January. The Oasis receives an average of 9.6 mm of rainfall per year. On an average there are 7.5 days per year with more than 0.1 mm of rainfall. The monthly relative humidity varies from 33% in May to 60% in December. Wind speed varies from 13.7 km/h in April to 8.6 km/h in November and the average value of wind speed is 10.6 km/h.

Geomorphology of the Siwa Oasis had been studied by many investigators (Abu El-Izz, 1971; Zakaria, 1972; Harga et. al., 1975; Gomaa, 1976 and Abdel-Sammie, 2000). On the basis of these studies, Siwa Oasis can be divided into three main geomorphic units. It displays numerous characterizing landforms as follows:

- **The northern elevated plateau:** It occupies wide tract to the north and east of Siwa depression, with a surface underlain by resistant middle Miocene limestone. It is dissected by shallow dry drainage lines, which are directed southwards to the depression (Figure 2). Some of these drainage lines are filled with young Holocene deposits formed by successive accumulation of down washed deposits. The elevation of the plateau reaches about 100 m above sea level with gentle escarpment slopes.

Fig. 2 : The Northern dissected Plateau and Siwa Lake

- **The southern elevated plateau:** This plateau is located in the southern part of the piedmont plain about 60 m above sea level. It is underlain by Miocene Limestone and shale. The surface is occasionally covered with drift sand accumulations, mainly developing into sand dunes of the barchans types (Figure 3).

Fig. 3 : Orchards close to the Southern plateau of Siwa Oasis

- **Depression floor:** It is the transition from the foot of the northern elevated plateau to the southern elevated plain. The lowest point in this plain reaches about 17 m below sea level. Degradation and aggregation processes in this plain are presently still operating. The surface of this plain could be differentiated into piedmont slope, sabkhas, isolated hills, arable lands, sand dunes and lakes (Figure 4).

Fig. 4 : Encroachment of sand dunes on the southern border of Siwa Lake

The people of Siwa Oasis are mixture of Bedouin, Barbar and Sudanese races. They are not of Egyptian origin, being more closely connected with North Africans than with the Nile Valley. The total population of the Oasis amounted to 13,000 persons (El-Attar, 1998) reaches to 21,482 in 2006 (CAPMAS, 2006). Also it shows that 75.3 % of population lives in Siwa city followed by El-Maraqi 11.3 % and Aghormi 8.9%. The main activities of the population are agriculture together with some small industries such as olive oil extraction, olive pickling, date drying, carpet manufacture and natural water industry. The cultivated area is continuously increasing owing to the surplus of water and land resources as well as population. In 1962 the cultivated area was 800 ha only. It reached 1500 ha in 1977 according to the Arab Organization for Agriculture Development (AOAD, 1977). At present it is estimated at about 4200 ha according to the Desert Research Center (2003). Olive and palm orchards occupy 85% of the cultivated area (Figure 5). Few areas are cultivated with Citrus, Apricot, Pomegranate and Vegetables, while alfalfa is usually intercropped with the newly cultivated orchards. Basin irrigation (4x4 or 5x5m) prevails in the Oasis. Irrigation intervals vary from one to two weeks and the depth of water applied was estimated as 7.5 cm (Shatanawi, 1991). This indicates that the amount of water applied throughout the year reaches about 26,000 m^3/ha which is extremely higher than the irrigation water requirement of the cultivated crops in the Nile valley, even under saline water irrigation and climatic conditions approximately similar to that in the Oasis. This situation results in excessive deep percolation with consequent water-logging problems.

Fig. 5 : Olive and palm orchards

Water Resources

Ground water is the only water source in Siwa Oasis. The hydrological studies (Mudallal, 1991; Abdel-Mogheeth et. al., 1995; Abdel-Fatah, 1998 and Gad, 1999) showed that the ground water comes from three main aquifers; which are under artesian head and their waters are naturally flowing and can be classified as:

- **The upper limestone aquifer:** It is about 200 - 250 m thick, belongs to the middle Miocene period with water salinity of 2500 to 7000 ppm.
- **The lower limestone aquifer:** It lies directly under the Miocene rocks with 350 m thickness characterized by water salinity of 6500 to 8000 ppm.
- **The Nubian sandstone aquifer:** This aquifer is separated from the above described aquifers by a thick section of shale and marl related to the upper cretaceous age. This aquifer which has an average thickness of 120 m is recently tapped by some deep wells (1000-1200m). It is characterized by high water salinity of 200 ppm and high rate of discharge (450 $m^3/h/$ well).

Soils

The soils of the Oasis show considerable variations in their physical and chemical properties. According to Harga et. al (1975), the Oasis includes the following groups:

- Shallow sandy soils: These are mainly found at zero contour line in the form of the crescent shape. The soil depth is limited mainly by water table and ranges between 50 and 75 cm accordingly the high salinity is expected.
- Deep sandy soils: These soils are mainly located between the southern scarp and the main depression. These soils are slightly-calcareous where the $CaCO_3$ content is general less than 10%.
- Saline soils: It is occupying an appreciable portion of the depression and partly covered with patchy cultivations. Their surfaces are covered with a thick salt crust locally named *Korchief.*
- Calcareous soils: This type is mainly located in the eastern and western parts of the Oasis. They are shallow to medium in depth, and highly salt-affected. Calcium carbonate content varies between 10 to 50% of the total soil matrix.
- Hydromorphic saline sandy soils: This group is mainly located at the shore-line of the old lakes. Profiles of such soils are shallow and extremely salt-affected.

The properties of these soils are mainly a function of their parent materials. There is a variation with regard to their texture and calcium carbonate contents. Soils which are formed as a result of weathering processes on limestones and transported by water differ in texture, whereas those which are mostly due to wind action are sandy in texture. Generally, the salinity levels of arable land soils are much lower than those of uncultivated ones due to continuous irrigation and leaching. Soils are in most cases mildly alkaline. Organic matter content is generally low although they may, in few cases, reaches 29 % (Gomaa, 1976 and Sherif, 1979). Concerning the soils classification, the *Typic Haplosalids* sub-great group is the dominant soil type (163.52 Km^2) representing 51.1% of soils. These soils are characterized by a shallow soil profile depth, water logging and high to very high soils salinity level (Figure 6). The sub-great group *Typic Torripsamments* has a significant coverage, particularly at the south west of the oases, where it cover 57.28 Km^2 representing 17.9% of soils. Less coverage of sub-great groups *Typic Torriorthents, Typic Calciorthids* and *Typic Haplogypsids,* exist in patches clustered in the north and middle of the oases. They cover areas of 18.56, 8.64 and 4.16 km^2, representing 5.8, 2.7 and 1.3% respectively. The rock land and sand dunes cover an area of 23.6 Km^2 i.e. 7.4% of the total area. Regarding to soil capability, it is found that the soil salinity, soil depth, texture, CEC and drainage condition are the main limiting factors in Siwa Oasis. The Oasis has no capable soils (class II) due to the high severity of limiting factors, specially salinity and drainage condition. The moderately capable soils represent 11.89% of the total area. Low capable soils dominate the Oasis as they represent 75.49% of the total area. The low capable soils are associated with the *Aridisols* soil order (ASRT, 2009).

Fig. 6 : Water logged areas in the middle of Siwa depression

Irrigation and Drainage

Traditional surface irrigation system is dominant in Siwa Oasis. Farmers construct reservoirs around wells and springs to collect waters which are of slow discharge to use them afterwards in irrigation. Flood irrigation is applied at 7-16 days intervals with a rate of 46 m^3/acre/day. There are a number of aspects with regard to wells and spring water, the daily discharge rate is about 1,90,000 m^3/day, out of which, about 77,000 m^3 is highly saline (<8000 mg/l). The discharge is nearly constant all the year round, and the maximum need for water is during the summer months. During the winter there is excess water, part of which is used for leaching saline soils and the other flows to the lakes in the lower parts of the depression. Flowing springs have discharge rates which vary from one region to anther and even within one region (AOAD, 1977 and Gad, 1999). In the past irrigation water was provided by 200 natural springs with water flowing freely on the land surface. About 2000 acres were under irrigation in 1960s but due to the rehabilitation of these springs more water becomes available, and by 1980s the irrigated areas increased to about 3500 acres (Arar, 1991 and Desert Research Center, 2003). During the eighties more people came to Siwa Oasis and new land was put under irrigation. In the present time it is estimated that the irrigated areas have approached 12000 acres. This development depended on water source from about 1000 hand dug wells (20-30 m deep) with an average discharge of 20- 25 m^3/hr and about 300 drilled wells (70-130m deep) with an average discharge of 40 m^3/hr. At present about 235 million cubic meter/year (Mm^3/year) is being used for irrigation, of which 60 Mm^3/year from springs, 105 Mm^3/year from hand dug wells and 70 Mm^3/year from the drilled wells. This indicates that about 23500 m^3/acre are being used for irrigation which is very high compared to the actual amount of irrigation requirements. Actually, Siwa is a closed depression and there are no outlets to discharge excess water outside it. As in the past, the excess drainage water and irrigation tail water were diverted to the lakes (Abdel-Mogheeth et. al, 1995).

The agricultural activities are fixed around springs, where each spring serves a group of farms, called *Hettya*. The area of the Hettya varies from one to about 180 acres according to the spring discharge. The Hettya is surrounded by an interceptor open ditch (Figure 7) which acts as surface and subsurface drain where surface drainage includes runoff and tail end losses of irrigation canals. The drainage network in the Oasis comprises of arbitrary made open field drains (sub-main) and main drains which in turn deliver the drainage water to the lakes. The main and sub-main drains carry out the drainage water from ditches, including the surface and subsurface drained water to the four main drainage lakes which are natural outlets to the irrigated areas. These lakes are Al-Maraqi in the west, Aghormi and Siwa in the middle and Al-Zietun in the east. Need to mention that in the far east of Siwa depression, there are other two lakes, Timera and El-Ma'asser. Nevertheless, the agricultural activities around these two lakes are very limited. The total number of the main and sub main drains in Siwa approaches 65 drains. Gad (1999) estimated the discharge of main drains which pour their water in the four lakes. He found that Maraqui Lake receives drainage water from 13 drains, with total discharge 83,492 m^3/ day during the winter and 35358 m^3/ day during the summer, meanwhile Siwa Lake receives drainage water from 25 drains with total discharge 223,784 m^3/ day during the winter and 139,684 m^3/ day during the summer (Figure 8). Aghormi Lake receives drainage water from 17 drains with total discharge 1,01,280 m^3/ day during the winter and 34,939 m^3/ day during the summer, Al-Zietun Lake receives drainage water from 8 drains, with a total discharge of 201,313 m^3/ day during winter and 42,683 m^3/ day during the summer.

Fig. 7 : Traditional drainage system (open ditch)

Fig. 8 : Siwa lakes : Natural outlets to the irrigated areas

Soil Degradation under Poor Drainage Condition

Several studies have been conducted on the effect of water table depth on soil properties. Van Hoorn (1958) reported that the soil texture is one of most important soil properties which degrade under the poor drained soils. Deterioration of soil structure may be due to the change in the soil-water-air ratio where percentage of soil air have decreased in the sub-irrigated treatment. The structure factor and soil aggregation have slightly increased while the dispersion ratio decreased after drainage (Abdel-Mawgoode, 1979 and El-Melegy, 1990). Bulk density is an important determinant of soil's potential to support plant growth. Soils with high bulk densities may restrict root growth, water movement, and gas exchange. Chieng and Hughes-Games (1995) reported the bulk density on poor drained plots averaged 1.45 Mg/m^3 and bulk density on drained plots averaged 1.35 Mg/m^3.

Carter and Hedstorm (1983) and Fouss et. al. (1987) have shown that subsurface drainage installed on a USDA sugarcane field research site improved soil aeration, increased depth of the root system, enhanced the crop growth and yield and allowed the resumption of farming activities to begin soon after rainfall compared to fields with surface drainage only. Apichai et. al. (2003) suggested that soil drainage and soil moisture were directly related to soil bulk density and soil porosity except at 30 cm depth of sandy clay loam. Bornstein and Hedstorm (1982) evaluated the drainage effectiveness and trafficability response using tensiometers and cone penetrometer measurements. They stated that improved trafficability developed more rapidly when subsurface drainage of the soil profile was provided. Abdel-Ghany et. al. (1990) reported that the value of drainable water volume and drainable porosity increased by increasing water table depth as a result of installing drainage system. Gupta (1990) found that drainable pore space increased with increasing depth of water table. He also added that the drainable pore spacing is a very important factor in calculating drain spacing using means of un-steady state flow equations. Abu-Sinna (1991) revealed that the drainable porosity was significantly higher in the drained soils compared to the non-drained ones. This in turn, increases the aeration porosity in the drained soil by 50%. Wahdan *et al.* (1992) showed that the total porosity increased slightly in the drained soils compared to the non-drained ones. Oosterbaan (1994) reported that the percentage of large pores increases with increasing depth of the water table. Mohamedin (2002) in his study on the effect of tile drainage on soil physical properties at the north of El-Fayoum Governorate (lake of Karoun) reported that soil drainable porosity, quickly drainable pores were improved upon installing tile drainage system in the area. Naguib (1987) found that sub-surface drainage increased the quickly drainable pores of > 28.8 μ. He also found that the soil aeration and drainable pores of > 8.62 μ have also increased. Naguib (2001), found that there was a highly positive correlation between drainable porosity and drain installation. However, many investigations have paid special attention to the impact of drainage on drainable porosity. El-Melegy (1990) concluded that the aeration porosity increased by 5.24% after drainage. El-Melegy (1990), Faltas *et al.* (1991), Ragab (2000) and Naguib (2001) found that the hydraulic conductivity values increased after installing drainage system and continued increasing yearly. The higher hydraulic conductivity values were found in the soil of 20 m drain spacing. Ritzema and Braun (1994) stated that installation of drainage system has two direct effects i.e. it reduces the amount of water stored on or in the soil and allows water to flow through the drainage system. They also added that these two direct effects are usually causing many indirect effects and these are often the true reasons for drainage .These reasons are the removal of excess surface or ground water, to achieve better soil aeration, leading to higher productivity of crops and leaching for salinity control to prevent increases in soil salinity, sodicity and to prevent a build-up of salinity in the root zone. Moustafa et. al. (1990) studied the effect of a tile drainage system on the properties and productivity of clayey soils in Fayoum Governorate. They found that tile drainage reduced salinity and alkalinity and increased productivity. Wenberg (1990) reported that subsurface drainage must be adequate to permit the necessary leaching, hold the water table to sufficient depth and prevent the upward movement of salty capillary water to reach the root zone. Abdel-Dayem and Ritzema (1990) noticed in many areas of the Nile delta, especially in the northern parts, that the heavy clay soils are affected by salinity and sodicity hazard. He found that drainage is an important factor in the reclamation process of these soils. Abu-Sinna (1991) found that subsurface concrete drains removed 52 and 82% of initial salts after 7 and 20 years of land drainage, respectively. He found that the desalinization has happened more rapidly in the surface than in the subsurface soil layers. He added that soil desalinization is associated with the decrease in SAR and ESP values due to the high leach-ability of Na^+ compared to that of Ca^{+2} and Mg^{+2}. Kandil (2000) reported that installing sub-surface drainage system in Nubaria area resulted in appreciable reduction in salts in all treatments of drain spacing (25, 40, 70, 90 and 150 m). The 25 m drain spacing was the best one for decreasing soil salinity. On his study on some agricultural drainage problems in Siwa Oasis, Rafie (2008) concluded that:

- The Oasis is characterized by a moderately shallow to moderately deep profiles where the depth to water-table varies, on average, from about 50 to 100 cm.
- The values of the hydraulic conductivity vary from 0.25 to 4.3 m/day indicating that the ability of such soils to transmit water ranges from moderate to rapid.

- The depth to water- table varies, from 73 to 120 cm in winter and from 83 to 135 cm in summer.
- The values of the saturated soil layer vary from 40 to 150 cm.
- The flow direction of ground water is perpendicular to the equipotential lines and generally towards the drainage lakes.
- The appropriate spacing should be 30 m in areas around Siwa Lake and 20 m in the low-lying areas which is still without drainage network.

Based on climate, soils, vegetation cover and land management environmentally sensitive areas to desertification in Siwa Oasis were defined. It is found that the high sensitive areas for desertification are found in the edges of Oasis representing 48.9% of the total area, where the soil quality, climatic quality and management quality are low. The areas of moderate to low sensitivity for desertification exhibit the middle parts of Siwa as it represents 51.10% of the total area. The environmental sensitivity to degradation in the Oasis are mostly affected by land management, soils and climatic condition (NARSS, 2010)

Conclusion

Soils of Oases are mainly ascribed to the climatic conditions, topography and human activities. So, these soils may show differences with regard to their texture, mineralogy content, depth to water table or bedrock and several types of pedological features such as accumulation of salts, carbonate, gypsum and iron oxides. According to soil taxonomy, various soil subgroups were identified in Siwa Oasis (e.g. *typic haplosalids, typic torripsamments typic torripsamments, typic torriorthents, typic calciorthids* and *typic haplogypsids*). The soils of the Oasis have numerous limitations that reduce their agricultural productivity and could induce land degradation processes. These limitations could be aggregated as rough topography, shallow soil depth, hardpans, salt accumulation, coarse textured soils and inappropriate land use and management. Overuse of groundwater combined with the inferior qualities of the prevailing soil source, physiography and human induced mis-management practices caused serious salinity problems. So, the assessments of the soil and water resources in quantity and quality have to be conducted for the entire Oasis.

References

Abdel-Dayem, S. and. H.P Ritzema, 1990. Verification of drainage design criteria in the Nile Delta, Egypt. *Irrigation and Drainage Systems*, 4, 2, pp. 117 - 131.

Abdel-Fatah, M. A., 1998. "The horizons of Agricultural development in Siwa Oasis" Symp. Fac.of Agric. Alex. Univ. (Saba Basha) May 19-20, 1998. P, 6.

Abdel-Ghany, M.B., M.S., Adel-Dayem, H.M. Kanndil and M.A. Abu Sinna, 1990. Investigation of physical and chemical soil properties for subsurface drainage design. Symposium on land Drainage for Salinity Control in Arid and Semi-arid Regions. Feb. 25th - March 2nd, *Drainage Research Institute* Vol. (2), Cairo, Egypt.

Abdel-Mawgoode A.S.A., 1979. Studies on tile drainage in sugar cane growing areas in Upper Egypt. B.Sc. Thesis, Fac. of Agric., AL-Azhar univ., Egypt.

Abdel-Mogheeth, S.M., M.S. EL-Hakim, and M.Y. Afifi, 1995. Wet lands Management for suitable development in Siwa oasis. Proc . of the Conf. on "Versatility of wet lands" FL.U.S.A. PP 745-755.

Abdel-Sammie, M. K. A., 2000. Classification and evaluation of Siwa oasis Soils. Ph D. Thesis. Fac. of Agric., Ain shams univ. Cairo Egypt.

Abu El-Izz., M. S., 1971. Landforms of Egypt. The Amer. Univ., Cairo, Egypt.

Abu-Sinna, M.A., 1991. Efficiency of tile drainage for providing good air- water balance in soil. Ph.D. Thesis, Fac. of Agric., Cairo Univ., Egypt.

Al-Attar, H., 1998. Soil and Water Resources in Siwa Oasis. Symposium on "The horizons of Agricultural development in Siwa oasis". *Fac.of Agric. Alex. Univ.* (Saba Basha) May 19-20, 1998. P, 6.

AOAD, 1977. Feasibility Study in Economic and Agricultural Development in Siwa. Arab Organization for Agricultural Development Internal Report 457p.

Apichai, T., P., D. Bumrung and W. Bummungket, 2003. The Effect of soil physical properties on the soil water drainage. The Globe Learning Expedition Conference Sibeanik Crooti, June 29 - July 4, 2003.

Arar, A., 1991. Water-logging and salinity problems in Siwa. Int. Workshop on Rehabilitation of Siwa oasis; organized by FAO and the Development Authority of the North-West Coast Egypt. Matruh, 4 Nov., 1991.

ASRT, 2009. Preparation of land resources data base for agriculture usage. Final report, Academy of Scientific Research and Technology (ASRT), Cairo, Egypt.

Bornstein, J., and W. E. Hedstorm, 1982. Trafficability factor in a silty clay loam soil. *Transactions of the ASAE* 25 (5): 1240 - 1244.

CAPMAS, 2006. Statistical Yearbook 2006. Central Agency for Public Mobilization And Statistics, (CAPMAS), Cairo, Egypt.

Carter, C. E., and W. E. Hedstorm, 1983. Subsurface drainage of an alluvial soil increased sugarcane yields. *Transactions of the ASAE* 26 (2): 426 - 429.

Chieng, S. T. and G. A. Hughes-Games, 1995. Effects of sub-irrigation and controlled drainage on crop yield, water table fluctuation and soil properties. p. 231-246. In H. W. Belcher and F. M. D'Itri, (eds.) Subirrigation and Controlled Drainage. Lewis Publishers, Boca Raton, Florida.

Desert Research Center, 2003. Situation of the agricultural development in Siwa depression. Final report, Water resources and Desert soils branch, DRC, Cairo-Egypt.

Egyptian Meteorological Authority, 1996. Climatic Atlas of Egypt. Published report, Ministry of Transport, Arab Republic of Egypt.

EL-Melegy. A. M., 1990. Drainage criteria in rice fields. M.Sc. thesis, Fac. of Agric. Al-Azhar Univ., Egypt.

Faltas, R.L., B.H., Naguib, and M.M. Selem, 1991. Changes of physico-chemical soil properties and crop yield under different drain spacing. *Egypt. J. Appl.* Sci., 6 (12): 374 - 38.

Fouss, J. L., R. L. Bengston and C. E. Carter, 1987. Simulating subsurface drainage in the Lower Mississippi Valley with Drain mod. *Transactions of the ASAE* 30 (6): 1679 - 1688.

Gad, M. M., 1999. Water budget analysis for the top most aquifer for some cultivated areas in the central part of Siwa oasis, western desert, Egypt. Ph.D. thesis, Geology Dept. Fac. of Science Al-Azhar Unvi.

Gomaa, A.E., 1976. Physical characterization of Siwa soils. M.Sc. Thesis Fac. Agric., Cairo Univ., Egypt.

Gupta, S.K., 1990. Subsurface drainage system design for semi-arid climate in India. In Symposium on Land Drainage for Salinity control in Arid and Semi-arid Regions. Feb. 25th - March 2nd, *Drainage Research Institute* Vol. (2), Cairo, Egypt.

Harga, A. A., M. A. Hammad, and M.A. Abdel Salaam, 1975. The soils of Siwa Oasis. Desert Inst. Bull., *ARE*, 25(1&2):173-186.

Kandil, N. F., 2000. Drain spacing assessment in calcareous sandy loam soil in Egypt. *Minufiya, J.* Vol. 5 NO. 6: 1525 - 1539.

Mohamedin, A.A., 2002. Tile drainage efficiency as affected by soil and hydrological properties at the North of EL-Fayoum Governorate. Ph.D. Thesis Fac. of. Agric., Ain Shams. Univ ., Egypt.

Mudallal, O.H., 1991. Availability of water resources in Siwa oasis; Egypt .Int. Workshop on Rehabilitation & Development of Siwa oasis; Organized by FAO and Development Authority of the North-West Coast Egypt. Matruh, 4 Nov., 1991.

Mustafa, F.B., F.M. Abraham, and S. Fikry, 1990. Drainage efficiency and soil productivity in Fayoum Governorate. Agric. Res., 68: 253-261.

Naguib B. H., 2001. Evaluation of tile drainage efficiency through its effect on soil hydro - chemical properties and productivity. *Egypt J. Appl. Sci* 16: (7), 728 - 747.

Naguib, B. H., 1987. Tile drainage in relation to soil characteristics and plant growth Ph.D. Thesis Fac. of Agric Moushtohor, Zagazig Univ. (Benha Branch), Egypt.

NARSS, 2010. Assessment and Mapping Desertification Sensitivity Aspects of Some Western Desert Oases, Egypt, Using Remote Sensing and GIS. Final report, National Authority for Remote Sensing and Space Science (NARSS), Cairo, Egypt.

Oosterbaan, R. J., 1994. Agricultural drainage criteria. H.P. Ritzema (Ed), Drainage Principles and Applications, ILRI Publication 16, second edition, Wageningen, The Netherlands.

Rafie, M. R., 2008. Studies on some Agricultural Drainage Problems in Siwa Oasis. Ph.D. Thesis, Fac. of Agric., Al-Azhar Univ., Cairo, Egypt.

Ragab, M. A., 2000. The effect of calcium carbonate content efficiency and crop yield under different designs of tile drainage system. Ph.D. Thesis, Fac of Agric., Ain – shams University, Egypt.

Ritzma, H. P., 1994. Sub-surface flow to drain in drainage principles and Applications. ILRI., Pub. 16 Second Addition Wageningen, The Netherlands.

Shatanawi, M. R., 1991. Irrigation development in Siwa oasis, Egypt. Workshop on Rehabitilitation & Development of Siwa oasis; organized by FAO and Development Authority of the North-West Coast Egypt. Matruh, 4 Nov., 1991.

Sherif, M. A., 1979. Evaluation for fertility status in soils of Siwa Oasis. M.Sc. Thesis, Fac. Agric., Ain Shams Univ. Egypt.

Van Hoorn, J. W., 1958. Results of ground water level experimental field with arable crops on clay soils. *Netherlands J. of Agric. Science*, 6:1-10.

Wahdan, A. A., A. M. Helalia and H. M. Nasr, 1992. Impact of subsurface drainage on some soil properties of Upper Egypt. Proceeding of the 5th-International drainage Workshop, Lahore, Pakistan, ICID, *IWASRL*, 3: 671- 679.

Wenberg, D. R., 1990. Drainage for salinity management. In Sym. On land Drainage for Salinity control in Arid and Semi- Arid Regions. Feb 25th - March 2nd, DRI, Cairo Egypt.

Zakaria, M. H., 1972. Geological and Hydro-geological Studies of Siwa Oasis. M.Sc. Thesis, Fac. Sci., Alex. Univ.

19

Economic Evaluation of Precast Concrete Lining in Desert Areas

Tahir M.A., Tarig Elgamri and Isam M.A. Magid

Abstract

Three seepage tests were carried out for an earth canal at Rawakeeb Research Station (RRS)-west of Omdurman and an average seepage rate of 3.4 m/day was obtained. This value suggests that irrigation canal at RRS should be lined. In addition to that preliminary economic evaluation showed that canal lining is imperative in RRS. A concrete slab form was well designed for lining of RRS canal. Precasting technique and construction technique provided the required water tightness. Economic evaluation was done for RRS canal to investigate the extent of occurrence of supposed advantages, and to estimate costs and benefits and rate of return to lining. The capital expenditure (C), net present worth (NPW) and benefit/cost ratio were found to be 2600.4, 1961.9 and 1.3 respectively which offers a good indicator to adopt the technique.

Introduction

Life is tied to water as it is tied to air and food. Of 20 African countries faced food emergencies in recent years, half are already stressed by water shortage (Ahmed et al, 2009). Sudan is the largest country in Africa with an area of 2,506,000 km^2 and estimated population of 31 million people. Two-thirds of Sudan are located within arid and semi-arid regions. Recently and especially during the latter half of the previous century, the arid and semi arid regions of the country were subjected to different forms of land degradation. West Omdurman such as RRS, located within the arid region, experienced severe land degradation such as sand encroachment and deforestation in the past few decades, as a result of the influx of displaced people who entered the area fleeing from famine in the drought stricken Kordofan state and other parts of the country (El Gamri, 2004).

With the present rate of population growth of 2.9% per year, Sudan is below the water poverty line of 1000 m^3 per capita per year. This fact necessitates adoption of different ways and means of water conservation including lining of irrigation canals (Ahmed et al, 2009). Various types of water losses occur in an irrigation stream. Attempts are made to reduce these losses to achieve an efficient use of water. Moreover, canal lining proved to be highly adaptable to different physical, economical and social conditions.

In many locations around the globe unlined irrigation channels are characterized by very high water losses (Ahmed et al, 2008). Only a portion of the water delivered to a farm is beneficially used by the crops and about one-third of all the water diverted is lost in the conveyance and distribution system in the form of seepage, evaporation, operational losses and tail water losses (run off). Out of these losses, seepage from the bottom and sides of irrigation canal is most important (Ahmed *et al.*, 1984).

Seepage losses from irrigation canals have widely been identified as environmentally critical for the resulting ground water accessions and associated drainage problems (Raiz and Sen, 2005). Significant amounts of water can be saved by improving water management at the farm (Ahmed, 2007). According to Hill (2000), the main types of linings are:

(a) Paved or hard-surface.

(b) Exposed membrane.

(c) Buried membrane.

(d) Earth or conditioned earth.

(e) Soil sealants and stabilizers.

Rawakeeb Research Station (RRS)-west Omdurman is characterized by sandy clay loam soil and hence the need to reduce water losses and improve irrigation efficiencies is evident. Hill (2000) described a methodology to evaluate construction costs, water value, drainage problems, protection from failure and increased capacity in a reliable manner. The lining of a canal is economically viable if the extra cost of providing the lining is less than the benefit resulting from it. In other words, lining is financially justified if the benefit-cost ratio is greater than unity (Arora, 2002). In estimating and budgeting benefits and costs to be used in an evaluation, existing data was used in conjunction with information obtained from professionals and researchers from the field of engineering, agriculture and with farmers who daily deal with the application of water to their land (F.C, 1994). In the pursuit of this objective canal lining was adopted using precasting technique suggested by El Gamri (2005). Concrete more or less meets all the requirements for lining than any other material (Ahmed *et al.*, 1984).

Materials and Methods

Seepage measurement at RRS earth canal

Using two 50 mm Parshal flumes (Plate No.1) seepage from the earth canal of RRS was measured by following the inflow-outflow method. Three seepage tests were carried out and the length of the test reach was 40 m. Two flumes were positioned at the upstream and the downstream ends of the canal then the discharges were observed from the standard flume calibration chart as shown in Fig 1. Seepage rate was calculated using equation 1 as suggested by Weller *et al.* (1994).

$$P_s = Q_1 - Q_2 - Q_A - Q_i / W_p L \quad (1)$$

Where,

P_s = seepage loss per unit area of the wetted perimeter ($m^3/m^2/s$)

Q_1 = input discharge to the test reach (m^3/s).

Q_2 = output discharge from the test reach (m^3/s).

Q_A = the discharge abstracted from the test reach a long its length (m).

Q_i = the discharge abstracted into the test reach a long its length (m).

W_p = the width of the wetted perimeter (m).

L = length of the test reach (m).

Economic viability

Lining of a canal is economically viable if the extra costs of providing the lining are less than the benefits resulting from it. Economic viability was done using the methodology as suggested by Arora (2002):

1. The annual benefits (B) can be estimated using equation 2 as follows:

$$B = a + b + M \quad (2)$$

Where,

B = annual benefits (SD G).

a = annual saving in value of water likely to be lost by seepage (SD G).

b = other annual benefits (SD G).

M = annual saving in operation and maintenance of the channel after lining (SD G).

Plate 1 : 50 mm Parshal Flume

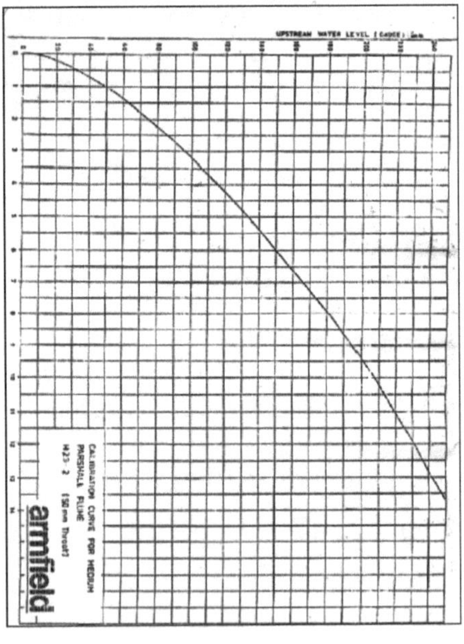

Fig. 1 : Standard Parshal flume calibation chart

2. Annual saving in the value of water (a) can be expressed as presented in equation 3

$$a = p\,l\,s\,W - P\,l\,S\,W \qquad (3)$$

Where,

a = Annual saving in value of water (SD G).

p = wetted perimeter of unlined channel (m).

P = wetted perimeter of lined channel (m).

s = seepage losses in unlined channel (m^3/ m^2 of wetted surface /day).

S = seepage losses in lined channel (m^3/ m^2 of wetted surface / day).

l = length of channel (m).

W = value of water (SD G per m^3).

3. The annual saving in operation and maintenance (M) can be expressed as shown in equation 4

$$M = p\,L\,c_1 - P\,L\,c_2 \qquad (4)$$

Where,

M = annual saving in operation and maintenance cost (SD G).

p = wetted perimeter of unlined channel (m).

P = wetted perimeter of lined channel (m).

L = length of channel (m).

c_1 = operation and maintenance cost of unlined channel (SD G/ m^2).

c_2 = operation and maintenance cost of lined channel (SD G/m^2).

4. Additional capital expenditure on the construction of lined canals (C) is given by equation 5

$$C = c\,T\,L \qquad (5)$$

Where,

C = capital expenditure on construction of lined canals (SD G).

c = cost of lining (SD G per m^2).

T = total perimeter of lining (m).

L = length of the channel (m).

5. The net present worth (NPW) of the total annual benefits (B) over the life of lining is determined from the relation presented in equation 6

   $N P W = B [(1+i)^{N-} 1 / i (1+i)^{N}]$ (6)

Where,

N P W = net present worth of total annual benefits.

B = total annual benefits.

i = prelevant rate of interest.

^{N} = life expectancy of concrete lining.

6. The additional cost of the lined channel should be less than the net present value (NPW) of the annual benefit. Thus

   $C \leq N P W$

7. The benefit/cost ratio is given by equation 7

   B/ C ratio = N P W / C (7)

Information relevant to RRS

The data used for evaluation was collected from irrigation studies at RRS and farmers.

1. According to El Gamri (1998) average seepage rate of earth canal (s) at RRS = 4.8 m/day.
2. According to Deacon (1984) average seepage rate for concrete made from various local aggregate (S) = 4.6 * 10 -7 m/s.
3. According to El Gamri (1998), wetted perimeter (P) for the earth canal at RRS = 0.66 m.
4. Based on canal hydraulic design, wetted perimeter for lined channel (p) is assumed to be = 0.8 m.
5. According to F.C. (1994), value of water (W) as farmers willingness to pay is assumed to be = 0.44 SD G/m^3.
6. With reference to local market up to date estimates of the costs (c) of the cement, gravel, sand, bricks and other costs were found to be 7.92, 0.88, 0.88, 12.197 and 1.76 (SD G/m) respectively.
7. According to Arora (2002), life expectancy for concrete N is assumed to be = 50 years.
8. According to The Bank of Sudan (2004), interest rate (i) is 10%.

9. Annual maintenance cost of earth canal (c_1) with reference to RRS = 2.96 SD G/m^2.
10. Annual maintenance cost of lined canal (c_2) with reference to RRS = 0.88 SD G/m^2.
11. Length of the channel to be lined is assumed to be = 100 m.

Canal construction in RRS

Precasting of concrete slabs

The form

Form was fabricated and developed to produce concrete slabs of 1 m x 0.6 m x 0.06 m (length, width, thickness) as shown in Plate 2. As suggested by EL Gamri (2005; Pers. Comm.), the slabs are to be constructed with acute angle of 45^0 at the forge and back in order to provide the jointing system. Three sides of the form were provided with 10 x 10 cm hinges with sliding pins of 5 mm. The materials used to construct the form were (Ahmed et al, 2009):

1. Angle bars of 1¼″side thickness (3.15 cm).
2. Angle bars of 1½″ side thickness (3.81 cm).
3. Angle bars of 3″ side thickness (7.72 cm).
4. Hot rolled (HR) of 3 x 6 cm rectangular pipes.
5. Cold rolled (CR) of 3 x 6 cm rectangular pipes.
6. Hot rolled (HR) of 4 x 8 cm rectangular pipes.
7. Cold rolled (CR) steel sheet of 0.9 mm thickness.
8. Removable pins.

Construction of RRS Canal

RRS lined canal was constructed in a rectangular section as follows:

1. The soil and gravel mixture used in subgrade preparation was collected from the vicinity of RRS.
2. Two parallel rows of bricks were laid transversely on the compacted subgrade. The gaps between bricks were filled with local coarse aggregates and well compacted again.
3. Transverse joints which were made using the same slab form were positioned every 10 m along the canal to support the walls.

4. Two straight walls of (30-35 cm) height were constructed using bricks and a cement mortar of 1:4 as shown in Fig (2).

Plate 2 : Concrete slab form

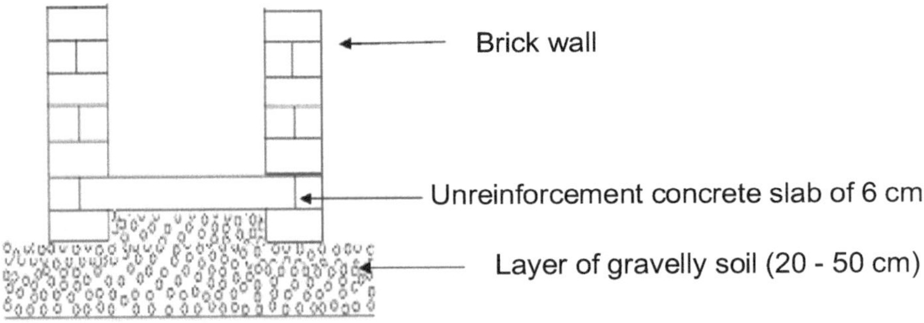

Fig. 2 : RRS canal section

Results and Discussions

Seepage measurement at RRS canal

Figures (3), (4), (5) show the graphical presentation of each seepage test. Seepage rate was calculated using equation (1) with detailed calculations. The average seepage rate of 3.4 m/day was obtained at RRS. The general practice in Pakistan is to adopt canal lining if the seepage rate is equal to or more than 2.94 m/day. Since this value is less than the average value obtained in this work, canal lining is highly advisable at RRS. The average seepage rate obtained from these tests was found to be less than that is reported by El Gamri (1998) of 4.8 m/day and this may be due to the length of the test reach (40 m) which was greater than that used by El Gamri (1998) of 15 m.

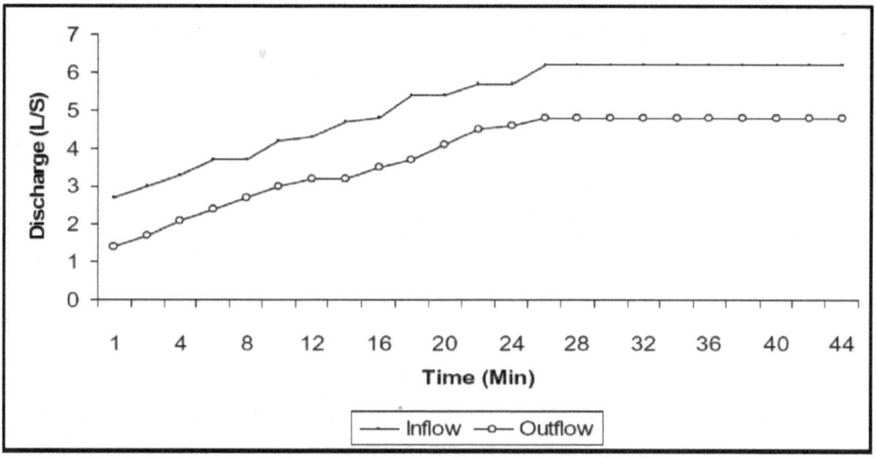

Fig. 3 : Graphical presentation of seepage test (1)

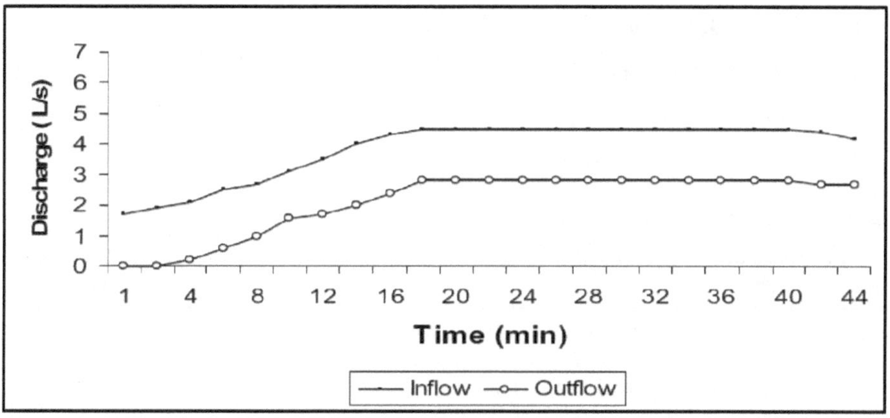

Fig. 4 : Graphical presentation of seepage test (2)

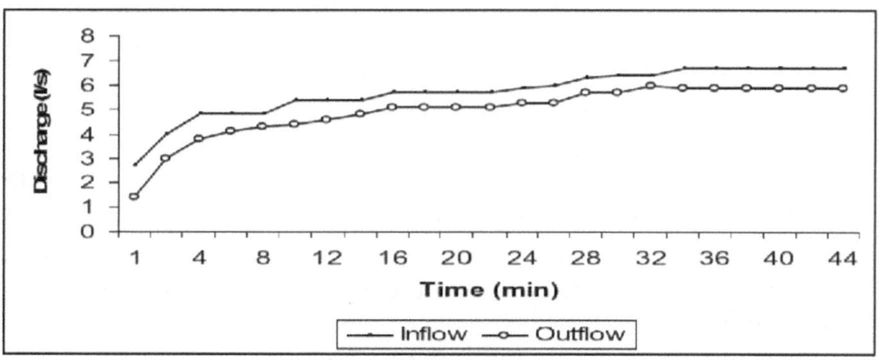

Fig. 5 : Graphical presentation of seepage test (3)

In view of the above assessment and due to the wind blown sands and heterogeneity of RRS soil, it may be concluded that seepage rates obtained from these tests are sufficiently reliable and representative of RRS canal losses.

Economic Evaluation

Canal lining had been practiced before at RRS but no detailed economic evaluation (i.e. B/C ratio) was done. Table 1 shows the results of the economic evaluation. It has clearly appeared that B/C ratio is greater than unity by about 1.3 times. The values for NPW, B/C obtained from equations (6) and (7) respectively are acceptable. The value of B/C ratio shows that the benefits resulting from canal lining at RRS is about 1.3 which offers a good indicator to adopt the precasting technique. According to Khan (1994), this form of canal lining was adopted in the mountains and remotel areas of Balochistan (Pakistan). Nevertheless, adoption of this form of canal lining in RRS is considered beneficial for the following:

1. Considerably high durability.
2. Ease of construction.
3. Ease of maintenance.
4. Use of locally available materials.

Table 1 : Results of economic viability

S.No.	Item	Value (SD G)
1.	Annual savings in the value of water (a)	137.9
2.	Annual saving in operation, maintenance of the channel after lining (M)	122.1
3.	Annual benefits (B)	260.04
4.	Capital expenditure (C)	1961.9
5.	Net present worth (NPW)	2600.4
6.	Benefit/cost ratio (B/C)	1.33

Conclusions

1. There are substantial seepage losses in RRS which were found to be greater than the permissible limits.
2. Once B/C ratio was found to be 1.3, concrete lining using precasting technique can be adopted at RRS and is economically viable.
3. RRS canal was constructed efficiently.
4. The final decision to line or not to line an irrigation canal must be made after consideration of individual and specific site

factors and not relying solely on formulas or experience from elsewhere.

References

Ahmed, T.M. and El Gamri, T. and Magid, I.M. 2009. Use of Precast Concrete Slabs and Bricks for Lining Irrigation Canals. Journal of Science and Technology. Sudan University of Science and Technology. Pp. 199-213.

Ahmed, T.M. and El Gamri, T. and Magid, I.M. 2008. Impacts of Canal Lining on Some Irrigation Efficiencies and Crop Productivity. Albuhuth Journal, The National Center for Research. Pp. 70-82

Ahmed, M and Dutta, S.C. and Abul kair. 1984. Development of low cost indigenous technology to minimize water losses due to seepage in irrigation canal. Journal of Agricultural Mechanization in Asia, Africa & Latin AMA. Pp. 77.

Ahmed, T.M. 2007. Canal Lining and its Impacts on Irrigation Efficiencies. M. Sc. Thesis, Council of Engineering Research and Industrial Technologies, Sudan Academy of Sciences-SAS. (Unpublished).

Arora, K.R. 2002. Irrigation Water Power and Water Resources Engineering. 5th Edn Standard Publisher Distributaries, Nai Sarak, Delhi.

Bank of Sudan. 2004. Financial and Supply Policies 1983-2004. Bank of Sudan, Khartoum. (Unpublished). Pp. 212-213.

Deacon, N.H.G. 1994. Seepage and Durability of Irrigation Canal Linings: A Review of Published Data. Country Technical Note OD/TNS, Hydraulic Research Institute. 1994.

El Gamri, T. 1998. Development and Evaluation of Low Cost Building Materials for Lining Irrigation Canals. M. Sc. Thesis, Faculty of Agriculture, University of Khartoum (Unpublished).

El Gamri, T. 2004. Prospects and Constrains of Desert Agriculture: Lessons from West Omdurman. Environmental Monitoring and Assessment Journal Kluwer Academic Publishers. Netherlands. 99:57-73.

El Gamri, T. 2005. Desertification Research Institute. Water Resources Department. personal contact.

Felicity Chancellor (FC.). 1994. Economics of Canal Lining : An Evaluation Method. Proceeding of the International Workshop in Canal Lining and Seepage. Lahore, Pakistan. Pp. 437-451.

Hill, R. 2000. How well does Your Irrigation Canal Hold Water? Does it Need Lining? Electronic publishing, Utah State University Extension in cooperation with the U.S. Department of Agriculture, Cooperative Extension Service, Utah State University, Logan, Utah. (EP/03 2000/DF).

Khan, A. S. 1994. Balokhistan Minor Irrigation and Agricultural Development project-Approach to Channel lining. Proceeding of the International Workshop in Canal Lining and Seepage, Lahore, Pakistan. Pp. 121-127.

Raiz, M. and Sen, Z. 2005. Aspects of Design and Benefits of Alternative Lining Systems. EWRA European Water Resources Association. E.W. Publications 11/12: 17-57.

Weller, J.A. and Mc Ateer, P. 1994. Seepage Measurement Techniques and Accuracy. Proceeding of the International Workshop in Canal Lining and Seepage, Lahore, Pakistan. Pp. 171-196.

20

Images are an Outstanding Evidence of Rapid Development : A Perfect Example from United Arab Emirates (UAE)

Colonel Mahash S. Al Hameli and Major Ali M. Al Shehhi

Abstract

Before 1960's, UAE was a poor developing country and almost there were no real signs of civilization. During the 1970's two important things happened: H. H. Sheikh Zayed Bin Sultan Al Nahyan, the president of UAE, and the other Emirates rulers united together and established the United Arab Emirates and considerable oil findings were discovered. The UAE Leadership has proved to have deep future vision; in three decades the country jumped from a relatively backward scattered emirates to a well developed and civilized country.

Development is very obvious : from tents and sandy tracks to tower buildings and highways, from desert to green belts of palm trees and vegetables in all parts of the country, and most important, education, healthcare and all other services became the best in the region. Urban development was so sharp; someone who left the place 10 years ago can hardly know his location and services in all sectors became the best at the international level. The oil money was properly invested in development.

UAE Air Force Space Reconnaissance, used satellite and aerial images for the past 20 years and successfully utilized images from various sources including aerial photographs, Russian satellites images, SPOT and Landsat images and recently, in late 2000, established a National Ground Station with direct reception of IKONOS, IRS and KOMPSAT satellites imagery data and with nationals controlling all stages of data reception, processing, analysis and production.

The satellite images are an obvious outstanding proof of the rapid development which has been achieved in the past few decades in all life aspects. It reflects size of efforts and plans that have been steadily put in action for the benefit of the people and their future.

Introduction

Motivation and Aim

Being in a satellite receiving station and continuously acquiring images of UAE and surrounding countries and because UAE is undergoing extensive development for the past three decades, we decided to express this development revolution by using different kinds of imagery, as a proof of rapid development. This will enhance the application of high-resolution imagery, particularly in planning, mapping and research purposes.

Location and general information

UAE is located in SW Asia, east of Saudi Arabia, with about 600 km of coastline of the Arabian Gulf (figure 1) and on the east it has 100 km coastline on the Indian Ocean. UAE area is about 83,000 Sq. Km and population is ~3.75 millions (2002 census). The sand and gravel desert dominates most of the country, with steep mountains in the north and northeastern parts. Apart from Alain city, all major cities lie on the coast including Abu Dhabi, the capital of UAE. The weather is warm and sunny in winter, and hot and humid in summer. UAE's confirmed oil reserves is 97.8 billion barrels; ~ 9% worldwide, with revenues $ US 15-20 billions/year.

Summary of recent UAE history

In 1966 H.H. Sheikh Zayed Ben Sultan Al Nahyan, the present president was chosen as ruler of Abu Dhabi and later in 1971, H.H. Sheikh Zayed Bin Sultan Al Nahyan, along with HH Sheikh Rashid Almaktoum and other Emirates rulers came together and initiated the Federation of the seven Emirates, hence United Arab Emirates was established; Seven Emirates One Nation. A new era of development started.

Some development started in the late 1960s but was remarkable in the mid-1970's with more oil discoveries and price jump, attracting

100,000's of workers and technicians from many parts of the world who shared in the development.

UAE Development Facts

The rapid development of UAE's towns and villages from small oasis in the 1950's to a large modern and fully functional cities required a heavy investment in infrastructure. Roads, airports, seaports, housing, sewage, electricity, water and telecommunication networks have all been developed to meet the demands of a high-tech society (UAE Yearbook, 2003). Tens of billions of $US have already been invested in infrastructure at both federal and emirate levels. In housing, 1000's of villas, low income houses and land parcels have been granted to nationals, residential complexes that have a high standard services. Agriculture development is very obvious, with an area of 270,942 hectares incorporating more than 40 million palm trees, vegetables, crops and fruits; e.g. Liwa farms (figure 1).

The paved highways exceeded 4000 km, with first class and fully lightened and serviced highways linking the major cities. Major recreation and housing projects such as the expansion of Abu Dhabi Korniche (2003-2005) and the Palm Project of Dubai are going on. Jabal Ali Free Zone has more than 2300 companies registered (UAE Year Book 2003). UAE is classified as one of the best in information and technology applications; ranked as top among Arab countries and 21st worldwide with more than 25% of the population subscribed to internet services (Alsuweidi, 2004). In education, number of schools jumped from 20 with 4000 students in year 1962 to 747 with 556,500 students in year 2001/02, with a budget of $ US 1.4 billions for 2001/02. Schools changed from huts with students sitting on the sand in the 1950's (Ronald Codari, 1990) to first class modern ones. UAE women are sharing in the workforce by 41% in education sector and 34% in heath sector.

On international level UAE is becoming a regional and global trade and tourism hub and a center for important exhibitions such as International Defense (IDEX), IT (GITEX) and Dubai Air Show. UAE give $ US 10's millions of aid yearly to poor countries and has funded many projects in over 50 developing countries worth of $ US 6.2 billions. H. H. Sheikh Mohammad Bin Zayed Al Nahyan initiated the UGS project and it was operational in year 2000, with the help of Space Imaging, the leading world company in high resolution imaging satellites business.

Fig.1 : Location of the United Arab Emirates (UAE), the seven united Emirates are named

The electrical energy generated exceeded 8000 MW/hr and water generation is around one billion gallons/day. The country has 6-international airports (e.g. Dubai; fig. 2).

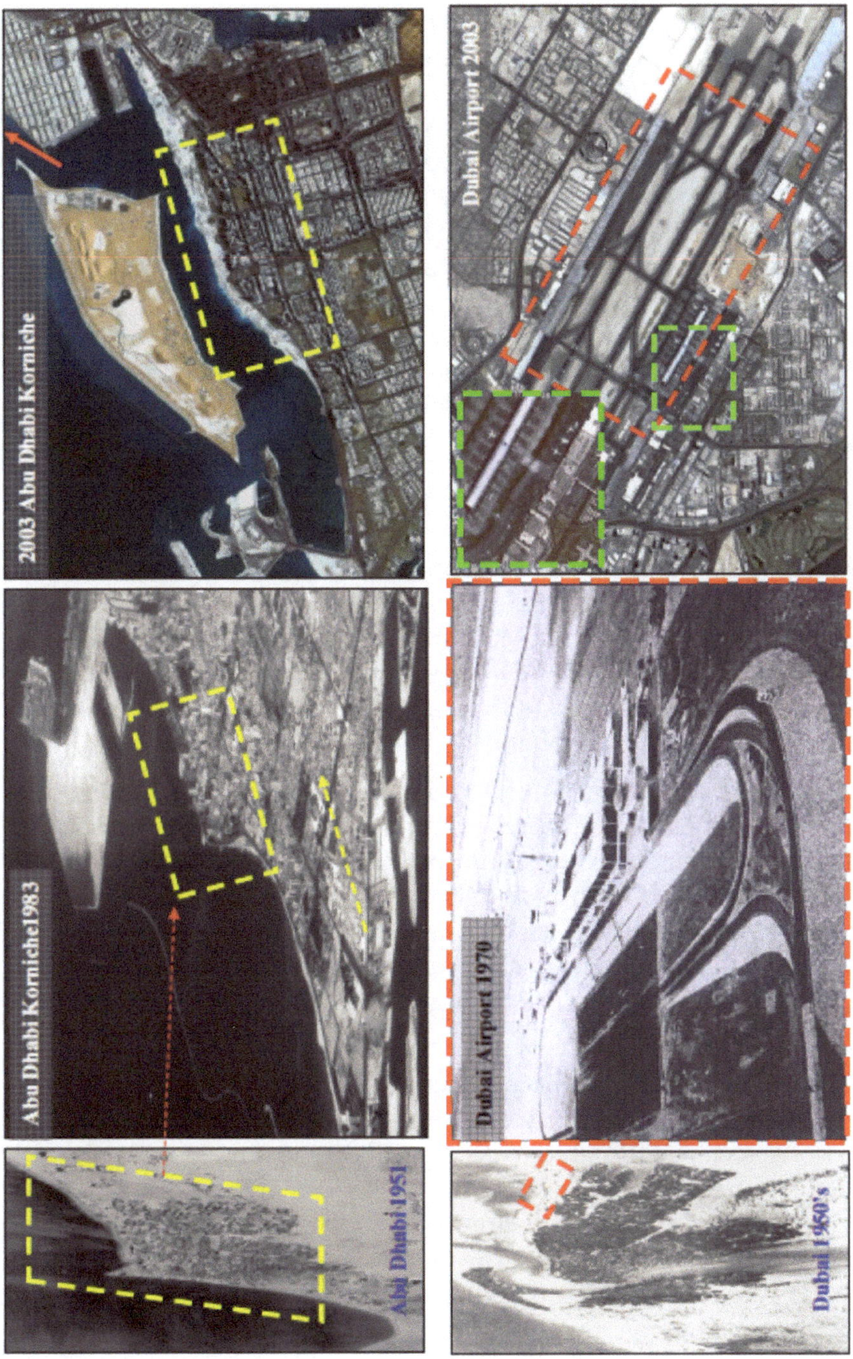

Fig.2 : Historical development of Abu Dhabi & Dubai cities from images.

Methodology and Input Data

The main aim of the study is to emphasize the rapid development, which has been achieved in UAE during the past three decades with the use of images. The study involves first collection of imagery data and reports and documents. Internet surfacing of UAE government sites was useful. Five Landsat images covering the major cities of Abu Dhabi, Dubai and Alain dated 1973, 1984, 1990 and 1995 were collected and IKONOS images collected in 2003 by UAE, UGS were used. The built-up parts of both Abu Dhabi and Dubai were vectorized from images and areas were determined. In addition, municipality maps were rectified, classified and used for correlation purposes. ERDAS Imagine software was utilized for image processing, enhancement, vectorization and classification. A number of images and maps were scanned using a large format scanner.

Data Analysis and Results

The study involved acquisition of historical Landsat MSS (1973) and TM images (1984, 1990 and 1995) and compilation of references on UAE development. Landsat MSS and TM bands 4,3 and 2 false color composite (figures 3 and 4) and composite of TM bands 7, 4 and 2 parallel with recently acquired IKONOS images were used for vectorization of built up areas in Abu Dhabi and Dubai cities.

Generally speaking Abu Dhabi city is calm, well organized and most recent development is devoted to outside the island in the E, SE and NE directions. The island is almost fully developed. The Lulu island are somewhat small islands and these have been recently established just west of Abu Dhabi island, on which remarkable recreational and residential/commercial projects will be developed. Dubai city has largely expanded, particularly along the SW coast and to the east, with more emphasis on tourism/recreational and industry.

Using an old aerial photograph of Abu Dhabi dated 1951 (figure 2), the built-up area which was simple and scattered huts, tents and some houses were estimated to be around one sq. km. In 1973, the built-up area (figure 3, table 1) was about 15 sq. km and in 11 years time it was doubled about 5 times rising up to 82.9 sq. km. In 1984. However, the 1973 Landsat MSS image provide a rough estimation and many vacant areas must have been included. In the years 1984-1995 a sort of steady expansion was there, with an average expansion of 10 sq. km / year; ~ 8-12%. Landsat TM images permitted better estimate of urban expansion than MSS, but still some exaggeration might be there.

Table 1 : UAE cities expansion estimates from images

Year/image type	Abu Dhabi area Sq. Km	Dubai area Sq. Km
1951 Air photo	0.9	NA
1973 Landsat MSS	15 (6/yr)	13 NA
1984 Landsat TM	82 (6.2/yr)	82 (6.5/yr)
1990 Landsat TM	128 (7.6/yr)	112 (5/yr)
2003 IKONOS	262 (12.75/yr)	238 (13.75/yr)

IKONOS images of Abu Dhabi and Dubai collected by UAE UGS in late 2003 were enhanced and built-up areas were accurately vectorized (figure 3). The area of Abu Dhabi city and associated surrounding towns such as Alshahamah, Musaffah, Almafraq and Baniyas have considerably expanded in the past 8 years from 160 to above 260 sq. km. Still some parts of these towns are not included because they fall outside the sample area selected (figure 3). These towns are fairly recent and have been undergoing rapid expansion especially during the past two years.

Dubai city on the other hand has developed rapidly and the built-up area (only in the sample area; figure 3) jumped from ~13 sq. km. in 1973 to 82 sq. km. in 1984, 128 sq. km. in 1995 and reached 238 sq. km. in 2003(table 1). However, two large areas are not included, the Emirates road and Jabal Ali, parts of which has undergone a rapid expansion, particularly in the last three years. It should also be in mind that there is vertical growth in the city center, Sheikh Zayed Street and Aljumeirah areas. The Musaffah industrial area has doubled four times; from 11 sq. km in 1984 to 16 sq. km in 1990 and from 18 sq. km. in 1995 jumped to 40 sq. km in 2003. It is very obvious from Landsat TM bands 7, 4 and 2 image that a real revolution in agriculture is there, particularly in the middle of the desert, namely in Liwa, Alain and SE of Ras Alkheimah (figure 1).

Scanning, rectification and classification of city expansion maps (UAE National Atlas, 1993), enables to make absolute areas of city expansion samples from Abu Dhabi and Dubai. Abu Dhabi built-up area was 11.2 sq. km in 1969, 15.5 sq. km in 1974 increased to 34.9 sq. km in 1980 and 42.9 Sq. km in 1990. This implies rapid growth in the late 1970's after the jump in oil prices and slow growth in the 1980's. Similarly in Dubai the city area was doubled from 25.3 sq. km in 1976 to 58.2 sq. km and in 1988 it reached 67 sq. km. It should be clear that major cities in other Emirates have also undergone considerable development and although may not be of the same rate as Abu Dhabi

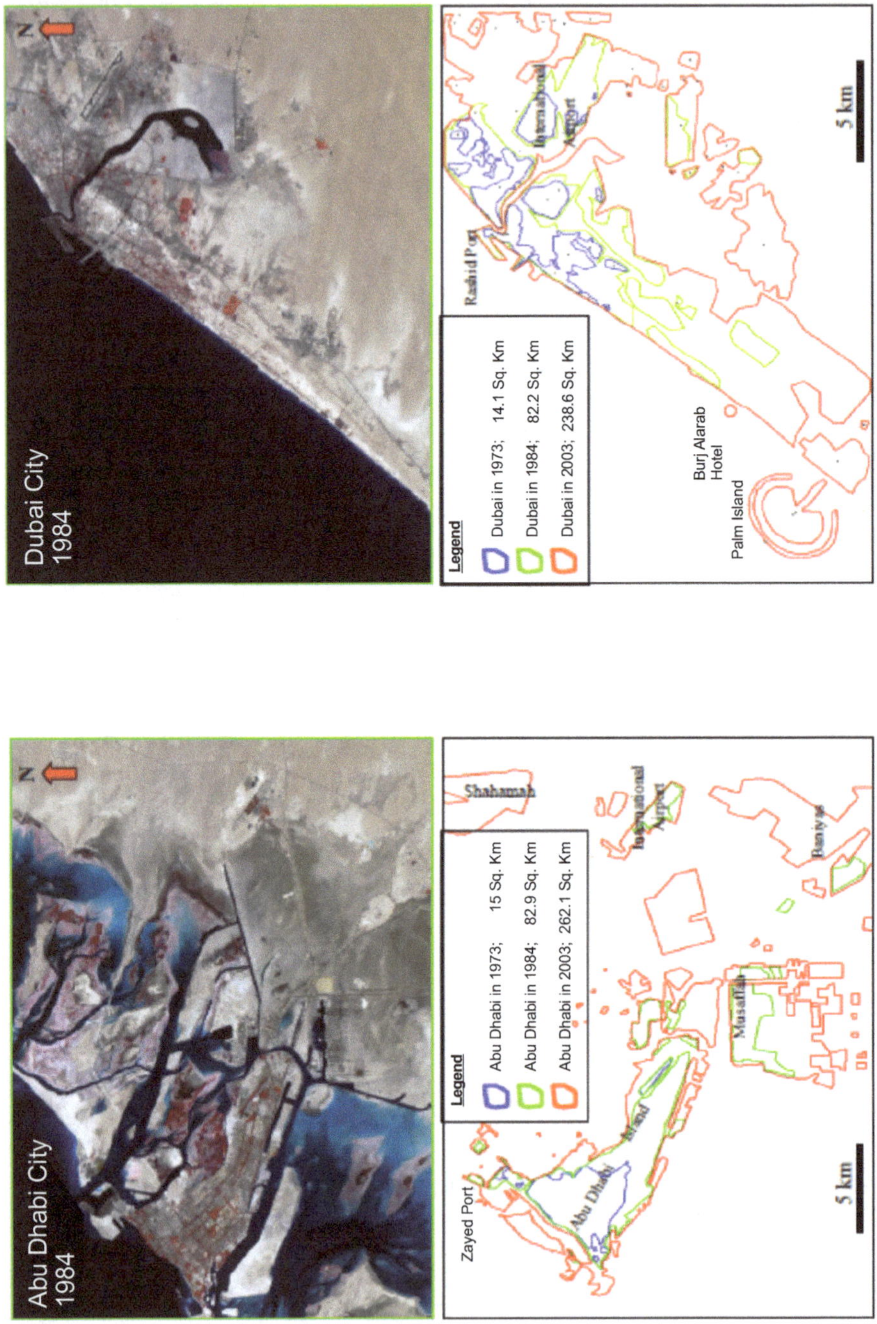

Fig.3 : Expansion of Abu Dhabi and Dubai cities in thirty years, (using Landsat and IKONOS satellities imagery)

and Dubai, but still all Emirates have equal basic services of education, health, electricity, etc.

Conclusions

- Images of different kinds stand as a historical record of facts and changes, which help current and future generations to have an honest, look at development rates. UAE proved to be an excellent case study with rapid development in the past 30 years.
- The wise policy of H. H. Sheikh Zayed Ben Sultan Al Nahyan, the UAE president and other UAE Emirates rulers combined with proper utilization of the oil money was behind this great achievement; a move from a poor non-developed to a highly civilized society was a kind of miracle.
- IKONOS imagery were superior in delineation and identification of urban development classes, whereas Landsat MSS image was poor and only shown major roads and building clusters, with fair estimates from Landsat TM images.
- Availability of images is a crucial factor in all studies dealing with historical development, especially in the past about 50 years. In case of UAE, having a national Ground Receiving Station, parallel with good bilateral relations with the world community, facilitates the acquisition, processing and analysis of images.

Acknowledgements

Sincere thanks to H.H. Sheikh Mohammad Bin Zayed Al Nahyan, Deputy Crown Prince of Abu Dhabi and Chief of Staff of UAE Armed Forces, for the patronage of keeping the UAE Armed Forces up to date with high technology, represented here by initiating the establishment of the National ground station. Thanks also goes to Brigadier Khalid Abu Alainain, Commander of UAE Air Force and Air Defense for continuous support and encouragement. The assistance provided by the UGS staff is appreciated.

References

Alsweidi, E. B., 2004. The school as a total change center; an Arabic report, personal communications.

Codari, R. 1990. The Seven Sheikhdoms, Stacy International, London.

National Atlas of United Arab Emirates, 1993. The United Arab Emirates University. GEOprojects ltd. U. K.

UAE Yearbook 2003. UAE Ministry of Information & Culture.

Useful Websites

www.uae.interact.co, www.uae.gov.ae/mop, www.uae.gov.ae/intro www.uae.gov.ae/maf

21

Urban Sustainability Perspectives on Changes in Adaptation to Saudi Arabian Desert Environment: The Role of Geospatial Technologies

Yusuf Adedoyin Aina

Abstract

The desert environment is one of the most challenging environments for human habitation. Inhabitants have to deal with climatic factors such as high temperature, sandstorms, flash floods and lack of water due to low rainfall. The people of Saudi Arabian desert had developed various traditional adaptation approaches such as living in tents, wearing head dresses, settling close to source of water, using camel for transportation, irrigation farming and traditional land management (hema). In recent times, modern technologies for air conditioning, transportation and housing have replaced the traditional adaptation methods especially in the cities. The modern technologies have reduced the challenges faced by the traditional society but has thrown up new challenges to sustainability. The new issues include urban flooding, depletion of underground water, high energy demand and consumption and loss of natural habitat to urbanization. This chapter reviews the traditional and the modern adaptation techniques in Saudi Arabia especially in the urban areas. It examines the implication of the changes for urban sustainability and explores different approaches for reducing the negative impacts. The chapter also dwells on the role of geospatial techniques in adapting to the desert environment.

Introduction

Background to the study

The desert environment is one of the most challenging environments for human habitation. The environment is characterized by dryness and sparse vegetation cover as the amount of precipitation is too low to support abundant vegetation. About 20% of the world (land surface) gets less than 25 cm of rainfall per year and in addition to that, about 15% of the world is semi-arid (Watson, 1981). Generally, the amount of rainfall in a desert is less than 50 cm in a year. Lack of vegetation cover exposes a desert to water and wind actions. A large percentage of rain water flows away as flash flood with high intensity and speed. So, little water percolates into the soil to replenish underground water and the force of the run-off is great enough to erode most things on its path. Similarly, strong and unimpeded wind can drive large amount of sand to become heavy sandstorm and when the sand settles, it forms dunes. The major challenge of the desert environment is the extreme weather conditions. It can be very hot during the day and extremely cold, even freezing, at night.

Despite the vicissitudes of the desert environment, living organisms including human beings had found ways of adapting to the extreme environmental conditions. Some plants have small leaves and chlorophyll on their stems to reduce their water loss. They also develop long root system to reach underground water. Animals burrow into the ground to avoid the high temperature during the day. Some animals have shells and scaly skins to reduce loss of water. Also, human inhabitants had developed adaptation techniques for living in the desert. In Saudi Arabia, a country that covers most of the Arabian desert the people had developed traditional adaptation techniques like living in portable shelters (tents), settling close to source of water, using camel as means of transportation, irrigation farming and traditional land management (hema) (Watson, 1981; Vincent, 2008).

In recent times, after the unification of the country and the sedentarization of nomads, human adaptation shifted from the traditional ways to modern techniques and the settlement pattern was transformed (from tent to built settlement) (Al-Said, 1986; Cole, 1981). The developmental changes were catalyzed by the revenues from oil production. The oil wealth was used to build urban and industrial infrastructure and to provide social amenities. Technological means of adaptation such air conditioning system, salt water desalination and automobiles began to replace the traditional approaches especially in the cities. Thus, the cities became centers of attraction to the populace and quite a number of rural dwellers started migrating to the cities. Large urban population puts a lot of pressure on the fragile environment and also exposes a large number of people to natural disasters such as flooding. The demand for water and energy at the urban areas is increasing with increasing population. These issues of high energy and water demand, urban flooding and loss of natural habitat call to question the sustainability of Saudi cities.

Related works

Watson (1981) highlighted how people living in the deserts of the world had adapted to the environment and the problems and challenges posed by growing desert population He opined that "the growth of towns and cities has created problems. For instance, there is an increasing demand for water for household uses, and water generally must come from underground" (Watson, 1981). Increasing use of underground water could lead to the depletion of water in an aquifer. High population increase in the desert cities will increase water and energy demand and waste generation, which will affect the fragile desert environment especially the semi-arid and sub-humid regions at the edge of the deserts (Maukonen *et al.*, 2006). Moran (2008) also dwelt on human adaptability in arid regions and environmental sustainability of rapidly growing cities. The study highlighted the use of spatially explicit techniques in modern ecosystem studies.

In the context of Saudi Arabia, Mallakh (1982) examined the rapid development and transformation that occurred in the Kingdom in the 1970s. The study discussed the development of Saudi oil economy and industrialization including the establishment of industrial cities (such as Yanbu and Jubail). These rapid economic changes led to the development of infrastructure including desalination for the provision of water for domestic, agricultural and industrial uses. Some studies have examined the socio-economic (Cole, 1981) and environmental impacts (Vincent, 2008; Tolba *et al.*, 2009) of these changes including the effects on urban spatial pattern and sustainability (Al-Said, 1986; Eben Saleh, 1998; Al-Hathloul, 1996; Alshuwaikhat and Nkwenti, 2002; Choguil, 2006).

The studies by Al-Said (1986), Cole (1981), Eben Saleh (1998) and Al-Hathloul (1996) portrayed the transformations and social changes in the way of life of the Bedouins and also the changes in the urban settlement pattern. The old settlement pattern had been rapidly replaced by a new pattern. The courtyards had been replaced by villas and wider streets. In a view to highlight the environmental issues (that might be related to the socio-economic changes) facing Saudi Arabia and the Arab world, Vincent (2008) and Tolba *et al.* (2009) addressed challenges such as urban hazards, desertification, soil salinization, climate change, sea level rise and sustainable development while Alshuwaikhat and Nkwenti (2002), Choguil (2006) and Alshuwaikhat and Aina (2006) examined the issue of urban sustainability in arid regions and Saudi cities of Riyadh and Dammam. The authors agreed that environmental aspect of sustainability has not received adequate attention. Geospatial techniques, such as GIS, are regarded as viable tool for promoting urban sustainability and effective urban governance (Alshuwaikhat and Aina, 2006; Alshuwaikhat and Nkwenti, 2002).

Saudi Arabian Desert Environment

Saudi Arabia lies between longitude $34^{0}29' 30''$ and $55^{0}40' 00''$ east and latitude $16^{0}22' 46''$ and $32^{0}14' 00''$ north. It occupies the major part of the Arabian Peninsula and covers an area of about 2,000,000 km^2. It has border with Jordan, Iraq and Kuwait in the north; Qatar and United Arab Emirate (UAE) in the east; Oman and Yemen in the south and south east. Saudi Arabia has a coastline of about 3400 km mainly at the Arabian Gulf and Red Sea coasts (Saudi Geological Survey, 2003). Saudi Arabia is located in one of the arid regions of the world - The Arabian Desert (Figure 1). Rainfall is not more than 200 mm in most of Saudi Arabia and the temperature could be over 40 °C in summer and less than 20 °C in winter. There are a few sub-humid areas mainly in the south-western part of the country. These areas receive more rainfall and they have more vegetation cover than other areas.

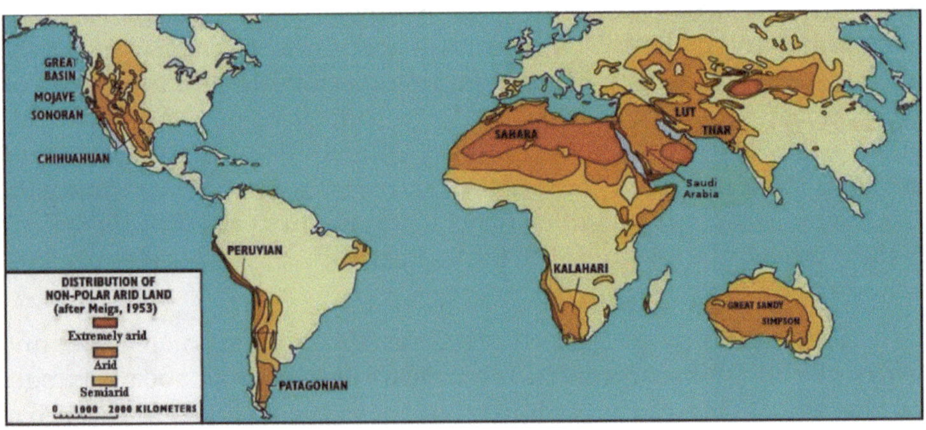

Fig. 1 : The location of Saudi Arabia (Source: Adapted from Mustapha, Lim, and Mat Jafri, 2012)

The physical terrain of Saudi Arabia can be divided into coastal lowlands, mountain ranges, volcanoes, plateaus and sandy areas (Saudi Geological Survey, 2003). There are no permanent overland water bodies. The coastal lowlands are found at the eastern and western coastline of Saudi Arabia. The western coastal lowlands are called *tihamah* (Vincent, 2008). The coastal plain runs from north to south along the *Red Sea* coast. The coastal plain contains marshlands and salt flats (*sabkha*). The eastern coastal plain runs along the Arabian Gulf coast. The mountain ranges lie to the east of the western coastal lowlands (Figure 2). The mountain ranges have varying elevations from 300 m to about 3,000 m at the Asir region in the south (Saudi Geological Survey, 2003). The mountain ranges are interspersed with valleys (wadi) such as Wadi Fatimah, Wadi Bisha, Wadi Dawaser and Wadi Ramah. The volcanoes (*harrah*) include Alharrah volcano in Jouf region in the northern part of Saudi, Uwayrid volcano close to Al-Ulla town, Khaybar volcano close to Madinah, Salma volcano and Rahat volcano. The plateaus include Najd upland located close to Riyadh (east of the mountain ranges) and about 600 km wide at its maximum (Vincent, 2008). Other plateaus are Hismaplataeu (found in the north west - east of Tabuk), Tabuk plain (average altitude of 800m) and Summan (narrow and mainly limestone) (Vincent, 2008).

Fig. 2 : Western coastal lowland with mountain ranges at the background

The sandy areas cover about a third of Saudi Arabia and include Jafurah, Nafud, Dahna, and Rub al Khali. Jafurah lies in the eastern part of Saudi Arabia. It contains dunes, sand sheets and Sabkha. The Nafud lies between Jauf and Hail and has average elevation of about 900m. It sand dunes are reddish due to the iron oxide content. Dahna is located south of Nafud and it is a narrow link between Nafud and Rub al Khali in the south. It lies in the central part of the Kingdom and extends to about 1300 km^2 Rub al Khali also known as empty quarters covers most of the southeast of Saudi Arabia. Sand dunes with height close to 250 m cover the terrain. Rub al Khali is rich in oil and it contains most of the oil fields including Shaybah oil fields.

According to 2010 census, Saudi population is about 27 million. About half of the population resides in just five cities; Riyadh (5.2 million), Jeddah (3.4 million), Makkah (1.5 million), Dammam (1.2 million) and Madinah (1.1 million). The concentration of the population in few cities put a lot of pressure on the urban infrastructure and has implications on sustainability of the cities. The government has made recent efforts to establish new cities to decentralize population distribution. Industrial and knowledge cities have been established in Thuwwal, Jizan, Hail and Ras Alkhair. The pressure on infrastructure and demands from the five cities will still be enormous even if the population increases in those cities are abated.

Traditional Adaptation Methods: The Nomads, Agricultural Settlers and Town Dwellers

The traditional inhabitants of Saudi Arabia can be grouped into nomads, agricultural settlers and town dwellers. The settlements composed of tiny villages distributed in about three belts; Hijaz-Asir belt, Najd belt and Ahsa-Qatif belt (Mubarak, 1992). Each group of settlers had means of adapting to the desert environment and sometimes the settlers changed their roles from nomads to agricultural settlers or to town dwellers. The nomads are hunters and gatherers and herdsmen who traverse from one oasis to another to get pasture and water. Due to their high mobility, the nomads have portable dwelling in the form of tents that protect them from the weather elements and can be easily moved. Animals such as camels and donkeys were helpful to the nomads in moving across the desert and transporting loads. Despite the high mobility of the nomads, their socioeconomic activities are within known territories controlled by tribal system (Nahedh, 1989). They regarded the resources within a tribal territory as communal and the resources are managed under the communal system to ensure preservation and reduce overgrazing.

The agricultural settlers settled where rainfall could support cultivation in the southwest region (*Asir*) and the southern coastal plain (*Tihamah*). Also, agricultural settlers could be found around the oases such as Alhasa oasis and Qassim region. In other farming villages, scattered all over the kingdom, farmers depended on irrigation from underground water. Ditches and surface tunnels were made to store and direct water to the farms. Sometimes salt water would be drawn into the farm from the wells and after evaporation the soil would become salt pan (*sabkha*). The problem of salinization has become more pronounced with the advent of modern technology that enables farmers to draw more volume of salt water unto the farms at a faster rate. The nomads and agricultural settlers had developed means of conserving vegetation through the traditional rotation grazing system (*Hema*) (Zahran and Younes, 1990).The system is based on "a set of regulations" that is used to control "the extent and intensity of utilization of resources"(Zahran and Younes, 1990). The hema system has about three levels of control that include tribal, village and individual levels. The system has been widely practiced in Saudi Arabia and Zahran and Younes (1990) noted more than 200 of hemas in the Hijaz region alone.

The town dwellers were those who settled in towns and urban centers such as the holy cities of Makkah and Madinah. The settlers engaged in small scale artisanship and commerce. The influx of pilgrims to the two holy places had boosted socioeconomic activities and urban population in the cities and surrounding areas. The growth of the port city of Jeddah had also been influenced by these activities since it serves as gateway to the holy sites. The development of the cities followed the traditional organic pattern with courtyards, narrow streets and dense urban form (Figure 3). The economy of these cities was not based on firm economy (Mubarak, 1992).The artisans generally got the skills from their families and the enterprises were run as "risk-minimizing rather than profit-seeking" for the expansion of productivity (Mubarak, 1992).

Fig. 3 : A typical traditional courtyard pattern

Modern Technological Adaptation: Nomads became Settlers and Movement to the Cities

After the establishment of the Kingdom in 1932 and the discovery of oil, all the aspects of traditional adaptation techniques were transformed by technological and socioeconomic changes. The nomads had been settled in agricultural or urban communities and the traditional local tribal system had been replaced with centralized national system of governance. The portable tents with passive cooling systems were replaced with static built environment that relies on energy for cooling. The transportation system has also become dependent on energy as 'beasts of burden' were replaced with automobiles (Al-Said, 1986).

Agricultural activities had expanded beyond the limit hitherto imposed by availability of water. Modern techniques of irrigation and extraction of underground water had been introduced to cater for agricultural needs. Prior to the introduction of modern techniques, only about 0.2% (Abo-Hassan, 1981) of 2 million km^2 was cultivated and the percentage grew in multiples (Mubarak, 1992) after modernization. Despite the huge growth in the agricultural sector, the sector lost a sizable number of workers to the booming oil industry and growing cities (Mubarak, 1992).

Rapid urbanization is one of the results of the transformations brought about by the oil economy and the introduction of modern technology. The percentage of urban population rose from 16% in 1950 to 49% in 1970 and to 77% in 1992 (Alkhedheiri, 2002).Urbanization in Saudi Arabia has been reported by Alshuwaikhat, Aina and Rahman (2006) and it has reached to 88.5% in 2006 and probably more than 90% by 2012 according to the general Gulf Cooperation Council (GCC) data (UNEP, 2012).Due to the fast pace of urban and technological development, urban built environment has been rapidly changed in form, pattern and structure. The mud brick and stone houses gave way to cement brick and concrete houses and the traditional organic pattern gave way to grid iron pattern with wide streets (Figure 4). It is noteworthy that the Saudi urbanization, as mentioned above, is polarized to five main cities (Riyadh, Jeddah, Makkah, Madinah and Dammam) (Table 1). The five cities are the residences for about 46% of the population. These transformations and rapid urbanization have implication on the urban sustainability of Saudi cities. The increasing energy and material consumption of the teeming urban population, waste generation, increasing pressure on infrastructure and resilience of the cities to urban hazards and climate change are factors that could impact on sustainability.

Fig. 4 : A typical wide street of the modern era.

Table 1 : Population growth in major urban centers in Saudi Arabia, 1974 - 2010

City	1974	1992	2004	2010
Riyadh	666,840	2,777,096	4,087,152	5,188,286
Jeddah	561,104	2,046,251	2,801,481	3,430,697
Makkah	366,801	956,697	1,294,168	1,534,731
Madinah	198,186	608,295	918,889	1,110,093
Dammam	127,844	482,321	744,321	1,243,512
Total (million)	1.921	6.871	9.846	12.507
National total population (million)	7.010	16.903	22.678	27.137
Percentage of national population	27.4	40.6	43.4	46.1

Source: Ministry of Economy and Planning, Saudi Arabia.

New Challenges of Modern Saudi Cities and Urban Sustainability

The challenges posed by rapid urbanization to urban sustainability are not only due to mere increase in population but also the fragility of the environment. Increasing urban growth enacts pressures on the already fragile environment especially the semi-arid areas and also impact on urban resilience to hazards. The main challenges to urban sustainability of Saudi cities include water and energy demand, urban and natural hazards, urban infrastructure, vegetation loss and desertification, urban ecological footprint and climate change.

Water and energy demand

The increasing demand for water and energy is one of the main challenges of rapid population and urban growth. Highest demands for water and energy are from domestic use in rapidly expanding cities, industrial activities and agriculture. The demands for water are met by underground water, desalination and waste water treatment. The utilization of underground water for agriculture has already resulted in depleting aquifer as the demand surpasses aquifer recharge rate (Al-Ibrahim, 1991).Also, the problem of salinization has been exacerbated with increasing use of saline underground water for irrigation especially at the coastal areas. For example, some of the farmlands at Wadi Fatimah (close to Jeddah) have been abandoned because of salinization. Saudi Arabia has become the largest producer of desalinated water in the world due to its efforts to meet water demands through desalination (Mezher, Fath, Abbas and Khaled, 2011). The Kingdom produces about 10.5 million m^3/day of desalinated water and the domestic water consumption in 2005 was about 2.1 billion m^3 (Mezher, Fath, Abbas and Khaled, 2011).The increasing production of desalinated water is considered to have environmental impact on the Arabian Gulf and Red Sea (Purnama, Al-Barwani and Smith, 2005).

The demand for electricity in the Kingdom is also high like the demand for water. It is used for industrial production and also for domestic use especially air conditioning during the hot and humid summer period. Electricity consumption is intricately linked with crude consumption as Saudi power supply is mainly generated using crude oil. In the 2000s, Saudi used less than 200,000 barrels/day which has risen to more than 450,000 barrels/day in 2009 (Husain, 2011). The total electricity consumption in 2000 was 109 billion megawatt and it has risen to 186 billion megawatt in 2009 (US-EIA, 2012).

Urban and natural hazards and urban infrastructure

The contingency plan and resilience of a city during urban and natural disasters impact on urban sustainability. Most of the cities of the world are faced with risks of one or more hazards. Hazards have impacts on available urban infrastructure if they result in the destruction of infrastructure such as road, water and sewage system. Lack of adequate infrastructure resulting from urban hazards due to damages to infrastructure exacerbates the impacts on urban populace. Effective disaster management goes a long way to reduce the impact of hazards and ensure resilience. In Saudi context, urban flooding poses a risk to the cities due to high intensity of desert rainfall and inadequate infrastructure in form of drainage to collect runoff. An example is the Jeddah flood that occurred in November 2009. The flood claimed over 100 lives and caused about a billion Saudi Riyal losses (Al-Zahrani, 2009). It was concluded that lack of disaster management planning contributed to the scale of the impacts of the disaster (Akeel, 2009). Some of the Saudi major cities such as Riyadh, Jeddah and Makkah also have pending problems on inadequate infrastructure especially sewer network, water distribution and housing for urban poor.

Vegetation loss and desertification

Vegetation loss, land degradation and desertification result from increasing human activities such as overgrazing, saline irrigation and urbanization. The study by Barth (1999) exemplified the case of desertification in Saudi cities using Jubail city as a case study. It is asserted that "land degradation was intensified by the rapid social and economic changes that occurred due to extensive oil exploration that led to a population increase and urban development. In the following decades, traditional husbandry has been transformed into a commercial ranging system. The provision of governmental subsidized concentrated foodstuffs replaced the dependence upon the open range, which led to over exploitation of range resources" (Barth, 1999).The study by Khan and Al-Homaid (2003) has also shown that the development of urban infrastructure has led to the loss of mangrove stands of about 260 hectares in Tarut Bay and Tarut Island from 1973 to 1997. In Riyadh, the vegetation cover has decreased from 40 km^2 in 1994 to 30 km^2 in 2007 due to encroachment by urban expansion (Masheeand Malthus, 2009).

Urban ecological footprint

The national ecological footprint of Saudi Arabia has surpassed the bio-capacity since 1979 and in 2008 there is a gap of about 3 global hectares per capita between Saudi bio-capacity and ecological footprint (GFN, 2012). Ecological footprint is the per-person resource demand while the bio-capacity is the capacity of the ecosystem to meet the demand. Thus, in the case of Saudi Arabia demand is greater in three folds than the ecosystem capacity. Since urbanization is around 90% in the Kingdom, the cities are the sources of the high demand on the ecosystem. Even at the city level, the ecological footprint of Riyadh is found to be higher than the footprint of other cities in neighboring countries (Choguill, 2006).

Climate change, global warming and sea level rise

There is a link between climate change and urban development through vehicle-mile-travel (VMT) and carbon emission (Ewing *et al.*, 2008). Sprawling urban development and automobile-dependence in Saudi cities contributes to the high CO_2 emission from petroleum consumption relative to surrounding countries. Saudi Arabia's emission is about one third of Middle East emission and about 3% of global emission (Table 2). Apart from the issue of its high contribution to Green House Gas emission (GHG), Saudi is also vulnerable to impacts of climate change and sea level rise. According to the projected climate change scenario; there will be an increase in temperature of about 4%, a minimum increase in precipitation of about 30% (except in the northern region, -4%) and a minimum increase in runoff of about 100% (Al Zawad, 2008).It means Saudi Arabia will be vulnerable to heat waves (especially in summer months) and flooding. A study by (Babu, Sivalingam and Machado, *in press*) has also found that places along the Arabian Gulf coast are vulnerable to coastal inundation due to sea level rise.

Table 2 : Saudi Arabia CO_2 from the consumption of petroleum (Million Metric Tons)

	1980	1990	2000	2010
Middle East	292	490.6	658.5	958.9
Bahrain	2.2	2.9	3.5	6.6
Kuwait	15.3	12.2	39.8	56.2
Oman	2.4	5.7	7.9	18.5
Qatar	1.5	3.4	5.2	16.8
Saudi Arabia	88.5	138.7	191.8	306.3
United Arab Emirate	10.6	40.1	51.8	78.8
World	8825.2	9118.0	10174.2	11174.6

Source : United States Energy Information Administration (EIA)

Approaches to achieving sustainable cities in Saudi Arabia

Quite a number of efforts have been made through different approaches to improve the sustainability of Saudi cities. The approaches include reducing the demand for energy and water through a more efficient utilization, exploring alternative sources of energy and water, introducing conservation zones to protect habitats and implementing environmental impact assessment regulations and early warning systems. The study by Rahmanand Khondaker (2012) has highlighted the different initiatives (including policies, plans, programs and projects) for reducing GHG. It has also presented the efforts by Saudi Aramco at lowering CH_4 emissions.

Similarly, a study on conservation exemplified that protection against grazing and human impacts has led to an increase in vegetation abundance and species richness at the eastern coast of the Kingdom (Shaltout, El-Halawany and El-Kady, 1996). Programs of the National Commission for Wildlife Conservation and Development (NCWCD) such as the one at Farasan Archipelago are aimed at conserving biodiversity and protecting the Kingdom's wildlife which are threatened by urban expansion and related anthropogenic activities (Mutairi *et al.* 2012; Abuzinada, 2003). In other to ensure sustainability, some of the urban development authorities in the Kingdom have started including sections on environment in their plans. An example is the inclusion of environmental development plan for Wadi Hanifa in the Riyadh City development scheme by High Commission for Development of Riyadh (ADA) (Arab News. 2011).Also, there are plans to implement light rail, public transport system, flood drainage system and housing projects (Abuzinada, 2003).

There is the need to complement these efforts with a better implementation of environmental impact assessment. A study by van der Merwe,Lattemann and Amy*(in press)* asserted that non-transparent and inaccessible data and ambiguous policies militate against having a thorough environmental impact assessment of discharge from desalination plants. In line with the inadequacies of environmental impact assessment in the Kingdom, there is a need for an effective institutional framework (for the coordination of private and public stakeholders) and a strong political will in the drive towards sustainable cities. "A comprehensive approach needs to be adopted for the planning and management of urban centers in order to enhance systems performance, human interaction, and contained developments" (Alshuwaikhat and Nkwenti, 2002). Also, efforts must be made to develop adequate and effective methods and tools for baseline studies and measuring sustainability. Geospatial techniques are very relevant in developing the needed tools and methodologies.

The Role of Geospatial Techniques in Sustainable Urban Development in Saudi Arabia

The study by Alshuwaikhat and Nkwenti (2003) has argued for the adoption of emerging technologies such as geospatial techniques in managing urban development. In the same vein, Aina (2012) posited that "the myriad of challenging issues facing the world today ranging from urban growth and biodiversity to climate change have spatial dimension. Geospatial technologies are needed in addressing these challenges". The application of geospatial techniques in managing urban development is not new in Saudi Arabia. The studies by Masheeand Malthus (2009) and Aina *et al.*(2008) has applied geospatial techniques in studying urban expansion in Riyadh. Also, Alshuwaikhat andAina (2006), Al Zawad (2008) and Babu *et al.*(*in press*) has used GIS to study urban sustainability, highlight different climate change scenarios and depict the impacts of sea level rise.

Apart from individual researchers, different municipalities and urban planning authorities such as Jeddah, Dammam, Madinah, Riyadh, Royal Commission for Jubail and Yanbu (RCJY) and Ministry of Municipality and Rural Affairs (MOMRA) have developed urban planning systems that are based on GIS and other geospatial techniques. Even some of the municipalities, such as Jeddah Municipality, have developed urban observatories for monitoring urban development. However, these efforts are individual in nature and there is no strong link between these urban planning systems. A program has been started by GIS Committee at Eastern province and also Riyadh Authority to implement a wider network of collaborative system but it has not been fully implemented. The study by Alshehri (2011) affirmed that "there is lack of knowledge and no clear framework describing the optimal way for stakeholders, users, providers or administrators, to collaborate effectively in establishing a National Spatial Data Infrastructure (NSDI). Moreover, the complex, multi-layer and multi-jurisdiction system of government leads to competing interests and mandates in coordinating spatial activity". It highlighted the problems of lack of unified data standard and appropriate legislations.

In order to fully reap the benefits of applying geospatial techniques, there is a need to develop the needed institutional framework and regulations for implementing adequate and effective national spatial data infrastructure. The collaborative framework will promote efficient use of data by reducing data redundancy and ensuring data sharing. Also, the accessibility to data by the public will be ensured without jeopardizing confidentiality. Currently, data sharing and public accessibility to data is haphazardly implemented. It depends on the discretion of the possessor of the data and even the possessors of data sometimes do not know which data can be shared with the public or other government establishments.

Concluding Remarks

This article has highlighted the different transformations that have occurred in Saudi Arabia since the establishment of the Kingdom and the discovery of oil especially those related with urban development. It has enumerated the different challenges posed by these changes to sustainable urban development. Though several efforts have been made on different aspects of sustainability to reduce the negative environmental impacts of rapid urban development, there is need for further actions in establishing adequate institutional framework through a strong political will in the drive towards sustainable urban development. In the drive towards urban sustainability, the role of geospatial techniques is underscored by the need for environmental baseline information, regular monitoring, early warning system and cross-boundary nature of environmental impacts. Saudi Arabia needs to implement a national spatial data infrastructure to ensure adequate data accessibility and collaboration between different stakeholders.

References

Abo-Hassan, AA. 1981. Rangeland management in Saudi Arabia. Rangelands, 3 (2), pp. 51-53.

Abuzinada, AH. 2003. The role of protected areas in conserving biological diversity in the Kingdom of Saudi Arabia. *Journal of Arid Environment*, 54, pp. 39-45.

Aina, YA. 2012. Applications of Geospatial Technologies for Practitioners: An Emerging Perspective of Geospatial Education. In SJ.Miah. (Ed.): Emerging informatics - Innovative concepts and applications. InTech, Croatia, pp. 1-20.

Aina, YA., Van der Merwe, JH. andAlshuwaikhat, HM. 2008. Urban Spatial Growth and Land Use Change in Riyadh: Comparing Spectral Angle Mapping and Band Ratioing Techniques. Proceedings of the Academic Track of Free and Open Source Software for Geospatial (FOSS4G) Conference, 29 September - 3 October, Cape Town, South Africa.

Akeel, M. 2009. Lack of disaster-management planning led to present crisis. Arab News, Available from: http://www.arabnews.com/node/330638, Accessed on June 25, 2012.

Al Zawad, FM. 2008. Using GIS Technology to assess the Impact of Climate Change on Water Resources in Saudi Arabia. Proceedings of the Third National GIS Symposium in Saudi Arabia, 7-9 April 2008, Al-Khobar, Saudi Arabia.

Al-Hathloul, SA. 1996. The Arab-Muslim city: Tradition, continuity, and change in the physical environment. Dar Al-Sahan, Riyadh, Saudi Arabia.

Al-Ibrahim, AA. 1991. Excessive use of groundwater resources in Saudi Arabia: Impacts and policy options. *Ambio*, 2 (1), pp. 34-37.

Alkhedheiri, AA. 2002. The Role of Secondary Cities in the National Development Process of Saudi Arabia. Riyadh, Saudi Arabia.

Al-Said, FAM. 1986. Bedouin settlements: their origin, physical pattern and growth (Dukhnah, Al-Qassim, Saudi Arabia). M. Arch & MCP thesis, The University of Pennsylvania, USA.

Alshehri, SAF. 2011. National spatial data infrastructure collaboration for the Kingdom of Saudi Arabia.Unpublished PhD thesis, University of Nottingham, UK.

Alshuwaikhat, H., Aina, Y. and Rahman, SM. 2006. Integration of urban growth management and strategic environmental assessment to ensure sustainable urban development: The Case of Arabian Gulf Cities. *International Journal of Sustainable Development and Planning,* 1 (2), pp. 203-213.

Alshuwaikhat, HM. and Aina, YA. 2006. GIS-based urban sustainability assessment: The case of Dammam city, Saudi Arabia. *Local Environment,* 11 (2), pp. 141-161.

Alshuwaikhat, HM. and Nkwenti, DI. 2002. Developing sustainable cities in the arid regions. Cities, Pergamon, 19 (2), pp. 85-94.

Alshuwaikhat, HM. and Nkwenti, DI. 2002. Visualizing decision-making: perspectives on collaborative and participative approach to sustainable urban planning and management. *Environment and Planning* B, 29, pp. 513-531.

Alshuwaikhat, HM. and Nkwenti, DI. 2003. Collaborative Planning and Management Frameworks: Approaches to Effective Urban Governance by Adoption of Emerging Technologies. *International Journal of Management,* 20 (3), pp. 294-305.

Al-Zahrani, S. 2009. Damage may top SR1 billion. Saudi Gazette, Available from: http://www.saudigazette.com.sa/index.cfm?method=home.regcon&contentID=2009112755625, Accessed on June 25, 2012.

Arab News. 2011. Focus on environment as Riyadh unveils new projects. Available from: http://www.arabnews.com/node/375557, Accessed on June 25, 2012.

Babu, D., Sivalingam, S. and Machado, T. (in press). Need for adaptation strategy against global sea level rise: An example from Saudi coast of Arabian Gulf. Mitigation and Adaptation Strategies for Global Change, pp. 1-16.

Barth, HJ. 1999. Desertification in the eastern province of Saudi Arabia. *Journal of Arid Environments,* 43 (4), pp. 399-410.

Choguill, C. 2006. Assessing the urban sustainability of Riyadh. Paper presented at the Jeddah International Urban Forum and Exhibition, April 2006, Jeddah, Saudi Arabia.

Cole, DP. 1981. Bedouin and social change in Saudi Arabia. In JG Galaty and PC Salzman (Ed.): Change and development in nomadic and pastoral societies. Brill, Leiden, The Netherlands. Pp. 128-149.

EbenSaleh, MA. 1998. Life and death of traditional settlements of Southwest Saudi Arabia. *Journal of Architectural Education,* Wiley-Blackwell, 51 (3), pp. 177-191.

Ewing, R., Bartholomew, K., Winkelman, S., Walters, J. and Anderson, G. 2008. Urban development and climate change. Journal of Urbanism: International Research on Placemaking and Urban Sustainability, 1 (3), pp. 201-216.

Global Footprint Network (GFN). 2012. Country trends – Saudi Arabia. Available from: http://www.footprintnetwork.org/en/index.php/GFN/page/trends/saudiarabia/, Accessed on June 25, 2012.

Husain, SR. 2011. Growing Saudi consumption for its own crude cause for concern. Arab News, Available from: http://www.arabnews.com/node/385194, Accessed on June 22, 2012.

Khan, MA and Al-Homaid, N. 2003.Remote sensing study on mangrove depletion, Tarut Bay, Saudi Arabia. In A.S. Alsharhan, W.W. Wood, A.S. Goudie, A. Fowler and E.M. Abdellatif (Ed.): Desertification in the third millennium: Proceedings of an International Conference, Dubai, 12-15 February 2000, pp. 227-234, Swets&Zeitlinger B.V., Netherlands.

Mallakh, R. 1982. Saudi Arabia: Rush to development. Croom Helm, London, UK.

Mashee, HA.and Malthus, TJ. 2009. Object-oriented Image Analysis for Monitoring Urban Expansion in the City of Riyadh, Saudi Arabia. Proceedings of the Fourth National GIS Symposium in Saudi Arabia, 4-6 May 2009, Al-Khobar, Saudi Arabia.

Maukonen, T., Nagatani, K., Lee, M. and Ezcurra, E. 2006.Executive summary. In E Ezcurra (Ed.): Global deserts outlook. UNEP, Nairobi, Kenya.Pp. v-xii.

Mezher, T., Fath, H., Abbas, Z. and Khaled, A. 2011.Techno-economic assessment and environmental impacts of desalination technologies. *Desalination*, 266 (1-3), pp. 263-273.

Moran, EF. 2008. Human adaptability: An introduction to ecological anthropology. 3rd edition. Westview Press, Colorado, USA.

Mubarak, FA. 1992. Urbanization, urban policy and city form: Urban development in Saudi Arabia. Unpublished PhD dissertation, University of Washington, USA.

Mustapha, MR., Lim, HS. andMatJafri, MZ. 2012. Analysis of land cover classification in arid environment: A comparison performance of four classifiers. In Y. Chemin (Ed.): Remote Sensing of Planet Earth. InTech, Croatia, pp. 117-142. Available from: http://www.intechopen.com/books/remote-sensing-of-planet-earth/analysis-of-land-cover-classification-in-arid-environment-a-comparison-of-four-classifiers-, Accessed on June 22, 2012.

Mutairi, KA., El-Bana, M., Mansor, M., Al-Rowaily, S. and Mansor, A. 2012. Floristic Diversity, Composition, and Environmental Correlates on the Arid, Coralline Islands of the Farasan Archipelago, Red Sea, Saudi Arabia. *Arid Land Research and Management*, 26 (2) pp. 137-150.

Nahedh, M. 1989. The sedentarization of a Bedouin community in Saudi Arabia.Unpublished PhD dissertation, University of Leeds, UK.

Purnama, A., Al-Barwani, H. and Smith, R. 2005.Calculating the environmental cost of seawater desalination in the Arabian marginal seas. *Desalination*, 185 (1), pp. 79-86.

Rahman, SM. and Khondaker, AN. 2012. Mitigation measures to reduce greenhouse gas emissions and enhance carbon capture and storage in Saudi Arabia. *Renewable and Sustainable Energy Reviews*, 16 (5), pp. 2446-2460.

Saudi Geological Survey (SGS) 2003. An amateur guide to the land trips in Saudi Arabia (in Arabic). Saudi Geological Survey, Riyadh, Saudi Arabia.

Shaltout, KH., El-Halawany, EF. and El-Kady, HF. 1996. Consequences of protection from grazing on diversity and abundance of the coastal lowland vegetation in Eastern Saudi Arabia. *Biodiversity and Conservation*, 5 (1), pp. 27-36.

Tolba, MK. and Saab, NW. (Eds.) 2009. Arab environment: Impacts of climate change on Arab countries. Arab Forum for Environment and Development, Beirut, Lebanon.

United Nations Environment Programme (UNEP) 2012. GEO5 – Environment for the future we want. UNEP, Kenya.

United States Energy Information Administration (EIA) Independent statistics and analysis – International energy statistics. Available from: http://www.eia.gov/cfapps/ipdbproject/IEDIndex3.cfm, Accessed on June 22, 2012.

van der Merwe, R., Lattemann, S. and Amy, G. (in press). A review of environmental governance and its effects on concentrate discharge from desalination plants in the Kingdom of Saudi Arabia. Desalination and Water Treatment, pp. 1-11.

Vincent, P. 2008. Saudi Arabia: An environmental overview. Taylor & Francis/Balkema, Leiden, The Netherlands.

Watson, JW. 1981. Deserts of the world: Future threat or promise? Philomel Books, New York.

Zahran, MA. andYounes, HA. 1990. Hema system: Traditional conservation of plant life in Saudi Arabia. *Journal of King Abdulaziz University: Science*, 2, pp. 19-41.

MIX
Papier | Fördert
gute Waldnutzung
FSC® C083411

Zeitfracht Medien GmbH
Ferdinand-Jühlke-Straße 7
99095 Erfurt, Deutschland
produktsicherheit@kolibri360.de